A Prescription for
Healthy Living

A Prescription for Healthy Living
A Guide to Lifestyle Medicine

Edited by

Emma Short

Department of Cellular Pathology,
Division of Cancer and Genetics,
Cardiff University, University Hospital of Wales,
Cardiff, United Kingdom

ELSEVIER

ACADEMIC PRESS
An imprint of Elsevier

Academic Press is an imprint of Elsevier
125 London Wall, London EC2Y 5AS, United Kingdom
525 B Street, Suite 1650, San Diego, CA 92101, United States
50 Hampshire Street, 5th Floor, Cambridge, MA 02139, United States
The Boulevard, Langford Lane, Kidlington, Oxford OX5 1GB, United Kingdom

Notices
Knowledge and best practice in this field are constantly changing. As new research and experience broaden our
understanding, changes in research methods, professional practices, or medical treatment may become
necessary.

Practitioners and researchers must always rely on their own experience and knowledge in evaluating and using
any information, methods, compounds, or experiments described herein. In using such information or
methods they should be mindful of their own safety and the safety of others, including parties for whom they
have a professional responsibility.

To the fullest extent of the law, neither the Publisher nor the authors, contributors, or editors, assume any
liability for any injury and/or damage to persons or property as a matter of products liability, negligence or
otherwise, or from any use or operation of any methods, products, instructions, or ideas contained in the
material herein.

Library of Congress Cataloging-in-Publication Data
A catalog record for this book is available from the Library of Congress

British Library Cataloguing-in-Publication Data
A catalogue record for this book is available from the British Library

ISBN: 978-0-12-821573-9

For information on all Academic Press publications visit our
website at https://www.elsevier.com/books-and-journals

Publisher: Andre Gerhard Wolff
Acquisitions Editor: Kattie Washington
Editorial Project Manager: Barbara Makinster
Production Project Manager: Punithavathy Govindaradjane
Cover Designer: Matthew Limbert

Typeset by TNQ Technologies

Contents

SECTION 2 Mental health and wellbeing

CHAPTER 20   Skin health: what damages and ages skin? Evidence-based interventions to maintain healthy skin

Lorna Jeng and Anjaly Mirchandani

CHAPTER 21   Western medical acupuncture

Carolyn Rubens

SECTION 4 Nutrition and healthy eating habits

Contributors

Arfa Ahmed
General Practitioner, National Health Service, Manchester, United Kingdom

Ekua Annobil
General Practitioner, Sydney, NSW, Australia

Aria Campbell-Danesh
Associate Fellow, Division of Clinical Psychology, British Psychological Society, London, United Kingdom

Caroline Deodhar
Senior Resident Medical Officer Obstetrics and Gynaecology, Westmead Hospital, Sydney, NSW, Australia

Adam Douglas
Department of Cellular and Anatomical Pathology, Derriford Hospital, Plymouth, United Kingdom

Emmajane Down
General Practitioner, National Health Service, London, United Kingdom

Ellen Fallows
General Practitioner, The British Society of Lifestyle Medicine, London, United Kingdom

Liz Forty
School of Medicine, Cardiff University, Cardiff, United Kingdom

Farah Gilani
General Practitioner, Ayrshire Medical Group, National Health Service, Scotland, United Kingdom

Laura Gush
General Practitioner, National Health Service, Bridgend, United Kingdom

Athanasios Hassoulas
Programme Director MSc Psychiatry, School of Medicine, Cardiff University, Cardiff, United Kingdom

Saba Jaleel
Psychiatrist, Change Grow Live, Birmingham, United Kingdom

Lorna Jeng
Radiologist, Whittington Hospital, London, United Kingdom

Alexandra J. Kermack
School of Human Development and Health, Faculty of Medicine, University of Southampton, Southampton, United Kingdom

Liza Kirtchuk
General Practitioner and Clinical Lecturer, King's College London, London, United Kingdom

Emma Ladds
Academic Clinical Fellow, Nuffield Department of Primary Health Care Sciences, University of Oxford, Oxford, United Kingdom; General Practitioner, National Health Service, Oxford, United Kingdom

Devina Leopold
General Practitioner, National Health Service, Cwmbran Village Surgery, Cwmbran, United Kingdom

Nita Maha
General Practitioner, Primary Care, Bristol, United Kingdom

Hayley S. McKenzie
Medical Oncology, University Hospital Southampton NHS Foundation Trust, Southampton, United Kingdom

Anjaly Mirchandani
General Practitioner, Northfields Surgery, London, United Kingdom

Gemma Newman
General Practitioner, National Health Service, Ashford, United Kingdom; Advisory Board Member, Plant-Based Health Professionals UK, United Kingdom

Alka Patel
Lifestyle Medicine Physician, General Practitioner and Health/Lifestyle Coach, Lifestyle First, London, United Kingdom

Venita Patel
Community Paediatrician, Guy's & St Thomas NHS Trust & Registered Nutritional Therapist, London, United Kingdom

Thom Phillips
General Practitioner, National Health Service, Cwmbran Village Surgery, Cwmbran, United Kingdom

Carolyn Rubens
General Practitioner and Medical Acupuncturist, Lighthouse Medical Practice, Eastbourne, United Kingdom

Sonal Shah
General Practitioner, National Health Service, London, United Kingdom

Laura Sheldrake
General Practitioner, National Health Service, Southampton, United Kingdom

Emma Short
Department of Cellular Pathology, Division of Cancer and Genetics, Cardiff University, University Hospital of Wales, Cardiff, United Kingdom

Ailsa Sita-Lumsden
Medical Oncology, Guy's and St Thomas' NHS Foundation Trust, London, United Kingdom

Claire Stansfield
General Practitioner, National Health Service, West Yorkshire, United Kingdom

Ann Wylie
Lecturer, King's College London, London, United Kingdom

About the editor

Dr. Emma Short, BMBCh (Oxon) MA (Cantab) PhD MRCSEd PGCMEd
Instagram: @dr_emmashort

Dr. Emma Short studied pre-clinical medicine at Cambridge University and clinical medicine at Oxford University. She lives in Cardiff, with her GP husband and two young daughters. She completed her basic surgical training in Devon, before moving to Wales to specialise in histopathology. She has a PhD from Cardiff University in cancer genetics: Genetic Mechanisms in Colorectal Polyposis, 2018.

Dr. Short has published extensively in the scientific literature and has an active role in medical education. She is interested in the interaction between the mind and the body, and the impact that mental well-being and social connections have on health. Dr. Short is a qualified meditation teacher and is a great advocate of showing kindness in all spheres of life. She loves exploring different aspects of holistic well-being and has diplomas in Mindful Nutrition and Shinrin Yoku. She is a keen runner, having completed an ultramarathon, two marathons and many half-marathons, and is passionate about health promotion and disease prevention. She set up and runs a not-for-profit community running group in Cardiff, Sirius Running, and is a qualified personal trainer.

Preface

Being healthy means different things to different people. The World Health Organisation (WHO) defines health as *'a state of complete physical, mental and social well-being and not merely the absence of disease or infirmity'*. Good health allows an individual to thrive and to achieve their full potential.

Health is determined by a complex interaction of genetic, environmental and socioeconomic factors, which begin before conception and continue throughout life, being influenced by broader constructs such as politics and cultural norms. Good health is a basic human right.

The developed world is currently facing a major health crisis, with dangerously high rates of conditions such as obesity, diabetes, cardiovascular disease and cancer, many of which are preventable. This has a profound and negative impact on the wellbeing of individuals, on societies and on the economy. Whilst disease prevention and health improvement absolutely require funding and interventions from multiple levels, including from the government, education system, social care system and industry, there are many lifestyle modifications which individuals can make to take charge of their wellbeing and to improve their health.

Numerous long-term health problems stem from obesity. Across the globe, obesity has nearly tripled since 1975 [1]. Moreover, 39% of adults were overweight in 2016, and 13% were obese. It is disturbing that unhealthy habits are being passed on to future generations: around the world, 41 million children under the age of 5 are overweight or obese, as are 340 million children and adolescents from ages 5 to 19 [1]. Statistics from the House of Commons in the United Kingdom (UK) state that 26% of adults in England are obese, and a further 35% are overweight [2]. In the United States of America (USA), it is reported that the prevalence of obesity is 39.5%, which amounts to 93.3 million adults [3]. Being overweight increases the risk of cardiovascular disease, musculoskeletal disorders, type 2 diabetes and certain cancers, including endometrial cancer, breast cancer and colorectal cancer [1,2].

There are currently approximately 3.8 million people in England with diabetes, which is around 9% of the adult population [4]. In the USA, approximately 9.4% of the population are diabetic, equating to over 30 million adults [5].

Cardiovascular disease is the leading cause of mortality throughout the world, responsible for nearly a third of all deaths [6]. There are more than seven million people in the UK with cardiovascular disease [7], and in the USA, 610,000 people die every year from the disease [8]. In the USA, over 1.7 million people are diagnosed with cancer annually [9], and in the UK almost a thousand people will be diagnosed with cancer every day [10]. Around the world, one in every four individuals will suffer from a mental health or neurological problem [11].

It is, therefore, vitally important to be aware that a significant proportion of these chronic diseases can be prevented and that individuals can be empowered to take control of their own health and wellbeing. The WHO states that 80% of premature heart disease, stroke and diabetes can be prevented [12] and Cancer Research UK report that 40% of cancer is avoidable [13]. Lifestyle modifications and behavioural changes do not need to be complicated, time consuming or expensive. Significant health gains can be achieved through simple measures such as being physically active, minimizing the time spent sitting, eating a healthy and balanced diet, maintaining a healthy weight, not smoking, moderating alcohol intake and maintaining social relationships. This approach to healthcare has recently gained global popularity and is described as, 'Lifestyle Medicine'.

In this book, medical doctors, research scientists and healthcare professionals from a variety of backgrounds will provide informed advice on how to encourage patients to take charge of their health and future. The book shall give an evidence-based overview of a diverse range of Lifestyle Medicine and health-related topics. It shall address the impact that society and the environment have on the health of a population and shall explore the psychology of health-related behavioural change.

We hope that this book will inspire and encourage you to empower your patients to make simple behavioural changes which will have a large impact on their physical and mental wellbeing. All healthcare professionals have a role to play in supporting their patients to decide what lifestyle modifications are important to them and in helping them to plan well to make changes that they will enjoy and sustain.

Let your patients know that they are in charge of their future and they can help themselves to change it for the better.

Dr. Emma Short

References

[1] https://www.who.int/news-room/fact-sheets/detail/obesity-and-overweight.
[2] House of Commons Briefing paper 3336, March 2018.
[3] https://www.cdc.gov/obesity/data/adult.html.
[4] https://www.gov.uk/government/news/38-million-people-in-england-now-have-diabetes.
[5] http://www.diabetes.org/diabetes-basics/statistics/.
[6] https://www.who.int/en/news-room/fact-sheets/detail/cardiovascular-diseases-(cvds).
[7] https://www.bhf.org.uk/what-we-do/our-research/heart-statistics.
[8] https://www.who.int/en/news-room/fact-sheets/detail/cardiovascular-diseases-(cvds).
[9] https://www.cancer.gov/about-cancer/understanding/statistics.
[10] https://www.cancerresearchuk.org/health-professional/cancer-statistics/incidence#heading-Zero.
[11] https://www.who.int/whr/2001/media_centre/press_release/en/.
[12] http://www.who.int/chp/chronic_disease_report/part1/en/index11.html.
[13] https://www.cancerresearchuk.org/about-cancer/causes-of-cancer/can-cancer-be-prevented.

Society, health and behaviour

Social and environmental determinants of health

1

Liza Kirtchuk[1] and Ann Wylie[2,*]

[1]*General Practitioner and Clinical Lecturer, King's College London, London, United Kingdom;* [2]*Lecturer, King's College London, London, United Kingdom*

Introduction

During the early part of the 20th century, the medical profession and evolving healthcare systems focused on scientific discoveries. New knowledge led to new interventions and treatments becoming available for many conditions, which had not previously been curable. Notable developments included open-heart surgery and the emergence of pharmaceuticals for a range of complex health issues. Morbidity could be relieved, and mortality rates improved. It was in the late 20th century that this progressive trajectory in health and medical sciences started to plateau and research investment costs became disproportionate to potential health gains.

At that time, observational studies started to identify social and environmental factors which seemed to underlie the variant health status of some communities and populations. However, many of these studies were not published in high-impact journals. Furthermore, some publications were initially suppressed as they were politically challenging. Many academics were developing their research paradigms, with public health and health promotion practitioners taking a proactive interest.

In the latter part of the 20th century, the World Health Organization (WHO) became increasingly engaged in concepts such as 'Health for all' and the role of the political and social determinants of health. A key event, held in 1978, was the international conference on Primary Health Care in Alma Ata. The seminal declaration arising from the conference stated that there was a:

> *need for urgent action by all governments, all health and development workers, and the world community to protect and promote the health of all the people of the world.* [1]p1

The declaration outlined a target as follows:

> *A main social target of governments, international organizations and the whole world community in the coming decades should be the attainment by all peoples of the world by the year 2000 of a level of health that will permit them to lead a socially and economically productive life. Primary health care is the key to attaining this target as part of development in the spirit of social justice.* [1]p5

---

*To be elected Fellow of the Royal College of General Practitioners May 2021.

A Prescription for Healthy Living. https://doi.org/10.1016/B978-0-12-821573-9.00001-1

Since this landmark event, the WHO has continued to work to identify the social determinants of health, which are defined officially as:

> ... *the conditions in which people are born, grow, live, work and age. These circumstances are shaped by the distribution of money, power and resources at global, national and local levels. The social determinants of health are mostly responsible for health inequities - the unfair and avoidable differences in health status seen within and between countries.* [2]

The Ottawa Charter for Health Promotion is an international agreement signed at the First International Conference on Public Health, organised by the WHO in 1986. It was explicit in its arguments that the social and environmental determinants of health (SEDH) were integral to health and needed to be explicitly recognised and addressed. More recent international work now argues that the causes of morbidity should be addressed by tackling social and environmental factors, and research is being performed to support these endeavors [3,4].

Table 1.1 outlines the key concepts relating to SEDH as set out by the WHO.

This chapter outlines the importance of SEDH. It describes the emerging literature and research paradigms, data sources and their limitations, and health-related social interventions and the efficacy of such interventions. It shall consider how knowledge and awareness translate to clinical practice and how this is relevant to addressing the health of patients.

Social and environmental determinants of health

The literature describing the SEDH is rich and varied. This chapter shall focus on recent and relevant research findings that can inform current approaches to addressing these determinants of health from a primary care perspective.

Table 1.1 WHO social determinants of health: key concepts.
WHO social determinants of health: key concepts
Employment conditions
Social exclusion
Public health programmes and social determinants of health
Women and gender
Early childhood development
Globalisation
Health systems
Measurement and evidence
Urbanisation
Adapted from World Health Organization, Social Determinants of Health webpage https://www.who.int/social_determinants/sdh_definition/en/, Accessed 7th January 2020.

An historical background

The WHO Ottawa Charter, 1986, was a key document which argued that five key areas were important for health promotion, as shown in Fig. 1.1:

- building healthy public policy;
- creating supportive environments:
- strengthening community action;
- developing personal skills; and
- reorienting health services, as shown in Fig. 1.1.

These principles remain relevant and important, over 30 years later [5].

In the United Kingdom (UK), the key papers that recognised the SEDH were first published in the late 1970s and during the 1980s [6]. These papers and reports made explicit **the link between deprivation and health inequality.** However, they prompted little action, and it was not until the

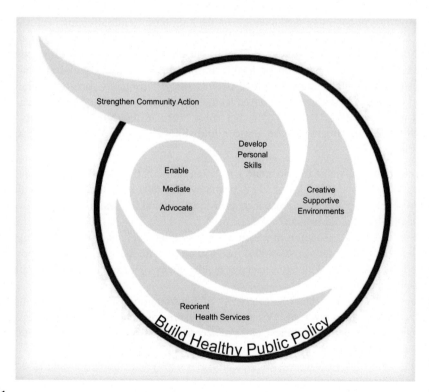

FIGURE 1.1

Recommendations of the Ottawa Charter 1986.

WHO Ottawa Charter for Health Promotion 1986 World Health Organization. Ottawa Charter for health Promotion; 1986. Geneva: World Health Organization.

mid-1980s that the Jarman indices were developed and used to assess levels of deprivation in different communities. The indices were utilised to adjust general practice payments for practices in areas of deprivation.

In 1997, the election of 'New Labour' led to the development of formal policies addressing the social determinants of health. Academics were funded to research such determinants [7,8], and the government produced and acted on a range of White Papers addressing the five areas of the Ottawa Charter. There was funding, collaboration, the commissioning of research, evaluation of pilot schemes, and resources for training and implementing a range of public policies. Across the many government departments, addressing inequalities per se was integral. One scheme established to help tackle social inequality was *Sure Start*, which aimed to provide support to parents and children under the age of 5 years who lived in areas of deprivation. Services were developed to support learning, positive health behaviors, well-being, social development and emotional development and were often delivered via children's centers. It has been reported that the Sure Start scheme reduced obesity levels, reduced hospital admissions and improved the physical health of the families who engaged in the programme [9,10]. Furthermore, it was associated with a reduction in parental use of harsh discipline and an improved home environment and life satisfaction for families.

The UK government commissioned a review on the long-term trends affecting health services, and in 2002, the Wanless report was published [11]. It was unusual that this innovative report came from the Treasury rather than the Department of Health, but the report explicitly acknowledged that the population's health depended not only on healthcare but on much wider social and environmental factors. The summary of the report described poor health determinants such as smoking, physical inactivity and obesity but additionally referred to the wider determinants of health such as **income, employment and education**. It was noted that **these were important determinants of chronic diseases such as chronic heart disease, cancer and diabetes**. Examples of the impact of SEDH include the following:

- Lower socioeconomic status is associated with increased emotional and developmental difficulties in childhood and a lower life expectancy. It is also associated with unhealthy eating habits, low levels of physical activity, increased rates of smoking and increased rates of alcohol-related hospital admission.
- Fuel poverty describes a situation in which a household has fuels costs which are higher than average, but that paying such costs would result in an income that was below the poverty line. Fuel poverty is associated with excess winter deaths, many of which are due to cardiovascular and/or respiratory disease.
- The physical and environmental characteristics of a community can have a major impact upon health. For example, high levels of UV radiation exposure are associated with an increased risk of the development of skin malignancies; a lack of access to clean water is associated with infectious diseases such as cholera; and air pollution can exacerbate respiratory disease.
- Poor workplace conditions can also affect health: Physically demanding jobs are associated with a risk of injury, while sedentary jobs increase the risk of cardiovascular disease.
- Educational attainment has a major impact on health literacy, as described in Chapter 4.

The impact of social and environmental factors on health in the 21st century

Social and environmental factors continue to have a significant impact on the health of individuals and communities in modern times. Examples illustrating this include refugee health and the obesity epidemic.

Refugee health

In 2015 nearly all European countries experienced exceptionally high number of refugees seeking safety, with similar phenomena occurring on a global scale [12]. The factors associated with this movement were predominately linked to war and conflict. Refugee health needs are significant and cumulative. Those making the decision to flee their home country may have already endured significant traumatic experiences and individuals with predisposing health conditions may have received suboptimal care.

Many refugees initially reside in camps in the countries to which they have fled, and healthcare is often very basic. Different camps have different structures and facilities, and they may have a number of environmental hazards, including a lack of clean water, a lack of warm shelters and inadequate toilet facilities, which can contribute to, or exacerbate, existing health problems.

A study addressing the healthcare needs of refugees from Syria, Afghanistan, Iraq, Pakistan, Nigeria and Somalia identified the main categories as being disabilities and injuries, mental health, pregnancy-related issues, infectious diseases, gastrointestinal disease, hydration and dental problems [12].

When refugees arrive at their ultimate destination, they often encounter challenges such as difficulties accessing the healthcare system and financial limitations if care requires payment. The refugee population and their experiences are diverse, and it is vital that their needs are met in a culturally competent and compassionate manner. Kang et al. [13] identified several factors that must be addressed to provide good healthcare for asylum seekers and refugees. These include tackling language barriers and suboptimal interpretation services, improving awareness of healthcare structure and services, recognising financial difficulties, providing adequate transport to clinics and addressing any perceived discrimination associated with race, religion or immigration status.

Obesity and the built environment

The obesogenic environment tends to predominate in higher-income countries. It describes an environment in which preprepared calorie-dense foods are readily available at low cost, and where physical activity is restricted or at low levels, for example due to a lack of green spaces, no sporting facilities or unsafe neighbourhoods. These factors are most frequently identified in areas of social deprivation. Such an environment limits the possibility of individuals consuming a balanced diet of fresh seasonal produce and is often associated with a lack of home preparation of meals and home cooking.

One of the global health challenges lies in tackling the obesity epidemic, and clinicians often need to encourage and support their patients in improving their diets. Research is being undertaken to determine the most effective means of achieving this, but it has been argued that there needs to be a focused collaboration between town planning and health strategies at a local level [14].

Neighbourhoods can promote physical activity through 'walkability' attributes, such as easy access to shops, provision of pavements and good public transport, as well as through provision of local recreational facilities such as playgrounds and cycling routes. Reducing the density of fast food outlets, particularly in areas of deprivation and near schools, is also important.

Food availability and physical activity have been addressed jointly through engagement with community gardens giving rise to fresh produce; benefits can extend further with evidence that gardening can also improve individuals' mental health [15]. With respect to factors affecting both food availability and physical activity levels, it is important to focus efforts on deprived communities, where the trend toward poorer facilities and higher density fast food outlets leads to 'deprivation amplification', further compounding the influence of SEDH [14].

The prevalence of obesity is very evident, yet while healthcare practitioners hone their skills for mediating behavior change and motivational interviewing using evidence-based techniques [16,17], environmental determinants will also need to be addressed to bring about significant change.

Addressing social and environmental determinants of health in the global north and high-income countries

In 2012, the European health policy framework, Health 2020, was adopted by the member states of the WHO European Region, and is committed to addressing health inequalities. While a number of European health departments have developed strategies to address SEDH and inequalities, these have not always been supported by economic policies influencing health outcomes [18]. It is hoped that the integration of the 2010 WHO Adelaide Statement on Health in All Policies into the legislation of European countries will be a strong driver for addressing SEDHs. The Adelaide Statement encourages cross-sectoral approaches to health equity [19]. For example, in Malmö, Sweden, the local government established a Commission for a Socially Sustainable Malmö, with recommendations for a range of sectors which were overseen by a cross-sector steering group. Interventions include improved access to culture and leisure activities, the employment of local people for new building work and financial investments specifically in social sustainability.

In North America, SEDH are high on the Canadian Government's agenda, both in terms of research and government policy. There is a *social determinants of health team* within the Public Health Agency of Canada and the Canadian Medical Association has demonstrated its commitment to this agenda. In the United States (US), the 2010 Patient Protection and Affordable Care Act had a big impact on placing disease prevention approaches on the national agenda. This resulted in a number of reforms relating to healthcare delivery and payment, which aimed to address inequity and SEDH. An example is the State Innovation Models Initiative, which provides funding to states to adopt healthcare models which promote coordination between healthcare providers, and to address population health needs. In Rhode Island, this funding was used to establish Community Health Teams which promote enhanced information-sharing between clinical and community services to address social determinants of health.

In Australia, there has been a particular focus on addressing inequities among the Aboriginal populations, and although this has dropped down the national agenda in recent years, South Australia has a Health in All Policies approach [20]. An example of a successful strategy was the focus on improving the health, sustainability and economic position of a particular region with various unmet health and social needs: The development of a Regional Atlas of Community Wellbeing focused on highlighting inequities and social determinants of health to inform government policymakers in areas such as economic growth, employment, education and healthcare provision.

The role of the healthcare provider

Julian Tudor Hart, a general practitioner (GP)/family physician working in Wales, UK, in the 1970s, coined the term 'inverse care law'. This was used to highlight the disparities in health outcomes across social class categories and to describe the way in which the provision of good medical care was inversely related to the needs of the populations being served [21]. He highlighted the role of the GP as a gatekeeper of precious resources which need to be responsibly distributed to the local community as a whole, with more than just a view on individual patient care. This broader role of GPs in addressing health inequalities was recognised by the Royal College of GPs (RCGP) in the United Kingdom. They formed the Health Inequalities Standing Group in the 1990s, which has worked with the Department of Health in England to identify the key areas in which GPs can address health inequalities. These include how GPs provide care as individuals and within the primary care team, commissioning at a local level, working in partnership with other relevant organisations and influencing the national agenda [22]. Primary care provision has also been heralded as a solution to local and global health inequalities by WHO.

Secondary care services also have an important role to play in addressing SEDH, and NHS England has developed a set of clinical and transformational indicators, some of which address health inequalities in hospitals. An example of how this has been implemented is the delivery of alcohol identification and brief advice by pharmacy technicians to all inpatients on medical wards within 48 h of admission at University Hospital Southampton NHS Foundation Trust.

The holistic consultation

In the United Kingdom, a 2010 King's Fund report 'Tackling inequalities in general practice' [23] highlighted the important role that primary care plays through generalism, advocacy and community- and population-level healthcare. However, it also highlighted the tension faced by GPs in addressing population-level concerns at the expense of responsive individualised care. An example of this tension includes alerts flashing up in a patient's electronic record to prompt the delivery of lifestyle counselling during consultations in which the patient's immediate agenda must be prioritised, such as the loss of a loved one. GPs are tasked with striking the delicate balance of population and individual level care but are in a uniquely privileged position to harmonise these by exploring the 'lived experience of inequalities at the individual level' [24] and by using a biopsychosocial model, in which an individual's health is approached as a dynamic interplay between biological, pathophysiological, psychological and socioenvironmental factors, to provide this holistic patient care.

Working with partner organisations

Social and environmental factors and health outcomes are inextricably linked. Where poor health is linked to nonclinical root causes, it is argued that there is great potential for social prescribing to form part of a clinical management plan [25]. Social prescribing is described in detail in Chapter 5. Primary care providers are also well placed to be advocates for change through proactive collaboration with local organisations, communities and public health [26]. One of the most notable collaborations of this kind has been the partnership between parkrun and the Royal College of General Practitioners (RCGP), UK. Parkrun is an increasingly international network of local volunteer-led runs which see over 350,000 people of varying backgrounds and abilities participating weekly. The emphasis is on inclusivity and community, and in 2018, parkrun partnered with RCGP to support practices to become 'parkrun practices', whereby links between the practice and local parkrun are strengthened, with a commitment to promoting their local parkrun and its associated health benefits to patients. This collaboration supports practices to engage with the local community and also provides a powerful vector for a number of health promotion messages. Parkrun expanded to the United States in 2012 and currently takes place across 42 locations; it is not formally partnered with any healthcare providers, although does receive sponsorship from the American Cancer Association.

Tomorrow's doctors

The medical profession, in particular Primary Care, not only acknowledges the links and issues around SEDH, but also largely acknowledges its role in addressing these. It is important that concepts such as SEDH, health equity and access to healthcare should be included as part of any medical education curriculum. Core medical undergraduate curricula in the United Kingdom, regulated by the General Medical Council (GMC), now have explicit learning outcomes related to SEDH [27–29] so that newly qualify doctors have a greater understanding of the nonbiological factors which impact the health and well-being of their patients. Similarly, in North America, SEDH are explicitly highlighted in the Canadian CanMEDS physician competency framework, in the subsection on the doctor as Health Advocate, and in the Association of American Medical Colleges report 'Behavioral and Social Science Foundations for Future Physicians' aimed at supporting medical schools to develop these areas of their curricula [30,31].

However, integrating teaching about equity, social justice and SEDH in medical undergraduate and postgraduate training is still in its infancy, and much work is needed to develop this further [32].

Conclusions

Social and environmental factors play a key role in determining the health of an individual and are being given increasing importance on the sociopolitical agenda. The WHO provides a prominent platform for global discourse about health promotion and SEDH, as highlighted by its seminal Ottawa Charter, 1986. There is a growing evidence base to support the links between SEDH and health outcomes, providing increased political capital for governments to address health inequalities through cross-departmental 'Health in all Policies' approaches, although achieving success remains challenging. SEDH represents a dynamic area, and contemporary issues have included refugee health needs and the obesogenic environment. Healthcare workers on the ground can address SEDH through recognition of their impact, developing the skills to address them within the clinical consultation, supporting provision of care that addresses unmet needs, advocating for the vulnerable in society, and influencing the political agenda. Increasingly healthcare organisations are recognising SEDH

and are extending their scope beyond a strictly biomedical model through the use of quality indicators that acknowledge health inequalities, and by engaging with social prescribing. Both under-
and postgraduate healthcare professional training bodies are acknowledging SEDH in their curricula,
paving the way for a future generation of clinicians better equipped to meet the exciting challenges
posed by SEDH.

Smoking case study

Sanjiv is a new GP at a practice in South London, UK, and has been tasked with improving smoking
rates for his practice population, based on work he did at his previous practice. In his old practice, he
worked with the practice healthcare assistant to set up a 'long-term conditions clinic' for patients with
conditions such as hypertension and diabetes. The emphasis was on supporting lifestyle change by
applying behavior change approaches: smoking cessation was one of the areas of focus. The
intervention was effective and practice prevalence figures for smoking went down from 7% to 4% over
the course of a year. Sanjiv has initiated a similar approach at the new practice, but results have been
disappointing, with no change to prevalence figures over the past 6 months: he has to face the reality
that this approach is not working.

He takes a look at the Public Health England National General Practice Profiles data set to gain a
better understanding of the differences in demographics and social determinants of health between the
two practice populations. Sanjiv identifies that the self-reported prevalence of long-standing health
conditions is lower in his new practice, and a greater proportion of the population are in paid work or
full-time education compared with his previous practice; this explains why a 'long-term conditions'
clinic during office hours is not attracting the target population. He also notes that the Index of
Multiple Deprivation (IMD) is much higher than in his previous area, and he ponders over the impact
that this might be having on the effectiveness of cessation approaches due to wider social factors and
challenges. This prompted him to do a literature search in which he identified that cessation services
are less successful in areas of deprivation and that these services are more likely to lose contact with
individuals in the 4 weeks following a quit date being set [33].

As a result of these investigations, he suggests that offering smoking cessation opportunities during
extended hour sessions is more likely to be accessible to the local population, and the practice makes
the necessary arrangements. He also suggests that the practice sets up text message reminders to
individuals who have set a quit date, to maintain contact with them. He recognises that smoking
prevalence rates may not be the only marker of 'success' and that bringing these down is likely to be
more challenging than at his old place of work. Instead, he focuses on measuring access to smoking
cessation opportunities as a marker of improvement, and there are future plans to explore perspectives
of the barriers to accessing smoking cessation services among the practice's patient participation
group.

Practice points

- Even within the same neighbourhood, SEDH can vary significantly, influencing the effectiveness
  of any public health or medical intervention.
- It is important to have an understanding of the local population demography and needs when
  designing any intervention.

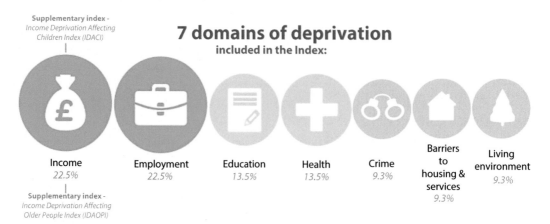

FIGURE 1.2

The English Index of Multiple Deprivation (IMD) 2015.

Department for Communities and Local Government Infographic: The Indices of Multiple Deprivation 2015 https://assets.
*publishing.service.gov.uk/government/uploads/system/uploads/attachment_data/file/464431/English_Index_of_Multiple_*
Deprivation_2015_-_Infographic.pdf.

- When measuring outcomes of interventions addressing SEDH, improvements to access may be as important as changes to disease prevalence or smoking status.
- Indices of deprivation for different geographic areas are important for supporting resource allocation decisions and addressing health inequalities. In England, the IMD is a local measure of deprivation and is calculated on the basis of seven differentially weighted domains as outlined in Fig. 1.2. New Zealand also has a well-established Index of Deprivation reflecting eight dimensions of socioeconomic deprivation, also used by policymakers in funding decisions. The United States has yet to develop a nationally agreed-upon and validated index, although there are some indices which show good potential for further development, such as the Robert Graham Center's Social Deprivation Index and the Social Vulnerability Index [34].

Travelling community case study

Maria is the director of a public health department and works with the local commissioning group who make funding decisions within the locality. She has been asked to consider inequalities faced by the local traveler community. Maria is aware that gypsies and travelers have an estimated life expectancy that is 10−25 years below the general population [35] and also represent the ethnic group who were least likely to rate their health as 'good' or 'very good' in the 2011 census in the United Kingdom [36]. To find out more, Maria accesses the most recent Friends, Family and Travellers (FFT) Annual Report [37], which highlighted barriers to accessing primary healthcare due to difficulties registering with a GP. This is mostly due to not being able to provide an address or proof of identity and is a difficulty affecting other groups such as refugees, homeless people and those fleeing domestic violence. The FFT Report also highlighted concerns about prejudiced and racist attitudes experienced when trying to access healthcare.

Maria contacted the practices within the locality and requested their registration policy and how these related to patients with no fixed abode or the traveler community. She also asked practices to share any difficulties experienced with respect to such registrations. One practice had a proactive policy regarding the traveler community and had overcome barriers cited by other practices, predominantly relating to effective communication with patients with no mailing address. Maria organised an opportunity for this practice to showcase the ways in which they were working to improve access at a local educational event, along with representatives from FFT who delivered cultural awareness training. Neighbouring practices were able to raise concerns, and a training need for frontline reception staff undertaking patient registrations was identified. She contacted practices again 3 months later: half had delivered registration training and 70% had developed a policy for registering individuals with no fixed abode.

Practice points

- The support of local commissioners and public health bodies is needed to advocate for vulnerable groups and quantify local unmet needs, as well as providing the funds to address these needs.

Summary

- SEDH play a significant role in health and well-being globally and should be addressed by governments in a pan-departmental manner, not solely in relation to healthcare policies.
- It is critical that clinicians are able to recognise the significance of SEDH in health and well-being and are equipped to elicit and address them as part of delivering nonjudgmental, holistic and contextual patient care.
- SEDH and health inequalities can be addressed by healthcare professionals through tailored individual patient care, at the level of a practice population, through local commissioning and through advocacy and influencing of the political agenda.
- There are ongoing developmental needs for clinicians in recognising and addressing SEDH, as well as applying behavior change approaches. This is increasingly recognised and is being addressed in undergraduate curricula.

References

[1] World Health Organization. In: Unicef. Primary health care: report of the international conference on primary health care, Alma-Ata, USSR, 6-12 september; 1978.

[2] World Health Organization. Social Determinants of health; Available from: https://www.who.int/social_determinants/sdh_definition/en/.

[3] World Health Organization. Ottawa charter for health promotion. Geneva: World Health Organization; 1986.

[4] Marmot M, Friel S, Bell R, Houweling TAJ, Taylor S. Closing the gap in a generation: health equity through action on the social determinants of health. Lancet 2008;372(9650):1661−9.

[5] Thompson SR, Watson MC, Tilford S. The Ottawa Charter 30 years on: still an important standard for health promotion. Int J Health Promot Educ 2018;56(2):73−84.

[6] Townsend P, Whitehead M, Davidson N. Inequalities in health: the black report and the health divide. London: Penguin; 1988.

[7] Marmot M. Fair society, healthy lives - strategic review of health inequalities in England post 2010. London: The Marmot Review; 2010.

[8] Marmot M, Allen J, Bell R, Bloomer E, Goldblatt P. WHO European review of social determinants of health and the health divide. Lancet 2012;380(9846):1011−29.

[9] Melhuish E, Belsky J, Leyland AH, Anning A, Hall D, Tunstill J, Ball M, et al. The impact of sure start local programmes on 5-year-olds and their families. Department for Education; 2010.

[10] Cattan S, Conti G, Farquharson C, Ginja R. The health effects of sure start. London: The Institute for Fiscal Studies; 2019.

[11] Wanless D. Securing our future health: taking a long-term view. London: HM Treasury; 2002.

[12] van Loenen T, van den Muijsenbergh M, Hofmeester M, Dowrick C, van Ginneken N, Mechili EA, et al. Primary care for refugees and newly arrived migrants in Europe: a qualitative study on health needs, barriers and wishes. Eur J Publ Health 2017;28(1):82−7.

[13] Kang C, et al. Access to primary health care for asylum seekers and refugees: a qualitative study of service user experiences in the UK. Br J Gen Pract 2019;69(685):e537−45.

[14] Townshend T, Lake A. Obesogenic environments: current evidence of the built and food environments. Perspect Public Health 2017;137(1):38−44.

[15] Papas MA, Alberg AJ, Ewing R, Helzlsouer KJ, Gary TL, Klassen AC. The built environment and obesity. Epidemiol Rev 1 January 2007;29(1):129−43.

[16] Aveyard P, Lewis A, Tearne S, Hood K, Christian-Brown A, Adab P, et al. Screening and brief intervention for obesity in primary care: a parallel, two-arm, randomised trial. Lancet 2016;388(10059):2492−500.

[17] Albury C, Stokoe E, Ziebland S, Webb H, Aveyard P. GP-delivered brief weight loss interventions: a cohort study of patient responses and subsequent actions, using conversation analysis in UK primary care. Br J Gen Pract 2018;68(674):e646−53.

[18] Donkin A, Goldblatt P, Allen J, Nathanson V, Marmot M. Global action on the social determinants of health. BMJ Global Health 2018;3(Suppl.1).

[19] Greszczuk C. Implementing health in all policies: lessons from around the world. The Health Foundation; 2019.

[20] Van Eyk H, Harris E, Baum F, Delany-Crowe T, Lawless A, MacDougall C. Health in all policies in south Australia—did it promote and enact an equity perspective? Int J Environ Res Publ Health November 2017; 14(11):1288.

[21] Tudor- Hart J. The inverse care law. Lancet 1971;297(7696):405−12.

[22] Ali A, Wright N, Rae M. Addressing health inequalities: a guide for general practitioners. 2008.

[23] Hutt P, Gilmore S. Tackling inequalities in general practice. The Kings Fund 2010;19.

[24] Popay J, Kowarzik U, Mallinson S, Mackian S, Barker J. Social problems, primary care and pathways to help and support: addressing health inequalities at the individual level. Part I: the GP perspective. J Epidemiol & Community Health 2007;61(11):966−71.

[25] Husk K, Elston J, Gradinger F, Callaghan L, Asthana S. Social prescribing: where is the evidence? Br J Gen Pract 2019;69(678):6−7.

[26] Allen LN, Barry E, Gilbert C, Honney R, Turner-Moss E. How to move from managing sick individuals to creating healthy communities. Br J Gen Pract 2019;69(678):8−9.

[27] General Medical Council. Outcomes for graduates 2018. GMC London; 2018.

[28] Wylie A, Leedham-Green K, Takeda Y. Engaging medical students and their teachers with the determinants of health: the approaches and impact of a curriculum development at one large UK medical school. MedEd Publish; 2014 [Accessed 7th January 2020].

[29] Wylie A, Leedham-Green K. Health promotion in medical education: lessons from a major undergraduate curriculum implementation. Educ Prim Care 2017:1−9.

[30] Frank J, Snell L, Sherbino J. The CanMEDS 2015 physician competency framework. Ottawa: Royal College of Physicians and Surgeons of Canada; 2015. http://canmeds.royalcollege.ca/uploads/en/framework/CanMEDS%202015%20Framework_EN_Reduced.pdf [Accessed 31 January 2020].

[31] Association of American Medical Colleges. Behavioral and social science foundations for future physicians. Washington, DC: Association of American Medical Colleges; 2011. https://www.aamc.org/download/271020/data/behavioralandsocialsciencefoundationsforfuturephysicians.pdf [Accessed 31 January 2020].

[32] Sharma M, Pinto AD, Kumagai AK. Teaching the social determinants of health: a path to equity or a road to nowhere? Acad Med 1 January 2018;93(1).

[33] Bauld L, Chesterman J, Judge K, Pound E, Coleman T. Impact of UK National Health Service smoking cessation services: variations in outcomes in England. Tobac Contr 2003;12(3):296–301.

[34] Phillips RL, Liaw W, Crampton P, Exeter DJ, Bazemore A, Vickery KD, Petterson S, Carrozza M. How other countries use deprivation indices—and why the United States desperately needs one. Health Aff 1 November 2016;35(11):1991–8.

[35] Parry G, Van Cleemput P, Peters J, Walters S, Thomas K, Cooper C. Health status of gypsies and travellers in England. J Epidemiol & Community Health 2007;61(3):198–204.

[36] ONS. 2011 Census analysis, what does the 2011 census tell us about the characteristics of gypsy and irish traveller in England and Wales, office for national statistics. Statistical Release; 2014.

[37] FFT. Friends. Families and aravellers annual report 2018–19. 2019.

How early childhood events impact upon adult health

2

Liz Forty

School of Medicine, Cardiff University, Cardiff, United Kingdom

Introduction

It is well established that physical and mental development during childhood is linked to health in adulthood [1]. When children have positive early experiences, their biological and psychological systems are strengthened, which has a positive impact on later health and well-being. Adversity plays an important role in normal childhood development. A degree of stress in the context of supportive relationships is beneficial for children and the development of a healthy stress response, for example, the first day at school. However, where stress is frequent, chronic and uncontrolled, it is harmful to the development of the child, particularly in the absence of supportive relationships and a safe environment.

Childhood adversity impacts on physiological, psychological and social development and can therefore have a significant impact on future physical and mental health. With childhood being a critical developmental period, it is also a critical period for risk prevention and intervention. Factors associated with conception and gestation also impact on health later in life; however, this chapter will focus on early childhood events (ECEs) that take place after birth.

Defining and measuring early childhood events

The term 'early childhood events' refers to events or conditions experienced by a child. ECEs, including both positive and negative events, may act as protective factors or risk factors in terms of child development. Most of the evidence around the impact of ECEs comes from studies focused on adverse childhood experiences (ACEs) or childhood trauma. There is great heterogeneity in definitions of ECEs, with substantial diversity in the types of exposure and the age ranges considered [2].

The term 'adverse childhood experiences' was originally developed in the United States (US), with the ACE's study. The ACE's study defines an ACE as exposure during childhood, before the age of 18 years, to:

- abuse: psychological, physical, sexual **and/or**
- household dysfunction: exposure to substance abuse, mental illness, violent treatment of mother and criminal behavior [3].

Since the original study, the list of ACEs has been expanded to include a broader set of experiences, for example, neglect, parental separation, exposure to violence outside the home, bullying and discrimination.

As the first 3—5 years of life are critically important for brain development, trauma researchers often focus on 'early' childhood trauma, usually referring to experiences occurring before the age of six years. Trauma researchers consider many types of trauma including, for example, those that are the result of intentional violence, such as physical or sexual abuse, or the result of natural disaster or accidents, such as the sudden loss of a parent or childhood injury.

Childhood trauma has been classified as type 1 or type 2 trauma [4]. Type 1 trauma refers to sudden, unexpected and isolated events, for example, a road traffic accident, rape, or terrorist attack. Type 2 trauma refers to traumatic events that are repeated and interpersonal, for example childhood physical or sexual abuse, or neglect, which is chronic and cumulative.

Table 2.1 provides examples of childhood experiences that are often described in studies of childhood trauma and adverse life events.

There is great variation across studies in how ECEs are measured. The majority of studies use self-report measures [5]. Some studies use retrospective recall, which may introduce recall bias whereby people with poorer health may retrospectively recall increased adversity, although the evidence suggests that false-positive reports of ACEs are rare [6]. Other studies collect data prospectively; however, this often relies on collecting data by proxy, for example via a parent or teacher, which introduces the potential for misclassification bias and underreporting. Both methods of study are valuable in contributing to understanding around the impact of ECEs on health outcomes.

Adverse experiences often cluster in children and young people's lives. ACEs research has tended to consider individual ACEs with equal weighting; however, it is important to note that not all events are 'equal', and ACEs have been shown to differ in terms of their impact [7—10]. It is therefore important to consider the type of childhood event, as well as the cumulative number of childhood

---

Table 2.1 Examples of adverse childhood experiences.

Neglect (physical, emotional)
Abuse (physical, emotional, sexual)
Bullying
Violence or coercion (domestic, gang related, victim of crime)
Life threatening accidents/injuries
Loss or separation from a parent or other loved one
Natural or manmade disasters
War or terrorist attacks
Food scarcity
Forced displacement or refugee status
Discrimination
Extreme poverty
Parental substance misuse or mental illness
Parental incarceration/imprisonment

events. Studies show that some ECEs having stronger associations with adult outcomes than others, for example childhood psychological abuse appears to have more of an impact on future mental health outcomes, compared with physical abuse [11]. Loss events appear to be associated with the development of mood disorders in adulthood, but not anxiety disorders [12]. Kessler et al. [12] also demonstrated that associations may attenuate with age, with childhood adversity being a stronger predictor of early onset mental disorders, in comparison with mental disorders with a later age at onset. In individuals exposed to a single adversity, the relative odds of onset of mental disorder were highest during the first few years following the adversity [12].

ACEs are relatively common, with between 45% and 65% of individuals experiencing at least one adverse childhood event during their lifetime [3,13–15]. Exposure to ACEs varies by gender [3,19–21] with some studies suggesting that women have higher rates of exposure to ACEs than men [3] and others finding higher rates of childhood adversity in men compared with women [19]. These differences in findings are likely to be due to variations in the specific ACEs considered and may also relate to cultural differences across countries. Felitti et al. [3] found that women were more likely to report sexual abuse or assault in childhood, whereas men were more likely to report physical abuse in childhood. Studies have consistently shown that exposure to childhood adversity varies according to race/ethnicity, with Black children having higher rates of specific adversities and total adversity counts, compared with White children. Black children are more likely to experience violence and discrimination for example, as well as parental incarceration [16–18].

Families living with socioeconomic hardship face challenges that increase parental stress and potentially increase the likelihood of adverse experiences. A key challenge for research focused on the effects of ECEs on health is how to separate the effects of ECEs and other contextual factors that are known to influence health. Some studies examining the impact of ACEs on health outcomes control for the confounding effects of social factors, such as education and employment, as these have been shown to be associated with both ACEs [22] and health outcomes [23]. Living in poverty may in itself be considered an adverse event in childhood [24].

Early childhood events, health behaviours and health outcomes

The ACE's study [3] was the first to clearly show an association between childhood adversity and numerous risk factors linked with adult mortality. Rates of health risk behaviours, physical illness and mental illness were associated with increased rates of exposure to ACEs. Experiencing more than one type of ACE was associated with multiple health risk factors in adulthood. The risk behaviours and illnesses included smoking, physical inactivity and increased numbers of sexual partners, ischaemic heart disease, cancer, lung disease, liver disease, depression, suicide, alcoholism and drug abuse.

These findings have been replicated extensively across the world [25–31].

Specifically, higher rates of respiratory problems [25], suicide [32], depressive disorders [33], cancer [34] and ischaemic heart disease [35] have been shown to be associated with childhood adversity. Green et al. [36] demonstrated that childhood adversities are associated, often subadditively, with mental illness throughout the life course. Early childhood adversity has also been linked to early mortality [21,37,38], a finding that is not solely related to health and social issues [38]. In addition to the impact of ACEs on the individual, health effects are also transmitted intergenerationally from parent to child [39].

Although the evidence clearly demonstrates the long-term impact of childhood trauma and adversity, research on ACEs and childhood trauma provides information on trends within the general population and therefore does not predict future health outcomes at the individual level. Adverse experiences affect people in different ways, and not all children exposed to ACEs will be negatively impacted as a result. For example, not all children who have a parent with a diagnosed mental illness will experience later adverse health outcomes. Exposure to some ACEs could have a protective effect on a child, for example parental separation in the context of domestic violence. Studies have also shown that for individuals who have experienced ACEs, the presence of positive childhood experiences can be protective and associated with positive health outcomes in adulthood [40,41]. This effect continues to be present even in those who have experienced three or more ACEs, although the effect is attenuated, suggesting that the protective effect of positive experiences is reduced in individuals who face a greater number of adversities in childhood [41].

How do early childhood events influence health?

The relationship between ECEs and adult health is complex with numerous pathways proposed involving the interaction of biological, psychological and social factors impacting on development. The impact of ECEs in each of these areas is considered below.

Biological development

Biological systems continue to mature throughout childhood and show profound changes during this developmental period. Studies show that ECEs can alter genetics and influence the development of several biological systems, including the immune, endocrine and nervous systems. These changes have been shown to persist into adulthood [42]. During this developmental period, the growth of a child's brain is influenced by genetic and environmental factors, including physical environment and social environment/relationships. Environmental factors, such as consistent and caring interactions with a parent, influence gene expression and therefore brain and physiological development.

It is well established that long-term activation of the stress response system has a significant physiological impact, particularly in early childhood. Studies have shown that exposure to ACEs can disrupt the body's ability to regulate its response to stress across the lifespan [43,44]. Chronic stress exposure is thought to lead to age-related disorders through its impact on allostatic systems including the brain, immune and endocrine systems. Long-term hypothalamic—pituitary—adrenal axis activation occurs in response to ongoing stress, for example long-term stress associated with ACEs, and influences glucocorticoid metabolism and immune function, resulting in increased susceptibility to infection and disease. The dysregulation of inflammatory response systems caused by long-term stress is thought to result in chronic 'wear and tear' on multiple organ systems [45].

Evidence indicates that exposure to high levels of stress, over prolonged periods of time, can influence the physical structure of the brain [42,46], for example, through effects on dendritic growth. Exposure to ACEs has been shown to affect areas of the prefrontal cortex, linked to executive function, and the limbic system, i.e. the hippocampus, amygdala, hypothalamus and thalamus, related to memory and emotion [47]. Underactivity in the prefrontal cortex results in issues with attention and learning. Overactivation of the amygdala leads to an overactive fear response, difficulties with feeling safe, calming down and sleep. Underactivity of the anterior cingulate cortex can result in difficulties with emotional regulation [42]. It is clear that childhood trauma and adversity can lead to long-term variations in brain structure and activity. Young people who have experienced trauma are likely to experience deficits across a range of cognitive functions [48].

Psychological development

ACEs may contribute indirectly to risk for poor health outcomes in adulthood through their impact on cognition, emotional regulation and social functioning that may increase the likelihood of an individual engaging with health risk behaviours. These intermediary factors play a key role in the relationship between childhood adversity and adult health outcomes [52]. Experiences of childhood trauma increase the likelihood that an individual will participate in 'risky' behaviours such as drinking, smoking, drug use or harmful eating behaviours (such as overeating, restricted eating), often as a way of coping and dealing with the emotional dysregulation that occurs as a result of trauma.

ECEs impact on attachment and attachment style with implications for health and well-being in adulthood [49]. Attachment refers to the emotional bond that is shared with a primary caregiver and influences an individual's ability to manage emotions and behaviour, relate to other people and manage social relationships. There is thought to be a biological basis to attachment, with infants having a predisposition to form a strong bond with a particular person because it aids their survival. A secure attachment in childhood supports the development of self-worth and social skills and is associated with self-regulation, self-esteem, better academic performance and fewer psychological and behavioural issues [50]. Insecure attachment, which can result from unmet needs or poor communication from a caregiver, is associated with externalising behavioural issues, for example aggression and lower levels of resilience, poorer executive function and poor emotional regulation.

The key developmental period for secure attachment is 0–5 years of age.

Individuals exposed to early childhood adversity, particularly trauma that involves or impacts on the primary caregiver, may develop an insecure attachment. Other relationships during childhood, for example, with a secondary caregiver such as a grandparent, or with a teacher in school, will also impact on attachment, positively and/or negatively.

Insecure attachment is associated with health risk behaviours such as aggression and alcohol use and substance use and has also been shown to impact on adult health outcomes. For example, a history of trauma in childhood has been associated with an anxious attachment style and increased risk for anxiety and depression [5,49]. An anxious attachment style has also been shown to be associated with chronic pain, stroke, heart attack and high blood pressure, independent of lifetime history of mental illness [51]. These findings suggest that risk of both mental and physical illness in adulthood may be increased in individuals with insecure attachment styles and experience of childhood trauma.

Social development

ACEs continue to be predictive of health outcomes even when demographic and socioeconomic characteristics have been taken into consideration. Social disadvantage also contributes uniquely to health outcomes [53]. This suggests that ACEs and social disadvantage need to be considered as separate but interacting factors in relation to adult health outcomes. The evidence also indicates that in those with a history of trauma, socioeconomic resources may have a protective effect in relation to adult health outcomes [54].

Poverty in childhood is associated with increased risk of maltreatment [55], poorer nutritional intake and poorer overall health. Children living in poverty are more likely to suffer from specific health problems such as anaemia, asthma and obesity [56–58], also influencing later health outcomes.

Traumatic events in childhood are linked to poorer academic performance, higher rates of school dropout and attendance issues and higher rates of behavioural problems [48], with lasting consequences in adulthood. Individuals with a history of violence and maltreatment in childhood have lower income levels in adulthood as a result of their lower educational and occupational performance [59], further perpetuating health inequalities.

ACEs can also lead to difficulties in interpersonal relationships in adulthood [60], and studies have suggested that at least some of the relationship between childhood trauma and adult health is mediated by social support and behavioural factors [11]. Socioemotional support has been shown to moderate the effects of ACEs on adult health [61,62].

There is robust evidence linking childhood trauma and future risk taking/health harming behaviour, for example alcohol use, drug use, smoking and unprotected sex [63–66]. Even when taking into account demographic characteristics, a dose response relationship remains between ACEs and health risk behaviours [67]. Clearly, a number of social and behavioural factors play a role in the relationship between ACEs and long-term health, providing key targets for preventative and interventional efforts to reduce the impact of childhood trauma and other adverse childhood experiences. The effects of ACEs can also be transmitted across generations, directly and indirectly, via complex interactions between the biological, psychological and social factors discussed above [68].

Prevention and intervention

Many factors interplay throughout childhood to influence development, with the personal encounters of the child with their environment and other people being of central importance. The research evidence clearly demonstrates that experiencing adversity in childhood can influence health outcomes in later life; however, it is also clear that there are factors that can support a child's resilience and that there are a number of ways in which these can be supported and developed.

Although the majority of individuals who experience childhood adversity will not go on to engage in health risk behaviours or develop adverse health outcomes [29], for some, experiences in early childhood can have negative impacts on health and behaviour that are lifelong. Raising awareness of the impact of childhood adversity is essential to address the potential health implications. There needs to be a focus on prevention, as well as interrupting the cycle of intergenerational transmission that perpetuates childhood adversity.

Preventing ACEs needs to be viewed within the broader context of addressing societal inequalities. Although childhood adversities are prevalent across the population, individuals living in areas of higher deprivation are more likely to be exposed to ACEs. ACEs can be seen as risk factors that can be addressed, and interventions aimed at reducing these risk factors in childhood could have widespread implications in relation to health promotion. Healthcare professionals have a role in preventing adversity where possible, and where it does occur, to intervene with the aim of compensating for the potentially harmful effects.

There are many evidence-based behavioural health interventions, for example, family therapy and parenting support, which can support children and families experiencing, or at risk of, adversities, as well as adults dealing with the effects of childhood adversity. Key areas of focus are considered in the following: (1) supporting parenting, (2) building relationships and resilience, (3) early identification of adversity and (4) responding to trauma [69].

(i) Supporting parenting

Positive parenting and parent–child relationships have a significant effect on child development and are key elements in facilitating children in overcoming adversity [61]. Interventions focused on parenting have been identified as key factors in the prevention of ACEs [70]. Supportive parenting can protect children from stress before it becomes harmful. The body's stress system can be influenced positively by the provision of safe, secure and nurturing relationships [71].

(ii) Building relationships and resilience

Although there is considerable variation in the adversity that children may experience, evidence suggests that outcomes associated with resilience may be related to specific individual, relational and school factors: Individual factors include emotion regulation and empathy; social factors include good relationships with caregivers and positive parenting approaches; school factors include a safe environment, positive relationships with teachers and student academic engagement, particularly in the context of poverty; community factors include community cohesion and cultural identity, including spiritual beliefs [72]. Factors such as these should be considered in public health interventions aimed at promoting resilience, preventing childhood adversity and reducing the impact of adversity where it does occur.

(iii) Early identification of adversity

Early identification of conditions that might contribute, or protect against, childhood adversity may enable early interventions, mitigating risk and potentially preventing disruption of the brain architecture during early development. There are a number of tools used to support early identification of trauma; however, routine enquiry about childhood adversity requires careful consideration due to the limited evidence around its feasibility, acceptability and impact [73]. Although the evidence for broad screening for ACEs is limited, healthcare professionals should continue to look for signs of childhood adversity in situations where there is potential for concern as this is a vital part of trauma-informed practice [74]. When screening for ACEs and childhood trauma, it is important to consider what exactly is being screened for, in which population/context, whether there are effective interventions available and whether there are any negative effects of screening that need to be considered [75].

(iv) Responding to trauma

Trauma-informed care involves understanding reactions and responses to trauma and adversity, including how trauma can affect patient presentation, engagement with services and healthcare outcomes. Trauma affects people differently according to individual characteristics, the nature of the trauma and the context and social environment. Trauma reactions are normal responses to abnormal circumstances. Trauma-informed care is based on trauma awareness, emphasises safety, provides opportunities to rebuild control and follows a strengths-based approach [76]. Recognition that provides care and treatment for an individual's current state of health, whether that be maladaptive behaviour, addiction, physical or mental illness, requires an understanding of how their current experiences are a consequence of their body's physiological and psychological response to stressors experienced earlier in life. Sweeney et al. [77] argue that trauma-informed care requires a cultural shift from 'What's wrong with you?' to 'What happened to you?' and to consider 'How has this affected your life?' and 'Who is there for you?'

Key to engaging patients who have been exposed to trauma is recognising that vigilance and suspicion are understandable, self-protective mechanisms that individuals may use to cope with trauma. Individuals exhibiting these characteristics, which are linked to experiences earlier in life, are less likely to engage with services and where they do, may face challenges building trust in healthcare services and providers. Stigma, associated with trauma, health risk behaviours and mental illness, may also influence an individual's willingness to seek support and engage with services. Improving understanding of ACEs and childhood trauma can help to reduce stigma and move the responsibility from individuals to community solutions.

Practical steps that can be taken to support trauma-informed care have been suggested including recognition of the impact of trauma on consultations; tailoring the length of consultations according to patient needs; providing trauma-focused therapy as well as physical healthcare and offering long-term, safe relationships with healthcare staff [78]. Service providers should focus on evidence-based practice, patients and providers should have access to trauma resources and continuity of care across services should be a priority. Healthcare practitioners should receive appropriate support and supervision and have a good understanding of their own emotional reactions and the principles of self-care, including connecting appropriately with others, for example via supervision and maintaining a balance across their professional and personal lives. For effective trauma-informed care, there must be recognition that healthcare practitioners themselves may have experienced trauma and adversity which may impact, positively and/or negatively, on their professional, as well as personal lives.

Conclusion

ECEs, both positive and negative, impact on child and adult health outcomes. As shown in Fig. 2.1, biological, psychological and social factors are influenced by ECEs and interact to impact on health

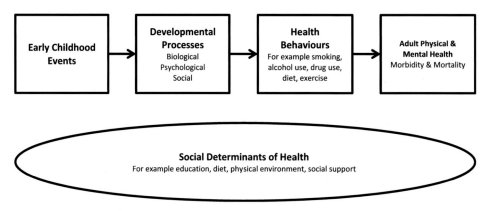

FIGURE 2.1

The relationship between early childhood events and adult physical and mental health.

behaviours and adult health outcomes. The social determinants of health are relevant through all stages of this process and have a significant impact on health outcomes.

The ACE's study has been quoted extensively in policy directed at addressing the impact of ACEs on health. However, it should be highlighted that the majority of individuals who are exposed to childhood adversity do not go on to experience poor health in adulthood as a result. The relevance of positive childhood events and factors associated with resilience must also be recognised. When interpreting the research evidence in relation to ACEs and adult health, consideration should be given to effect sizes, as well as the difference between absolute risk and relative risk.

Trauma-focused research has demonstrated that the ways in which an individual perceives what has happened to them, how they coped, and perceived their social environment and support are more correlated with future outcomes than the severity of the trauma itself [79]. In addition, research demonstrates that not all events are created equal. Experiencing one type of adverse event may not impact on an individual in the same way as another type of event. The extent of the impact also depends on the individual themselves and other contextual factors. It is also clear that adverse experiences do not tend to occur in isolation, rather they cluster in individuals. Attention should therefore focus on the importance of addressing these various factors, rather than limiting focus to any one type of event. Where resources are limited, these should be dedicated to those groups most likely to be exposed to multiple adversities in childhood.

Summary

1 Adverse childhood experiences are common, particularly in areas of social deprivation.
2 ECEs can impact on later adult health outcomes, both physical and mental health.
3 Not everyone who experiences childhood adversity will experience adverse health outcomes as a result.
4 A complex interaction of biological, psychological and social factors explain the relationship between ECEs and adult health outcomes.
5 Positive experiences and factors associated with resilience can mitigate the impact of early adversity on adult health outcomes.
6 Addressing the impact of childhood adversity requires interventions that focus on supporting parents, developing relationships and identifying adversity (or risk of adversity) at an early stage and care that is trauma informed.

References

[1] Delaney L, Smith JP. Childhood health: trends and consequences over the life-course. Future Child 2012; 22(1):43–63.
[2] Kelly-Irving M, Delpierre C. A critique of the adverse childhood experiences framework in epidemiology and public health: uses and misuses. Soc Pol Soc 2019;18(3):445–56.
[3] Felitti VJ, Anda RF, Nordenberg D, Williamson DF, Spitz AM, Edwards V, et al. Relationship of childhood abuse and household dysfunction to many of the leading causes of death in adults. The adverse childhood experiences (ACE) study. Am J Prev Med 1998;14(4):245–58.
[4] Terr L. Childhood traumas: an outline and overview. Am J Psychiatry 1991;148(1):10–20.
[5] Kalmakis KA, Chandler GE. Health consequences of adverse childhood experiences: a systematic review. J Am Assoc Nurse Pract 2015;27(8):457–65.
[6] Hardt J, Rutter M. Validity of adult retrospective reports of adverse childhood experiences: review of the evidence. J Child Psychol Psychiatry 2004;45:260–73.

[7] Fergusson DM, Boden JM, Horwood LJ. Exposure to childhood sexual and physical abuse and adjustment in early adulthood. Child Abuse Negl 2008;32(6):607–19.

[8] Schneider FD, Loveland Cook CA, Salas J, Scherrer J, Cleveland IN, Burge SK. Childhood trauma, social networks, and the mental health of adult survivors. Residency Research Network of Texas Investigators J Interpers Violence 2020;35(5-6):1492–514. https://doi.org/10.1177/0886260517696855.

[9] Lanier P, Maguire-Jack K, Lombardi B, Frey J, Rose RA. Adverse childhood experiences and child health outcomes: comparing cumulative risk and latent class approaches. Matern Child Health J 2018;22(3): 288–97.

[10] Cicchetti D, Rogosch FA, Gunnar MR, Toth SL. The differential impacts of early physical and sexual abuse and internalizing problems on daytime cortisol rhythm in school-aged children. Child Dev 2010;81(1): 252–69.

[11] Sheikh MA, Abelsen B, Olsen JA. Clarifying associations between childhood adversity, social support, behavioral factors, and mental health, health, and well-being in adulthood: a population-based study. Front Psychol 2016;7:727.

[12] Kessler RC, Davis CG, Kendler KS. Childhood adversity and adult psychiatric disorder in the US National Comorbidity Survey. Psychol Med 1997;27(5):1101–19.

[13] Rosenman SJ, Rodgers B. Childhood adversity in an Australian population. Soc Psychiatry Psychiatr Epidemiol 2004;39(9):695–702.

[14] Harbin Sacks V, Murphy D, Moore K. Adverse childhood experiences: national and state level prevalence. Child Trends 2014:2B.

[15] Merrick MT, Ports KA, Ford DC, Afifi TO, Gershoff ET, Grogan-Kaylord A. Unpacking the impact of adverse childhood experiences on adult mental health. Child Abuse Negl 2017;69:10–9.

[16] Maguire-Jack K, Lanier P, Lombardi B. Investigating racial differences in clusters of adverse childhood experiences. Am J Orthopsychiatry 2020;90(1):106–14.

[17] Strompolis M, Tucker W, Crouch E, Radcliff E. The intersectionality of adverse childhood experiences, race/ethnicity, and income: implications for policy. J Prev Interv Community 2019;47(4):310–24.

[18] Slopen N, Shonkoff JP, Albert MA, Yoshikawa H, Jacobs A, Stoltz R, Williams DR. Racial disparities in child adversity in the US. Interactions with family immigration history and income. Am J Prev Med 2016; 50(1):47–56.

[19] Almuneef M, ElChoueiry N, Saleheen HN, Al-Eissa M. Gender-based disparities in the impact of adverse childhood experiences on adult health: findings from a national study in the Kingdom of Saudi Arabia. Int J Equity Health 2017;16(90). https://doi.org/10.1186/s12939-017-0588-9.

[20] Leban L, Gibson CL. The role of gender in the relationship between adverse childhood experiences and delinquency and substance use in adolescence. J Crim Justice 2020;66. https://doi.org/10.1016/j.jcrimjus.2019.101637.

[21] Johnson J, Chaudieu I, Ritchie K, Scali J, Ancelin ML, Ryan J. The extent to which childhood adversity and recent stress influence all-cause mortality risk in older adults. Psychoneuroendocrinology 2020;111. https://doi.org/10.1016/j.psyneuen.2019.104492.

[22] Metzler M, Merrick MT, Klevens J, Ports KA, Ford DC. Adverse childhood experiences and life opportunities: shifting the narrative. Child Youth Serv Rev 2017;72:141–9.

[23] Daniel H, Bornstein SS, Kane GC. Addressing social determinants to improve patient care and promote health equity: an American college of physicians position paper. Ann Intern Med 2018 17;168(8):577–8.

[24] Hughes M, Tucker W. Poverty as an adverse childhood experience. N C Med J 2018;79(2):124–6.

[25] McKelvey LM, Saccente JE, Swindle TM. Adverse childhood experiences in infancy and toddlerhood predict obesity and health outcomes in middle childhood. Child Obes 2019;15(3):206–15.

[26] Merrick MT, Ford DC, Ports KA, Guinn AS. Prevalence of adverse childhood experiences from the 2011–2014 behavioral risk factor surveillance system in 23 states. JAMA Pediatrics 2018;172(11): 1038–44.

[27] Thompson R, Flaherty EG, English DJ, Litrownik AJ, Dubowitz H, Kotch JB, Runyan DK. Trajectories of adverse childhood experiences and self reported health at age 18. Acad Pediatr 2015;15(5):503–9.

[28] Copeland WE, Shanahan L, Hinesley J, Chan RF, Aberg KA, Fairbank JA, et al. Association of childhood trauma exposure with adult psychiatric disorders and functional outcomes. JAMA Netw Open 2018;1(7).

[29] Bellis MA, Hughes K, Leckenby N, Perkins C, Lowey H. National household survey of adverse childhood experiences and their relationship with resilience to health-harming behaviors in England. BMC Med 2014; 12:72. https://doi.org/10.1186/1741-7015-12-72.

[30] Ford K, Butler N, Hughes K, Quigg Z, Bellis MA. Adverse childhood experiences (ACEs) in Hertfordshire, Luton and Northamptonshire. Liverpool: Centre for Public Health; 2016.

[31] Hughes K, Bellis MA, Hardcastle KA, Sethi D, Bitchart A, Mikton C, et al. The effect of multiple adverse childhood experiences on health: a systematic review and meta-analysis. Lancet Public Health 2017;2: e356–66. https://doi.org/10.1016/S2468-2667(17)30118-4.

[32] Dube SR, Felitti VJ, Dong M, Giles WH, Anda RF. The impact of adverse childhood experiences on health problems: evidence from four birth cohorts dating back to 1900. Prev Med 2003;37(3):268–77.

[33] Chapman DP, Whitfield CL, Felitti VJ, Dube SR, Edwards VJ, Anda RF. Adverse childhood experiences and the risk of depressive disorders in adulthood. J Affect Disord October 15, 2004;82(2):217–25.

[34] Holman DM, Ports KA, Buchanan ND, Hawkins NA, Merrick MT, Metzler M, Trivers KF. The association between adverse childhood experiences and risk of cancer in adulthood: a systematic review of the literature. Pediatrics 2016;138(Suppl.1):S81–91.

[35] Dong M, Giles WH, Felitti VJ, Dube SR, Williams JE, Chapman DP, Anda RF. Insights into causal pathways for ischemic heart disease: adverse childhood experiences study. Circulation 2004;110(13):1761–6.

[36] Green JG, McLaughlin KA, Berglund PA, Gruber MJ, Sampson NA, Zaslavsky AM, Kessler RC. Childhood adversities and adult psychiatric disorders in the national comorbidity survey replication I: associations with first onset of DSM-IV disorders. Arch Gen Psychiatr 2010;67(2):113–23.

[37] Kelly-Irving M, Lepage B, Dedieu D, Bartley M, Blane D, Grosclaude P, et al. Adverse childhood experiences and premature all-cause mortality. Eur J Epidemiol 2013;28:721–34.

[38] Brown DW, Anda RF, Tiemeier H, Felitti VJ, Edwards VJ, Croft JB, Giles WH. Adverse childhood experiences and the risk of premature mortality. Am J Prev Med 2009;37(5):389–96.

[39] Lê-Scherban F, Wang X, Boyle-Steed KH, et al. Intergenerational associations of parent adverse childhood experiences and child health outcomes. Pediatrics 2018;141(6).

[40] Bethell C, Jones J, Gomboiav N, Linkenback J, Sege R. Positive childhood experiences and adult mental and relational health in a statewide sample: associations across adverse childhood experiences levels. JAMA Pediatr 2019;173(11). https://doi.org/10.1001/jamapediatrics.2019.3007.

[41] Crandall A, Miller JR, Cheung A, Novilla LK, Glade R, Novilla MLB, et al. ACEs and counter-ACEs: how positive and negative childhood experiences influence adult health. Child Abuse Negl 2019;96. https://doi.org/10.1016/j.chiabu.2019.104089.

[42] Danese A, McEwen BS. Adverse childhood experiences, allostasis, allostatic load, and age-related disease. Physiol Behav April 12, 2012;106(1):29–39.

[43] McEwen BS. Physiology and neurobiology of stress and adaptation: central role of the brain. Physiol Rev 2007;87:873–904.

[44] Lupien SJ, McEwen BS, Gunnar MR, Heim C. Effects of stress throughout the lifespan on the brain, behaviour and cognition. Nat Rev Neurosci 2009;10:434–45.

[45] Shonkoff JP, Garner AS. The lifelong effects of early childhood adversity and toxic stress. Pediatrics 2012; 129(1):e232–46. https://doi.org/10.1542/peds.2011-2663.

[46] McEwen BS, Gianaros PJ. Stress and allostasis induced brain plasticity. Annu Rev Med 2011;62:431—45.

[47] Luby JL, Tillman R, Barch DM. Association of timing of adverse childhood experiences and caregiver support with regionally specific brain development in adolescents. JAMA Netw Open 2019;2(9).

[48] Perfect M, Turley M, Carlson J, Yohanna J, Saint Gilles M. School-related outcomes of traumatic event exposure and traumatic stress symptoms in students: a systematic review of research from 1990 to 2015. Sch Ment Health 2016;8(1):7—43.

[49] Erozkan A. The link between types of attachment and childhood trauma. Univers J Educ Res 2016;4(5):1071—9.

[50] Mikulincer M, Shaver PR. Attachment patterns in adulthood structure, dynamics and change. New York: Guilford Press; 2007.

[51] McWilliams LA, Bailey SJ. Associations between adult attachment ratings and health conditions: evidence from the national comorbidity survey replication. Health Psychol 2010;29(4):446—53.

[52] Korotana LM, Dobson KS, Pusch D, Josephson T. A review of primary care interventions to improve health outcomes in adult survivors of adverse childhood experiences. Clin Psychol Rev June 2016;46:59—90.

[53] Nurius PS, Logan-Greene P, Green S. ACEs within a social disadvantage framework: distinguishing unique, cumulative, and moderated contributions to adult mental health. J Prev Interv Community 2012;40(4):278—90.

[54] Mock SE, Arai SM. Childhood trauma and chronic illness in adulthood: mental health and socioeconomic status as explanatory factors and buffers. Front Psychol 2010;1:246.

[55] Pelton LH. The continuing role of material factors in child maltreatment and placement. Child Abuse Negl 2015;41:30—9.

[56] Cook JT, Frank DA, Levenson SM, Neault NB, Heeren TC, Black MM, et al. Child food insecurity increases risks posed by household food insecurity to young children's health. J Nutr 2006;136(4):1073—6.

[57] Kirkpatrick SI, McIntyre L, Potestio ML. Child hunger and long-term adverse consequences for health. Arch Pediatr Adolesc Med 2010;164(8):754—62.

[58] Power C, Pinto Pereira SM, Li L. Childhood maltreatment and BMI trajectories to mid-adult life: follow-up to age 50y in a British birth cohort. PloS One 2015;10(3). https://doi.org/10.1371/journal.pone.0119985.

[59] Hyman B. The economic consequences of child sexual abuse for adult lesbian women. J Marriage Fam 2000;62:199—211.

[60] Luntz BK, Widom CS. Antisocial personality disorder in abused and neglected children grown up. Am J Psychiatry 1994;151(5):670—4.

[61] Bellis MA, Hardcastle K, Ford K, Hughes K, Ashton K, Quigg Z, Butler N. Does continuous trusted adult support in childhood impart life-course resilience against adverse childhood experiences — a retrospective study on adult health-harming behaviours and mental well-being. BMC Psychiatry 2017;17(1):110.

[62] Brockie TN, Elm JHL, Walls ML. Examining protective and buffering associations between sociocultural factors and adverse childhood experiences among American Indian adults with type 2 diabetes: a quantitative, community-based participatory research approach. BMJ Open 2018;8. https://doi.org/10.1136/bmjopen-2018-022265.

[63] Miller BA, Downs WR, Testa M. Interrelationships between victimization experiences and women's alcohol use. J Stud Alcohol 1993;11(Suppl. l):109—17.

[64] Widom CS, Ireland T, Glynn PJ. Alcohol abuse in abuse and neglected children followed-up: are they at increased risk? J Stud Alcohol 1995;56(2):207—17.

[65] Taha F, Galea S, Hien D, Goodwin RD. Childhood maltreatment and the persistence of smoking: a longitudinal study among adults in the US. Child Abuse Negl 2014;38(12):1995—2006.

[66] Wiehn J, Hornberg C, Fischer F. How adverse childhood experiences relate to single and multiple health risk behaviours in German public university students: a cross-sectional analysis. BMC Public Health 2018;18:1005.

[67] Garrido EF, Weiler LM, Taussig HN. Adverse childhood experiences and health-risk behaviors in vulnerable early adolescents. J Early Adolesc 2018;38(5):661—80.

[68] Lomanowska AM, Boivin M, Hertzman C, Fleming AS. Parenting begets parenting: a neurobiological perspective on early adversity and the transmission of parenting styles across generations. Neuroscience February 7, 2017;342:120—39.

[69] Di Lemma LCG, Davies AR, Ford K, Hughes K, Homolova L, Gray B, Richardson G. Responding to Adverse Childhood Experiences: an evidence review of interventions to prevent and address adversity across the life course. Wrexham: Public Health Wales, Cardiff and Bangor University; 2019, ISBN 978-1-78986-035-1.

[70] Lundahl BW, Nimer J, Parsons B. Preventing child abuse: a meta-analysis of parent training programs. Res Soc Work Pract 2006;16(3):251—62.

[71] Collishaw S, Pickles A, Messera J, Rutter M, Shearer C, Maughana B. Resilience to adult psychopathology following childhood maltreatment: evidence from a community sample. Child Abuse Negl 2007;31(3): 211—29.

[72] Gartland D, Riggs R, Muyeen S, Giallo R, O Afifi T, MacMillan H, et al. What factors are associated with resilient outcomes in children exposed to social adversity? A systematic review. BMJ Open 2019;9. https://doi.org/10.1136/bmjopen-2018-024870.

[73] Ford K, Hughes K, Hardcastle K, Di Lemma LCG, Davies A, Edwards S, Bellisa MA. The evidence base for routine enquiry into adverse childhood experiences: a scoping review. Child Abuse Negl 2019;91:131—46.

[74] Ko SJ, Ford JD, Kassam-Adams N, Berkowitz SJ, Wilson C, Wong M, et al. Creating trauma-informed systems: child welfare, education, first responders, health care, juvenile justice. Prof Psychol Res Pract 2008;39(4):396—404.

[75] Finkelhor D. Screening for adverse childhood experiences (ACEs): cautions and suggestions. Child Abuse Negl 2018;85:174—9.

[76] Hopper EK, Bassuk EL, Olive J. Shelter from the storm: trauma-informed care in homelessness services settings. Open Health Serv Pol J 2010;3:80—100.

[77] Sweeney A, Filson B, Kennedym A, Collinson L, Gillard S. A paradigm shift: relationships in trauma-informed mental health services. BJPsych Adv 2018;24(5):319—33.

[78] Brooks M, Barclay L, Hooker C. Trauma-informed care in general practice: findings from a women's health centre evaluation. Aust J Gen Pract 2018;47(6):370—5.

[79] Hyman SM, Gold SN, Cott MA. Forms of social support that moderate PTSD in childhood sexual abuse survivors. J Fam Violence 2003;18:295—300.

The psychology of health-related behaviour change

3

Aria Campbell-Danesh

Associate Fellow, Division of Clinical Psychology, British Psychological Society, London, United Kingdom

Introduction

Over the past century, global patterns of disease have changed significantly. The incidence and prevalence of infectious 'communicable' diseases has fallen, while the rates of 'noncommunicable' chronic diseases (NCDs) have risen dramatically [1]. The major NCDs throughout the developed world include cardiovascular disease, chronic respiratory conditions, cancer and type 2 diabetes.

The management of chronic diseases is one of the most pressing health challenges of the 21st century. NCDs kill 41 million people each year, equating to 71% of all global deaths [2]. Lifestyle factors such as poor diet, physical inactivity, smoking and alcohol use are all major risk factors for the development of NCDs [3].

Chronic diseases place substantial strain on health services: In developed countries, approximately 60%—70% of primary healthcare visits are related to lifestyle-based illnesses [4]. It is vital that measures addressing lifestyle factors are implemented to reduce premature mortality and morbidity and to lessen the current levels of demand on health and social services [5]. Prospective studies suggest that a high-quality diet, being physically active, maintaining a healthy weight and nonsmoking is associated with a reduced risk of mortality from chronic disease by 55%—60% [6,7]. Health-related behaviour change is a worldwide public health priority.

There is mounting evidence for the effectiveness of behaviour change practices in improving health [8]. Interventions have targeted lifestyle factors such as diet [9—12], physical activity [13—16], smoking [17—19] and alcohol consumption [20—23]. Encouragingly, small changes in health-related behaviours can result in significant health gains [24,25]. Changes in dietary intake and physical activity, resulting in a small amount of weight loss, can have a significant impact on cardiovascular risk factors. For each kilogram of weight an overweight or obese individual loses, there is a 16% reduction in the risk of developing diabetes [26] and a reduction in systolic blood pressure by 1.05 mm Hg [27]. However, research demonstrating long-term behaviour change is limited [28,29] and high relapse rates have been found following attempts to lose weight [30], reduce alcohol consumption [31] and stop smoking [32,33].

To increase the likelihood of positively influencing patients' lifestyle choices, it is important to have a sound theoretical understanding of the psychology underlying sustained behaviour change [34]. Applying behaviour change theories has been identified as a key intervention in improving healthcare and is an integral component of optimising patient motivation, supporting behaviour change and enhancing health-related outcomes on an individual level [35—37].

The most commonly used theories in health behaviour research in the past two decades include the transtheoretical model, the theory of planned behaviour and social cognitive theory [38,39]. However, there are many behaviour change theories, many of which overlap in terms of their core elements [40]. The task of a medical practitioner in knowing which theory to select and apply can therefore be difficult [41]. Furthermore, research has shown that primary care health practitioners avoid providing lifestyle advice due to self-doubts regarding their professional efficacy in this domain [42].

It is important that healthcare providers understand a simple set of evidence-based psychological constructs related to behaviour change and its maintenance. In light of this, experts have identified the key cognitive, behavioural and external factors involved in changing behaviours and maintaining behavioural changes over time and in different contexts [43].

This chapter describes the key elements of the psychology of behaviour change. It provides a simple evidence-based frame of reference to aid effective communication with patients about health-related behaviour change. The example of weight loss will be used throughout for simplicity and clarity. Advice shall be given regarding the practical application of psychological theories.

The psychology of behaviour change

There are three principal domains involved in the psychology of behaviour change, summarised in Table 3.1:

(1) Self-regulation
(2) Motivation
(3) Habits

Self-regulation

Self-regulation refers to the processes by which an individual creates, moves toward and achieves their goals. Long-term behaviour change is considered to result from continual self-regulation. Behaviour change approaches are frequently based on strategies that foster an individual's capacity to self-regulate their behaviour, for instance in the case of weight loss by focussing on changes to dietary intake and physical activity [44].

Self-regulation involves the processes of self-monitoring, self-evaluation and self-reinforcement [45]. These processes can be conscious and deliberate, as well as subconscious and automatic. Central features of self-regulation include setting goals, developing ways to attain those goals, assessing progress and revising goals and strategies accordingly [46]. Self-regulation can therefore be separated into two broad categories: goal *setting* and goal *striving* [47,48]. Goal setting involves creating a desirable goal and determining the criteria against which success is measured. Goal striving addresses the planning and execution of behaviours that promote the successful achievement of goals.

Table 3.1 Three main domains related to maintained behaviour change [43].		
Domain	**Description**	**Characteristics**
(1) Self-regulation	The processes involved in formulating, progressing toward and attaining goals. Successful behaviour change is the result of monitoring the new behaviour, assessing progress and revising goals and strategies accordingly.	Goal setting: - Outcome and process - Approach and avoidance - Challenging and easy Goal Striving: - Action planning - Coping planning
(2) Motivation	Behaviours are more likely to be maintained if they are inherently interesting, enjoyable and satisfying, chosen autonomously and aligned with an individual's goals, values and identity.	Intrinsic motivation Behavioural enjoyment Outcome satisfaction
(3) Habits	Behaviour change maintenance is optimised when behaviours become habits and are automatically triggered by specific cues.	Initiation Learning Stability

Goal setting
Goal definition

A goal is a cognitive representation of a desired outcome that an individual is trying to achieve [49,50]. Goals can drive thoughts, emotions and action in four ways [51] by:

(1) directing attention toward goal-relevant behaviours;
(2) raising the intensity of effort applied;
(3) increasing persistence; and by
(4) leading to the acquisition and use of goal-relevant knowledge and strategies.

Health-related goals can be subdivided into several categories based on goal characteristics [52].

Outcome and process goals

In sport and exercise settings, a useful distinction has been made between *outcome* and *process* goals. Outcome goals refer to a desired end result, such as winning a tennis competition. Process goals are centred on specific behaviours in which the individual will engage, for instance executing the optimal posture and racquet position when making a drop shot. Goals can also be conceptualised as *performance*, i.e. achieving a certain standard, or *mastery*-related, i.e. developing a skill.

With regards to weight loss, outcome or performance goals could be to lose 10 pounds, achieve a certain dress size or fit into an item of clothing. On the other hand, process or mastery goals focus on developing the skills and implementing the actions that move people closer to their desired outcome. Examples include walking for 30 min, keeping a food diary or learning how to cook healthy meals at home. In brief, outcome goals relate to the destination; process goals relate to the journey. Research suggests that both types of goals are effective in improving performance [53]. However, there is evidence that process goals lead to significantly higher levels of self-efficacy, cognitive anxiety control and concentration [54]. By focussing on learning or developing a skill, setbacks are likely to be evaluated more positively and utilised as feedback to improve, problem-solve and still engage with the goal [55].

Approach and avoidance goals

Approach goals involve moving toward a desired outcome, whereas avoidance goals are rooted in moving away from an undesired outcome [56]. For instance, someone attempting to lose weight may have the process goal of 'running twice a week' or 'avoiding sugary foods'. The former centres on implementing a concrete action, whereas the latter is based on avoiding a specific behaviour. Both relate to the outcome goal of losing weight, yet studies indicate that the way that goals are framed has psychological consequences [57,58]. Individuals who use approach rather than avoidance goals tend to experience more positive emotions. In contrast, those with more avoidance goals evaluate themselves more negatively on measures of self-esteem, optimism and depression. Such individuals also experience less satisfaction with their progress and are more likely to report that pursuing their goals decreased their sense of personal control and vitality.

Challenging and easy goals

Another crucial goal characteristic relates to level of difficulty. Clinicians may be aware of the SMART framework [59], which advocates for goals to be specific, measurable, attainable, realistic and time-sensitive. It is advisable that process goals are feasible and appropriately challenging for the individual.

There is growing evidence, however, that 'unrealistic' *outcome* goals may lead to better results for some people. A robust finding in organisational behavioural science research is that setting more challenging, specific goals yields the highest performance and effort, compared with easy goals, the instruction to 'try your best' or no goals. Performance levels flatten or decrease when ability levels are reached or commitment to the goal drops [51]. Goal commitment is impacted by the importance of the goal and self-efficacy — the belief that the goal can be accomplished. This shall be discussed later in this chapter.

In the weight loss field, traditional practice is to counsel people to let go of 'dream' weight losses and adopt 'realistic' goals. Weight loss outcome goals are typically defined as the amount of weight that an individual desires to lose. Setting realistic weight loss goals has been identified as one of seven obesity-related myths that persist despite contradictory research [60]. There is no consistent evidence that clinicians should advise patients to set realistic outcome goals. The general concern of the 'false hope syndrome' [61] is that individuals will be unable to attain highly ambitious targets, with eventual failure leading to distress and relapse. However, emerging evidence suggests that having high goals is not linked to psychological distress and failing to meet unrealistic weight loss expectations does not lead to weight regain [62,63]. A meta-analysis found that setting realistic goals does not lead to higher weight loss [64]. Theorists have proposed that the decision criteria that initiate behaviour change may differ from those that continue behaviour change: Individuals may be more motivated by favourable expectations but then maintain changes because of their satisfaction with the achieved progress [65]. A recent study investigating weight loss targets in more than 35,000 adults in a community sample [66] found that weight loss was greater for those who set targets; obese adults with a higher first weight loss target achieved better weight loss outcomes.

Goal striving
Goal striving relates to the planning and execution of behaviours that increase the likelihood of goal attainment.

Action planning
Action plans are a useful self-regulation tool to promote goal-directed actions and are concerned with the 'what', 'where', 'how often' and 'when' of target behaviours. For instance, an action plan may revolve around the goal: 'I will walk for 20 minutes by the river after lunch on a workday'. Researchers [67] recommend that

(1) patients determine their own action plans to heighten ownership;
(2) action plans are short-term, for example 1 week;
(3) patients rate their confidence in their ability to carry out the plan; and
(4) patients self-monitor the plan.

Self-monitoring is often considered to be the cornerstone of behaviour change [68]. Goal setting and self-monitoring are interlinked constructs: goals are set, specific behavioural plans to attain those goals are created, progress towards the goals is monitored and action plans can be modified in response to self-monitoring records.

Action plans can be created collaboratively relatively quickly and often within the time constraints of primary care visits. A study [69] in primary care found that 83% of patients made an action plan with their clinician. Goal setting discussions lasted 6.9 min on average, and 3 out of 4 clinicians rated the collaboration as equally or more satisfying than previous behaviour change attempts [70]. Furthermore, at least 53% of patients who made a plan reported implementing the behaviour change in the following 3 weeks.

Coping planning
While action planning is important for the initiation of behaviour change, coping planning is critical to the maintenance of behaviour change. Self-regulation involves coping with barriers, temptations, distractions and situations with a high risk of lapse. i.e. a single event that is incongruent with the desired goal, and relapse, i.e. a sequence of lapses. Coping planning involves prospectively identifying risks to goal-directed behaviour and managing those situations accordingly to protect against disruption of goal attainment. A systematic review [71] found that combining action plans with coping plans is more effective than using action plans alone.

Coping planning involves making contingency plans to provide solutions to anticipated challenges. An evidence-based tool is the use of *implementation intentions (IMPs)*, which are defined as conditional plans with the structure: 'If situation *x* arises, then I will do *y*, a goal-directed behaviour' [72]. For instance, when trying to reduce calorie consumption, a parent may be tempted to eat the leftover food of their children, resulting in unwanted caloric intake. A specific if-then plan could be 'if there is leftover food on my child's plate, then I will immediately throw it into the compost bin'.

While IMPs are often simple, they have been shown to be effective in adopting a low-fat diet, increasing physical activity levels, and reducing alcohol binging: a meta-analysis [73] showed that they had a positive impact on goal attainment, with a medium-to-large effect size (d = 0.65).

Motivation

Motives are central to the initiation and maintenance of behaviour. An individual's motivation for initiating behaviour change is often rooted in rational and long-term outcomes, for instance losing weight for aesthetic and physical health reasons [43]. For healthcare practitioners to support long-term behaviour change, an understanding of maintenance motives is crucial.

Intrinsic and extrinsic motivation

Intrinsic motivation relates to engaging in tasks that individuals find inherently interesting. Intrinsically motivating behaviours differ from those that are extrinsically motivated in that they do not rely on external variables, such as rewards/punishment or approval/disapproval from others. Intrinsic motivation is associated with improved learning, performance and persistence [74]. There is compelling evidence that intrinsic, self-driven interests are significantly more likely to lead to goal progress compared with external pressures [75,76].

Intrinsic motivation is hypothesised to be more stable and sustainable than extrinsic motivation [77]; self-sustained motivation is autonomous, i.e. it is personally endorsed and involves a sense of choice, as opposed to being linked to feeling a need to comply with external demands. Behaviour change is more likely to be lasting if the action is perceived to be personally meaningful, and it is aligned with an individual's values and sense of self. Behaviour change and identity is bidirectional: Individuals tend to self-regulate their behaviour in line with their beliefs about themselves [78,79], and behavioural changes can also lead to changes in self-concept [80].

Behavioural enjoyment and outcome satisfaction

Maintenance of behaviour change is more likely when reinforced by immediate, positive emotional and physiological outcomes as opposed to long-term, cognitive outcomes [43]. Positive motives are recommended, as the motivation to prevent negative health consequences is believed to be inadequate in sustaining behaviour that necessitates ongoing effort [81,82]. Individuals are more likely to continue engaging in behaviours that provide a sense of enjoyment, either from their execution or the immediate outcomes [83,84].

Habits

Initiating a goal-directed behaviour for the first time requires self-regulation, including planning and attention [65,85]. Coping with obstacles, for instance by inhibiting goal-disruptive thoughts, feelings and behaviours, requires cognitive capacity and motivational resources [86,87]. Dual process model theorists [86] propose that behaviour is determined by two interacting systems: a reflective and impulsive system. The reflective system is based on conscious thought and knowledge; the impulsive system influences behaviour more quickly through associative links and habits.

Research [88] suggests that psychological resources are limited; tasks that require conscious cognitive effort deplete an individual's self-regulatory capacity and lead to impaired self-control, lower effort, increased perceived difficulty, negative affect and fatigue. When cognitive resources are diminished, processes outside of conscious awareness predominantly influence behaviour [86]. Relying solely on conscious effort to protect goal-directed behaviour against derailment is a risky strategy. However, the need for self-regulation decreases as an individual repeatedly enacts a new behaviour and develops a habit [89]. Habitual behaviours reduce dependence on awareness, intentionality and motivational processes and thus are more likely to be engaged even when conscious resources are reduced [90]. Forming healthy habits therefore is a crucial component of long-term behaviour change.

Habits are actions that are automatically triggered by specific cues [89]. Habits emerge by repeating actions in stable contexts, thereby strengthening the association between a cue and an action. Habitual actions are accompanied by a psychological experience of automaticity [91]. Reinforcement is hypothesised to play a key role in habit formation [92,93]. However, whether external rewards are necessary is unclear. Research into the formation of habits in real-world settings [94] shows that intrinsically rewarding behaviours, in the absence of external rewards, can become habits. The length of time it takes a behaviour to reach a maximal level of automaticity on average is 66 days, with a range from 18 to 254 days. Simpler actions become second nature more quickly than complex ones. Habit formation can be divided into three phases [95]:

(1) Initiation phase: choosing the new action and the context in which it will be carried out.
(2) Learning phase: the context–behaviour association is reinforced by consistently repeating the action in the same context, resulting in automaticity.
(3) Stability phase: the habit is formed with the highest level of automaticity and the habitual behaviour continues with minimal effort.

Supporting patients in behaviour change: a practical application of psychological theory

A desirable outcome of many healthcare consultations is for a patient to be motivated and willing to engage in lifestyle modifications to improve their health and well-being. Healthcare professionals can support and encourage this process by adopting the following strategies:

Goal setting

Health professionals can support patients to form desirable goals and to develop criteria by which to determine success. Typically, goal setting precedes goal striving. However, during the pursuit of a goal, a patient may reevaluate and alter the goal or criteria for judging success, resulting in a cyclical process.

Clinicians can assist patients to form an overarching, long-term outcome goal, for instance to lose 20 pounds. Ambitious outcome goals do not have to be discouraged, as a high level of aspiration can be motivating.

It is advisable that outcome goals are supplemented with one or more immediate, actionable process goals. Failure to achieve a goal can be framed as feedback to reevaluate current strategies and problem-solve accordingly. This will increase the likelihood that the patient will still engage in goal pursuit when facing challenges, due to higher self-efficacy, and lower the risk that the goal is abandoned as a result of a sense of personal failure or incompetence. Process goals should be attainable, but challenging: the appropriate level of difficulty will depend on the individual's ability, goal commitment and self-efficacy.

Exploring the personally meaningful reasons behind setting a goal can enhance commitment. Commitment can also be boosted if patients make a public commitment to goals, for example sharing goals with family and friends, since people wish to be viewed as rational and consistent by others [96]. Self-efficacy can be raised by [97,98]:

(1) providing training in the necessary skills required for goal attainment;
(2) discussing or introducing the patient to role models or reference groups with whom she/he identifies that carry out the target behaviour; and
(3) positively communicating confidence that the patient is able to achieve the goal.

Assisting patients to create approach goals may be more beneficial than setting avoidance goals. One strategy is to convert avoidance goals into approach goals through a substitution; for instance 'eating fewer chocolate bars' can be reformulated into 'eating a yoghurt as a snack instead of a chocolate bar'.

The SMART framework can be applied to process, approach goals in the following way: 'I will eat a piece of fruit in the afternoon at work, instead of buying a packet of crisps from the vending machine, on Monday to Friday, for the next two weeks'.

Goal striving

Healthcare professionals can aid patients to develop action plans. As part of a collaborative process with their doctor, patients should specify **what** behaviour they wish to implement, for example running, **how** the behaviour will be performed, for example where to run and the duration/distance, **how often**, for example the number of times per week, and **when**, i.e. days and times.

Visualising performance of the target behaviour can assist people to decide on the how, when and where of goal-promoting actions, as well as anticipate potential challenges and obstacles. For instance, patients planning on taking part in an evening gym class may predict that they will be tired and tempted to sit down in front of the television after returning home. By anticipating this, alternative plans can be made: The patient might take athletic wear into work and go straight to the gym afterwards or attend a morning class instead.

Another potential obstacle is goal conflict. Health goals may compete with others in terms of time, psychological resources and money. When goal conflict occurs, individuals tend to prioritise those goals that are more important to them. The differing socioeconomic status (SES) of patients in clinical settings should be kept in mind: Goal conflict can arise when individuals are unable to afford certain health-related activities or if they are more concerned with their financial situation than their health. Research shows that individuals with higher SES are more likely to implement proactive coping strategies to offset ageing-related health problems in the future than those with lower SES [99]. Openly discussing the relative prominence of goals with patients can help them to be more aware of the level of priority of a health-related goal. Finding ways to consciously align goals can also improve outcomes [100]. For instance, clinicians could highlight research showing that exercise improves cognitive functioning and mental health and discuss with the patient ways in which prioritising exercise might benefit career prospects or save money on healthcare in the future.

Clinicians can ask patients to rate their confidence in their ability to carry out the plan, for example 'on a scale of 0–10, with a higher score representing higher confidence, how sure are you that you can elicit the behaviour?' If the confidence rating is below 7, the clinician should explore the challenges with the patient in more detail [67]. Asking what would need to happen for the individual to improve their confidence score to reach 7 might provide personal insights and prompt reevaluation and revision of the action plan. The healthcare practitioner can offer suggestions for the patient to consider. The patient should monitor execution of the plan and record observations or insights to make any necessary changes to the following week's plan. Continual feedback therefore leads to problem-solving, strategy refinement and tailoring of the plan to the individual. Fig. 3.1 illustrates an example of a health-related action plan.

Coping planning is a vital self-regulation strategy to shield goal-directed behaviour from derailment. Clinicians can facilitate long-term behaviour change by helping patients to create specific IMPs. For instance, in the context of performing high-intensity interval training (HIIT) on a Monday evening, an IMP could be 'if I miss the gym HIIT class because my work meeting overruns, then I will do a 30-minute YouTube HIIT workout at home'. The clinician can assist by guiding the patient to imagine future events and mentally rehearse the performance of the desired behaviour to self-generate potential challenges, by offering suggestions and by presenting lists of established barriers.

Motivation

Clinicians can facilitate lasting behaviour change by exploring the patient's own personally meaningful reasons to change. This may include asking about the personal relevance of a goal, why the person would persist and what the personal consequences of attaining the goal and carrying out goal-directed behaviours mean for the individual.

ACTION PLAN	
Date to commence	10th February
Action	Running for 20 minutes
Where	At the park next to the house
How often	3 times a week
When: days and times	Monday 7am; Thursday 6.30pm; Saturday 9am
Confidence in completing action	8/10
Observations or insights (including challenges or obstacles)	Too dark to run outdoors on Thursday night. Run at the gym instead.

FIGURE 3.1

An example of a health-related action plan.

Information can be provided about behaviours, such as exercise adherence or healthy eating, but care should be taken so that the patient does not feel pressure to comply with the clinician's expectations. Language such as 'you can', 'you might' and 'if you choose' is more likely to promote a sense of choice and volition than 'you should', 'you ought to' or 'you must' [74]. Similarly, the patient's language will provide clues as to whether intrinsically or extrinsically motivated goals are being formed.

Clinicians can discuss patients' values and support them to choose their own goals, as well as goal-promoting behaviours that are sources of spontaneous enjoyment. Interventions designed to boost patient autonomy in healthcare have been shown to have positive long-term behavioural outcomes [101,102]. Encouraging patients to notice and celebrate positive feelings and consequences of the outcomes of behaviour change, no matter how small or gradual, can augment satisfaction with outcomes.

A less tangible influence on motivation is identity. People interpret situations in line with active racial, ethnic and cultural identities and prefer identity-congruent actions [103]. Studies have shown that racial–ethnic minority identifying individuals are more likely to view health-promoting behaviours as white middle class and therefore are less likely to engage in health promotion [104]. Culturally sensitive attempts to encourage behaviours aligned with the patients' self- and social identities are advisable. Healthcare professionals can also aid patients to adopt positive self-concepts in line with their goals. For instance, an individual who takes up a new exercise activity can be encouraged to shift self-perceptions to include being an active or sporty person.

Habits

Promoting habit formation is a valuable strategy in behaviour change maintenance. Clinicians can assist patients to achieve clinically significant health outcomes through the formation of health-related habits by providing simple and easy advice [105]. Healthcare professionals can utilise the following steps [95] with patients to develop healthy habits:

(1) Personally choose a simple, health-related goal that you feel motivated to achieve, for example to eat more fruit.
(2) Decide on a simple action in line with the goal that can be implemented every day, for example to have a piece of fruit as dessert.
(3) Create a plan of where and when the action will be performed. Consistency is key: select the same place and time for each day if possible, for example 'after dinner at home I will eat a piece of fruit'.
(4) Perform the action at that place and time.
(5) Be patient. Forming a habit takes 9–10 weeks on average, but it will become second nature with time.
(6) Congratulate yourself each time the action is performed. Soon you will have formed a new healthy habit!

Healthcare professionals can also help patients to restructure their environment to remove cues that may trigger behaviour that is incongruent with their health goals [106,107]. For instance, studies have shown that making food less accessible by placing it further away reduces the probability and amount of snack intake [108].

Conclusion

Healthcare professionals have a unique opportunity to promote healthy lifestyle choices through their contact with individual patients. Given consultation time constraints and difficulty knowing *how* to support lasting behaviour change, clinicians may avoid offering advice on modifying behaviour. Providing simple and easy-to-implement evidence-based strategies that empower patients to live more healthily can reduce the risk of chronic diseases and promote beneficial health-related outcomes. The main elements of the psychology of long-term behaviour change include creating personally meaningful, enjoyable and satisfying goals in line with values and identity, developing concrete plans that are continually revised, fostering self-sustaining motivation and forming healthy habits that consistently move individuals towards their goals.

Summary

An individual's daily actions have a profound impact on short- and long-term health outcomes and quality of life.

Understanding the psychology of behaviour change and applying theory- and evidence-based tools are integral to sustain behaviour change and enhance health-related outcomes.

Healthcare professionals can support patients to achieve long-term health-related behaviour change by

- assisting in the formulation of ambitious outcome goals and attainable process goals that are specific, measurable, timely, approach-oriented and personally relevant;
- developing action plans and specific if-then plans with patients to promote and protect goal attainment;
- boosting sustainable motivation through helping individuals to choose their own personally meaningful, intrinsically enjoyable and satisfying target behaviours that are aligned with their values and identity;
- providing simple and easy advice to develop healthy habitual actions that are repeated in the same context every day.

References

[1] Lim SS, Vos T, Flaxman AD, Danaei G, Shibuya K, Adair-Rohani H, AlMazroa MA, Amann M, Anderson HR, Andrews KG, Aryee M. A comparative risk assessment of burden of disease and injury attributable to 67 risk factors and risk factor clusters in 21 regions, 1990–2010: a systematic analysis for the Global Burden of Disease Study 2010. Lancet December 15, 2012;380(9859):2224–60.

[2] World Health Organization. Global health estimates 2016: deaths by cause, age, sex by country and by region 2000–2016; 2018. World Health Organization; 2019.

[3] Danaei G, Ding EL, Mozaffarian D, Taylor B, Rehm J, Murray CJ, Ezzati M. The preventable causes of death in the United States: comparative risk assessment of dietary, lifestyle, and metabolic risk factors. PLoS Med April 28, 2009;6(4). e1000058.

[4] Australian Institute of Health and Welfare, Australian Institute of Health and Welfare. Chronic diseases and associated risk factors in Australia, 2006. Australian Institute of Health and Welfare; 2006.

[5] Abraham C, Kelly MP, West R, Michie S. The UK National Institute for Health and Clinical Excellence public health guidance on behaviour change: a brief introduction. Psychol Health Med January 1, 2009; 14(1):1–8.

[6] Knoops KT, de Groot LC, Kromhout D, Perrin AE, Moreiras-Varela O, Menotti A, Van Staveren WA. Mediterranean diet, lifestyle factors, and 10-year mortality in elderly European men and women: the HALE project. JAMA September 22, 2004;292(12):1433–9.

[7] Van Dam RM, Li T, Spiegelman D, Franco OH, Hu FB. Combined impact of lifestyle factors on mortality: prospective cohort study in US women. BMJ September 16, 2008;337:a1440.

[8] Michie S, West R. Behaviour change theory and evidence: a presentation to Government. Health Psychol Rev March 1, 2013;7(1):1–22.

[9] Ashton LM, Sharkey T, Whatnall MC, Williams RL, Bezzina A, Aguiar EJ, Collins CE, Hutchesson MJ. Effectiveness of interventions and behaviour change techniques for improving dietary intake in young adults: a systematic review and meta-analysis of RCTs. Nutrients April 2019;11(4):825.

[10] Fjeldsoe B, Neuhaus M, Winkler E, Eakin E. Systematic review of maintenance of behavior change following physical activity and dietary interventions. Health Psychol January 2011;30(1):99.

[11] Young C, Campolonghi S, Ponsonby S, Dawson SL, O'Neil A, Kay-Lambkin F, McNaughton SA, Berk M, Jacka FN. Supporting engagement, adherence, and behavior change in online dietary interventions. J Nutr Educ Behav April 27, 2019;51(6):719–39.

[12] Zhou X, Perez-Cueto FJ, Santos QD, Monteleone E, Giboreau A, Appleton KM, Bjørner T, Bredie WL, Hartwell H. A systematic review of behavioural interventions promoting healthy eating among older people. Nutrients February 2018;10(2):128.

[13] Buckingham SA, Williams AJ, Morrissey K, Price L, Harrison J. Mobile health interventions to promote physical activity and reduce sedentary behaviour in the workplace: a systematic review. Digit Health March 2019;5. 2055207619839883.

[14] Hobbs N, Godfrey A, Lara J, Errington L, Meyer TD, Rochester L, White M, Mathers JC, Sniehotta FF. Are behavioral interventions effective in increasing physical activity at 12 to 36 months in adults aged 55 to 70 years? A systematic review and meta-analysis. BMC Med December 2013;11(1):75.

[15] Howlett N, Trivedi D, Troop NA, Chater AM. Are physical activity interventions for healthy inactive adults effective in promoting behavior change and maintenance, and which behavior change techniques are effective? A systematic review and meta-analysis. Transl Behav Med February 28, 2018;9(1):147–57.

[16] Larkin L, Gallagher S, Cramp F, Brand C, Fraser A, Kennedy N. Behaviour change interventions to promote physical activity in rheumatoid arthritis: a systematic review. Rheumatol Int October 1, 2015;35(10): 1631–40.

[17] Carr AB, Ebbert J. Interventions for tobacco cessation in the dental setting. Cochrane Database Syst Rev 2012;(6).

[18] Lancaster T, Stead LF. Individual behavioural counselling for smoking cessation. Cochrane Database Syst Rev 2017;(3).

[19] Stead LF, Carroll AJ, Lancaster T. Group behaviour therapy programmes for smoking cessation. Cochrane Database Syst Rev 2017;(3).

[20] Alvarez-Bueno C, Rodriguez-Martin B, Garcia-Ortiz L, Gómez-Marcos MÁ, Martinez-Vizcaino V. Effectiveness of brief interventions in primary health care settings to decrease alcohol consumption by adult non-dependent drinkers: a systematic review of systematic reviews. Prev Med July 1, 2015;76:S33–8.

[21] Fergie L, Campbell KA, Coleman-Haynes T, Ussher M, Cooper S, Coleman T. Identifying effective behavior change techniques for alcohol and illicit substance use during pregnancy: a systematic review. Ann Behav Med October 31, 2018;53(8):769–81.

[22] Kaner EF, Beyer FR, Muirhead C, Campbell F, Pienaar ED, Bertholet N, Daeppen JB, Saunders JB, Burnand B. Effectiveness of brief alcohol interventions in primary care populations. Cochrane Database Syst Rev 2018;(2).

[23] Kelly S, Olanrewaju O, Cowan A, Brayne C, Lafortune L. Interventions to prevent and reduce excessive alcohol consumption in older people: a systematic review and meta-analysis. Age Ageing July 20, 2017; 47(2):175–84.

[24] Ezzati M, Lopez AD, Rodgers A, Vander Hoorn S, Murray CJ. Comparative Risk Assessment Collaborating Group. Selected major risk factors and global and regional burden of disease. Lancet November 2, 2002; 360(9343):1347–60.

[25] Swann C, Carmona C, Ryan M, Raynor M, Baris E, Dunsdon S, Kelly MP. Health systems and health-related behaviour change: a review of primary and secondary evidence. National Institute for Health and Clinical Excellence; 2010.

[26] Hamman RF, Wing RR, Edelstein SL, Lachin JM, Bray GA, Delahanty L, Hoskin M, Kriska AM, Mayer-Davis EJ, Pi-Sunyer X, Regensteiner J. Effect of weight loss with lifestyle intervention on risk of diabetes. Diabetes Care September 1, 2006;29(9):2102−7.

[27] Neter JE, Stam BE, Kok FJ, Grobbee DE, Geleijnse JM. Influence of weight reduction on blood pressure: a meta-analysis of randomized controlled trials. Hypertension November 1, 2003;42(5):878−84.

[28] Avenell A, Broom J, Brown TJ, Poobalan A, Aucott L, Stearns SC, Smith WC, Jung RT, Campbell MK, Grant AM. Systematic review of the long-term effects and economic consequences of treatments for obesity and implications for health improvement. Health Technol Assess May 31, 2004;8(21).

[29] Dombrowski SU, Knittle K, Avenell A, Araujo-Soares V, Sniehotta FF. Long term maintenance of weight loss with non-surgical interventions in obese adults: systematic review and meta-analyses of randomised controlled trials. BMJ May 14, 2014;348:g2646.

[30] Tsai AG, Wadden TA. Systematic review: an evaluation of major commercial weight loss programs in the United States. Ann Intern Med January 4, 2005;142(1):56−66.

[31] Moos RH, Moos BS. Rates and predictors of relapse after natural and treated remission from alcohol use disorders. Addiction February 2006;101(2):212−22.

[32] Hajek P, Stead LF, West R, Jarvis M, Hartmann-Boyce J, Lancaster T. Relapse prevention interventions for smoking cessation. Cochrane Database Syst Rev 2013;(8).

[33] Hughes JR, Keely J, Naud S. Shape of the relapse curve and long-term abstinence among untreated smokers. Addiction January 2004;99(1):29−38.

[34] Davis R, Campbell R, Hildon Z, Hobbs L, Michie S. Theories of behaviour and behaviour change across the social and behavioural sciences: a scoping review. Health Psychol Rev August 7, 2015;9(3):323−44.

[35] Campbell M, Fitzpatrick R, Haines A, Kinmonth AL, Sandercock P, Spiegelhalter D, Tyrer P. Framework for design and evaluation of complex interventions to improve health. BMJ September 16, 2000;321(7262): 694−6.

[36] Campbell NC, Murray E, Darbyshire J, Emery J, Farmer A, Griffiths F, Guthrie B, Lester H, Wilson P, Kinmonth AL. Designing and evaluating complex interventions to improve health care. BMJ March 1, 2007;334(7591):455−9.

[37] Glanz K, Bishop DB. The role of behavioral science theory in development and implementation of public health interventions. Annu Rev Publ Health April 21, 2010;31:399−418.

[38] Prestwich A, Sniehotta FF, Whittington C, Dombrowski SU, Rogers L, Michie S. Does theory influence the effectiveness of health behavior interventions? Meta-analysis. Health Psychol May 2014;33(5):465.

[39] Painter JE, Borba CP, Hynes M, Mays D, Glanz K. The use of theory in health behavior research from 2000 to 2005: a systematic review. Ann Behav Med July 17, 2008;35(3):358−62.

[40] West R, Brown J. Theory of addiction. John Wiley & Sons; August 14, 2013.

[41] Michie S, Johnston M, Abraham C, Lawton R, Parker D, Walker A. Making psychological theory useful for implementing evidence based practice: a consensus approach. BMJ Qual Saf February 1, 2005;14(1): 26−33.

[42] Lawlor DA, Keen S, Neal RD. Can general practitioners influence the nation's health through a population approach to provision of lifestyle advice? Br J Gen Pract June 1, 2000;50(455):455−9.

[43] Kwasnicka D, Dombrowski SU, White M, Sniehotta F. Theoretical explanations for maintenance of behaviour change: a systematic review of behaviour theories. Health Psychol Rev July 2, 2016;10(3): 277−96.

[44] Riekert KA, Ockene JK, Pbert L, editors. The handbook of health behavior change. Springer Publishing Company; November 8, 2013.

[45] Kanfer R. Motivation theory and industrial and organizational psychology. In: Handbook of industrial and organizational psychology, Vol. 1; November 1, 1990. p. 75–130. 2.

[46] de Ridder DT, de Wit JB, editors. Self-regulation in health behavior. John Wiley & Sons; June 5, 2006.

[47] Carver CS, Scheier MF. Control theory: a useful conceptual framework for personality–social, clinical, and health psychology. Psychol Bull July 1982;92(1):111.

[48] Mischel W, Cantor N, Feldman S. Principles of self-regulation: the nature of willpower and self-control. In: Higgins ET, Kruglanski AW, editors. Social psychology: handbook of basic principles. New York: Guilford Press; 1996. p. 329–60.

[49] Fishbach A, Ferguson MJ. The goal construct in social psychology. In: Kruglanski AW, Higgins ET, editors. Social psychology: handbook of basic principles. 2nd ed. New York: Guilford Press; 2007. p. 490–515.

[50] Locke EA, Shaw KN, Saari LM, Latham GP. Goal setting and task performance: 1969–1980. Psychological bulletin July 1981;90(1):125.

[51] Locke EA, Latham GP. Building a practically useful theory of goal setting and task motivation: a 35-year odyssey. Am Psychol September 2002;57(9):705.

[52] Mann T, De Ridder D, Fujita K. Self-regulation of health behavior: social psychological approaches to goal setting and goal striving. Health Psychol May 2013;32(5):487.

[53] Weinberg RS. Goal setting in sport and exercise: research and practical applications. Revista da Educação Física/UEM June 2013;24(2):171–9.

[54] Kingston KM, Hardy L. Effects of different types of goals on processes that support performance. Sport Psychol September 1, 1997;11(3):277–93.

[55] Elliott ES, Dweck CS. Goals: an approach to motivation and achievement. J Pers Soc Psychol January 1988; 54(1):5.

[56] Carver CS, Scheier MF. Origins and functions of positive and negative affect: a control-process view. Psychol Rev January 1990;97(1):19.

[57] Coats EJ, Janoff-Bulman R, Alpert N. Approach versus avoidance goals: differences in seff-evaluation and well-being. Pers Soc Psychol Bull October 1996;22(10):1057–67.

[58] Elliot AJ, Thrash TM. Approach-avoidance motivation in personality: approach and avoidance temperaments and goals. J Pers Soc Psychol May 2002;82(5):804.

[59] Latham GP. Goal setting: a five-step approach to behavior change. Organ Dynam 2003;32(3):309–18.

[60] Casazza K, Fontaine KR, Astrup A, Birch LL, Brown AW, Bohan Brown MM, Durant N, Dutton G, Foster EM, Heymsfield SB, McIver K. Myths, presumptions, and facts about obesity. N Engl J Med January 31, 2013;368(5):446–54.

[61] Polivy J. The false hope syndrome: unrealistic expectations of self-change. Int J Obes May 28, 2001;25(S1): S80.

[62] Fabricatore AN, Wadden TA, Womble LG, Sarwer DB, Berkowitz RI, Foster GD, Brock JR. The role of patients' expectations and goals in the behavioral and pharmacological treatment of obesity. Int J Obes November 2007;31(11):1739.

[63] Linde JA, Jeffery RW, Finch EA, Ng DM, Rothman AJ. Are unrealistic weight loss goals associated with outcomes for overweight women? Obes Res March 2004;12(3):569–76.

[64] Durant NH, Joseph RP, Affuso OH, Dutton GR, Robertson HT, Allison DB. Empirical evidence does not support an association between less ambitious pre-treatment goals and better treatment outcomes: a meta-analysis. Obes Rev July 2013;14(7):532–40.

[65] Rothman AJ. Toward a theory-based analysis of behavioral maintenance. Health Psychol January 2000; 19(1S):64.

[66] Avery A, Langley-Evans SC, Harrington M, Swift JA. Setting targets leads to greater long-term weight losses and 'unrealistic' targets increase the effect in a large community-based commercial weight management group. J Hum Nutr Diet December 2016;29(6):687–96.

[67] Lorig K, Laurent DD, Plant K, Krishnan E, Ritter PL. The components of action planning and their associations with behavior and health outcomes. Chron Illness March 2014;10(1):50−9.

[68] Lowe MR, Tappe KA, Annunziato RA, Riddell LJ, Coletta MC, Crerand CE, Didie ER, Ochner CN, McKinney S. The effect of training in reduced energy density eating and food self-monitoring accuracy on weight loss maintenance. Obesity September 2008;16(9):2016−23.

[69] Handley M, MacGregor K, Schillinger D, Sharifi C, Wong S, Bodenheimer T. Using action plans to help primary care patients adopt healthy behaviors: a descriptive study. J Am Board Fam Med May 1, 2006; 19(3):224−31.

[70] MacGregor K, Handley M, Wong S, Sharifi C, Gjeltema K, Schillinger D, Bodenheimer T. Behavior-change action plans in primary care: a feasibility study of clinicians. J Am Board Fam Med May 1, 2006;19(3): 215−23.

[71] Kwasnicka D, Presseau J, White M, Sniehotta FF. Does planning how to cope with anticipated barriers facilitate health-related behaviour change? A systematic review. Health Psychol Rev September 1, 2013; 7(2):129−45.

[72] Gollwitzer PM. Implementation intentions: strong effects of simple plans. Am Psychol July 1999;54(7): 493.

[73] Gollwitzer PM, Sheeran P. Implementation intentions and goal achievement: a meta-analysis of effects and processes. Adv Exp Soc Psychol January 1, 2006;38:69−119.

[74] Vansteenkiste M, Simons J, Lens W, Sheldon KM, Deci EL. Motivating learning, performance, and persistence: the synergistic effects of intrinsic goal contents and autonomy-supportive contexts. J Pers Soc Psychol August 2004;87(2):246.

[75] Koestner R, Lekes N, Powers TA, Chicoine E. Attaining personal goals: self-concordance plus implementation intentions equals success. J Pers Soc Psychol July 2002;83(1):231.

[76] Pelletier LG, Dion SC, Slovinec-D'Angelo M, Reid R. Why do you regulate what you eat? Relationships between forms of regulation, eating behaviors, sustained dietary behavior change, and psychological adjustment. Motiv Emot September 1, 2004;28(3):245−77.

[77] Deci EL, Ryan RM. The 'what' and 'why' of goal pursuits: human needs and the self-determination of behavior. Psychol Inq October 1, 2000;11(4):227−68.

[78] Bracken BA. Handbook of self-concept: developmental, social, and clinical considerations. John Wiley & Sons; 1996.

[79] Markus H. Self-schemata and processing information about the self. J Pers Soc Psychol February 1977; 35(2):63.

[80] Epiphaniou E, Ogden J. Successful weight loss maintenance and a shift in identity: from restriction to a new liberated self. J Health Psychol September 2010;15(6):887−96.

[81] Weinstein ND. The precaution adoption process. Health Psychology 1988;7(4):355.

[82] Weinstein ND, Sandman PM. A model of the precaution adoption process: evidence from home radon testing. Health Psychol 1992;11(3):170.

[83] Hall PA, Fong GT. Temporal self-regulation theory: a model for individual health behavior. Health Psychol Rev March 1, 2007;1(1):6−52.

[84] Stevens M, Bult P, de Greef MH, Lemmink KA, Rispens P. Groningen Active Living Model (GALM): stimulating physical activity in sedentary older adults. Prev Med October 1, 1999;29(4):267−76.

[85] Rothman AJ, Baldwin A, Hertel A. Self-regulation and behavior change: disentangling behavioral initiation and behavioral maintenance. In: Vohs KD, Baumeister RF, editors. Handbook of self-regulation: research, theory, and applications. New York: Guilford Press; 2004. p. 130−48.

[86] Strack F, Deutsch R. Reflective and impulsive determinants of social behavior. Pers Soc Psychol Rev August 2004;8(3):220−47.

[87] Baumeister RF, Bratslavsky E, Muraven M, Tice DM. Ego depletion: is the active self a limited resource? J Pers Soc Psychol May 1998;74(5):1252.

[88] Hagger MS, Wood C, Stiff C, Chatzisarantis NL. Ego depletion and the strength model of self-control: a meta-analysis. Psychol Bull July 2010;136(4):495.

[89] Verplanken B, Aarts H. Habit, attitude, and planned behaviour: is habit an empty construct or an interesting case of goal-directed automaticity? Eur Rev Soc Psychol January 1, 1999;10(1):101−34.

[90] Gardner B, de Bruijn GJ, Lally P. A systematic review and meta-analysis of applications of the self-report habit index to nutrition and physical activity behaviours. Ann Behav Med May 28, 2011;42(2):174−87.

[91] Wood W, Quinn JM, Kashy DA. Habits in everyday life: thought, emotion, and action. J Pers Soc Psychol December 2002;83(6):1281.

[92] Pavlov PI. Conditioned reflexes: an investigation of the physiological activity of the cerebral cortex. Ann Neurosci July 2010;17(3):136.

[93] Skinner BF. Science and human behavior. Simon and Schuster; 1965.

[94] Lally P, Van Jaarsveld CH, Potts HW, Wardle J. How are habits formed: modelling habit formation in the real world. Eur J Soc Psychol October 2010;40(6):998−1009.

[95] Gardner B, Lally P, Wardle J. Making health habitual: the psychology of 'habit-formation' and general practice. Br J Gen Pract December 1, 2012;62(605):664−6.

[96] Hollenbeck JR, Williams CR, Klein HJ. An empirical examination of the antecedents of commitment to difficult goals. J Appl Psychol February 1989;74(1):18.

[97] Bandura A. Self-efficacy: the exercise of control. Macmillan; February 15, 1997.

[98] White SS, Locke EA. Problems with the Pygmalion effect and some proposed solutions. Leader Q September 1, 2000;11(3):389−415.

[99] Ouwehand C, de Ridder DT, Bensing JM. Who can afford to look to the future? The relationship between socio-economic status and proactive coping. Eur J Publ Health March 23, 2009;19(4):412−7.

[100] Seijts GH, Latham GP. The effects of goal setting and group size on performance in a social dilemma. Can J Behav Sci/Revue canadienne des sciences du comportement. April 2000;32(2):104.

[101] Silva MN, Markland D, Carraça EV, Vieira PN, Coutinho SR, Minderico CS, Matos MG, Sardinha LB, Teixeira PJ. Exercise autonomous motivation predicts 3-yr weight loss in women. Med Sci Sports Exerc April 1, 2011;43(4):728−37.

[102] Williams GC, McGregor HA, Sharp D, Levesque C, Kouides RW, Ryan RM, Deci EL. Testing a self-determination theory intervention for motivating tobacco cessation: supporting autonomy and competence in a clinical trial. Health Psychol January 2006;25(1):91.

[103] Oyserman D, Destin M. Identity-based motivation: implications for intervention. Counsel Psychol October 2010;38(7):1001−43.

[104] Oyserman D, Fryberg SA, Yoder N. Identity-based motivation and health. J Pers Soc Psychol December 2007;93(6):1011.

[105] Lally P, Chipperfield A, Wardle J. Healthy habits: efficacy of simple advice on weight control based on a habit-formation model. Int J Obes April 2008;32(4):700.

[106] Prochaska JO, DiClemente CC, Norcross JC. In search of how people change: applications to addictive behaviors. Addictions Nursing Network January 1993;5(1):2−16.

[107] Thaler RH, Sunstein CR. Nudge: improving decisions about health, wealth, and happiness. Penguin; 2009.

[108] Maas J, de Ridder DT, de Vet E, De Wit JB. Do distant foods decrease intake? The effect of food accessibility on consumption. Psychol Health October 1, 2012;27(Suppl. 2):59−73.

Health literacy and how to communicate effectively with patients to elicit a long-term behavioural change

Sonal Shah

General Practitioner, National Health Service, London, United Kingdom

Introduction

Health is a universal right, an essential resource for everyday living, a shared social goal and a political priority for all countries [1]. Health literacy is a key determinant of health, with health literacy being a stronger predictor of an individual's health status than income, employment status, education level and racial or ethnic group [2].

Health literacy is defined by the World Health Organization as '*the personal characteristics and social resources needed for individuals and communities to access, understand, appraise and use information and services to make decisions about health*' [3]. Health literacy encompasses the dual nature of communication, referring to both the information that is being disseminated and how people are able to understand the information presented to them [4]. For professionals in a health setting, poor health literacy can be thought of as a 'risk factor' [5] for disease, and it may directly or indirectly impact an individual's health.

Improving communication between the healthcare professional and patient may help to address disparities in health literacy. Health literacy can be improved for individuals through the development of their skills and knowledge. This can enable patients to exert greater control over their health and the factors that shape their well-being.

It has been reported that low levels of health literacy are associated with a greater risk of long-term, life-limiting, health conditions and more difficulties in managing medications [6]. Low health literacy levels make functioning in a healthcare system challenging. This is especially true for those who are vulnerable or disadvantaged, as these individuals may not have the resources to enable them to be actively involved in making decisions regarding their own health. People with lower health literacy skills are also more likely to make adverse lifestyle choices [7].

Table 4.1 describes the adverse outcomes associated with low health literacy and lists the positive effects of improving health literacy.

Table 4.1 The consequences of low health literacy and the effects of improving health literacy.

Low health literacy is associated with Refs. [8−11]:

- Poor health outcomes, including increased risk of premature mortality
- Unhealthy lifestyle behaviours such as poor diet, smoking, lack of physical activity
- Lack of knowledge about medical conditions and therapies
- Lack of engagement with healthcare providers
- Reduced understanding of medical information
- Reduced use of preventative health services, increased use of emergency services
- Poorer self-reported health
- Increased hospitalisation
- Higher healthcare costs
- Late presentation of disease

Improvements in health literacy have been shown in Refs. [8,12−14]:

- Increase health knowledge
- Positively influence health behaviour
- Build resilience
- Encourage positive lifestyle changes
- Empower people to effectively manage long-term health conditions
- Reduce the burden on health- and social care services
- Reduce health inequalities
- Reduce disease and depression severity

A global problem

Poor health literacy is a problem throughout the world, with many countries experiencing a considerable mismatch between the complexity of health materials provided for the population and the population's knowledge and understanding. An observational study in England showed that between 43% and 61% of English working-age adults routinely do not understand health information [15]. Australian data suggest that only 41% of Australians aged 15−74 years had a level of health literacy that was at least adequate [16]. In the United States (US), the National Assessment of Adult Literacy (NAAL) showed that only one-third of individuals surveyed had the skills and abilities to read, understand, and demonstrate what is meant by a pill bottle label that says take 'two tablets by mouth, twice a day' [17].

Economics of poor health literacy

Poor health literacy is associated with significant financial burden for healthcare systems. In England, the economic cost is estimated to be between £2.95 and £4.92 billion every year [8]. In the United States, research suggests that the annual cost of poor health literacy is $8 billion, which equates to 3%−5% of the health budget [18]. At an individual level, health consumers with lower health literacy levels incur increased costs of between $143 and $7798 per person per annum compared with reference groups of health consumers with adequate health literacy levels [18]. The increased costs reflect the increased health needs of this specific patient cohort.

The quantification of health literacy and causes of disparities

As the understanding of the impact of poor health literacy has developed, it has led to advancements in methods to quantify the problem. In Europe, the European Health Literacy Survey (HLS-EU) was employed to assess healthy literacy in eight European countries [19]. The HLS-EU-Q questionnaire determined how individuals access or obtain information relevant to their health, how well they understand such information, their effectiveness at appraising or evaluating it and how they apply or use the information.

The information was correlated with income, social status, education and age. Health literacy was categorised into four tiers: insufficient, problematic, sufficient, and excellent. One in ten respondents showed insufficient health literacy, and almost one in two had limited health literacy, i.e. insufficient and/or problematic. There was a strong correlation between health literacy and self-assessed health, supporting the idea that poor health literacy leads to poor perceived health. Although there were some international variations, the HLS-EU confirmed that those with higher levels of education and those in higher socioeconomic groups have better health literacy. The survey also found that across all participating countries the amount of physical exercise people undertook was consistently and strongly associated with health literacy [20].

Similar results have been attained in the United States. The American Institute of Medicine reported that 90 million adults in America have trouble understanding and acting on healthcare information [21]. The National Assessment of Adult Literacy (NAAL), a comprehensive survey of literacy in America, found that adults who are socioeconomically disadvantaged, those belonging to ethnic minority groups and/or the elderly are disproportionately hindered by such literacy barriers [22]. In Australia, rates of literacy have also been found to be lower in those aged 45 years and over [23]. This may relate to differing expectations ofthe level of participation in healthcare among different generations, effects of cognitive decline on people's mental processing skills, the length of time since leaving formal education and the lower levels of formal education received by older generations [16]. Studies have also demonstrated an independent association between low levels of individual literacy and increased mortality among elderly people [24].

The majority of studies report that those at the greatest risk of poor health literacy are [15,25]

- disadvantaged socioeconomic groups;
- individuals from ethnic minorities;
- those whose first language is different to that of the national mother tongue;
- older people;
- individuals with long-term health conditions; and
- individuals with disabilities, including those who have long-term physical, mental, intellectual or sensory impairment.

One of the key challenges faced by policymakers and those trying to improve health is to ensure that the skills and abilities of the individual can be maximised, so that the demands and complexity of the healthcare and social care systems can be understood and navigated appropriately [4]. Evidence suggests that health literacy interventions at both system and practitioner level can impact positively upon health behaviours and health outcomes in those with low health literacy [26].

Interventions at the system level

In 1986, at the first international conference on health promotion, the Ottawa Charter identified the fundamental conditions and resources essential for health [27]. These included peace, shelter, education, food, income, a stable ecosystem, sustainable resources, social justice and equity. Several of these factors contribute to health literacy, in particular, education. It is evident that the causes of limited health literacy lie within social and cultural frameworks, the health and education systems that serve them and the interactions between these factors [21]. However, there are strategies that can be implemented at a system level to try to prevent and address inequalities in health literacy.

Nutrition

A good example of health literacy interventions at a system level can be seen when looking at the global approach to promoting healthy diets. In the first 2 years of a child's life, optimal nutrition and exercise fosters healthy growth and improves cognitive development [28]. Developing healthy eating habits from infancy and giving caregivers information to understand the relationship between diet and health is essential. It is therefore important to implement measures to strengthen health literacy from childhood as a means of reducing health inequalities [29].

Beginning in infancy, breastfeeding is considered the best form of nutrition for most babies and also has implications for the health of mothers. To ensure parents are provided with clear, unbiased information about breast-milk substitutes, without being influenced by health claims or marketing strategies, regulation of formula-milk advertising has been introduced by the World Health Organization (WHO). In 1981 the WHO instigated an international code for the marketing of breast-milk substitutes [30]. This code stipulates that the usual rules governing market competition and advertising should not apply to products intended for feeding babies. Rather, manufacturers should provide information to support parents make informed choices, which is particularly important for those with poor literacy who may be susceptible to marketing.

The WHO Collaborative Health Behaviour in School-aged Children (HBSC) survey found that health literacy is one of the main factors contributing to health differences and is associated with educational outcomes such as academic achievement and postschool aspirations [31]. The provision of education is a governmental responsibility, so having effective collaboration between health and education departments is vital.

Integrating health literacy into early years education is an effective means of establishing the foundations of good health. Implementation of an integrated, comprehensive core programme can promote health and nutrition literacy for school-aged children in schools and educational institutions [32]. The Australian National Framework for health-promoting schools [16,33], and the Schools for Health in Europe Network Foundation [34], among many other international agencies, have developed detailed guidelines to help support schools to health-promoting education. These include the development of health skills, knowledge, behaviour and communication through classroom teaching and whole-school approaches, taking into consideration the physical, social and emotional needs of all members of the school community.

Adolescents may have a poor understanding of nutrition. They may experience peer pressure, and they may be susceptible to the marketing of unhealthy foods. It is therefore important that a 'health-promoting' school environment embeds accurate nutritional information. Nutrition and food literacy knowledge can be undermined if there are conflicting messages in the settings where children gather. Schools, child-care and sports facilities therefore should support efforts to improve children's nutrition by making the healthy choice the easy choice and not providing or selling unhealthy foods and beverages [32]. A successful initiative is the partnership between museums in Victoria, Australia, and VicHealth, a health promotion foundation, which plans to phase out the sale of all sugary beverages [35]. Similarly, in France, a law has been passed banning the sale of any food or drink from automatic vending machines in schools [36].

Several governments have used political policies to guide and shape nutrition labelling to ensure consumers make informed food purchases and healthier eating choices [37]. The Codex Alimentarius Commission of the Joint Food and Agriculture Organisation of the United Nations (FAO) and WHO Food Standards Programme have also developed a set of international standards, guidelines and related texts for food products to protect consumer health and encourage fair practice in international food trade [38]. Providing nutritional information to consumers in a clear format, through the use of standardised terms, nutritional values, reference intakes and implementing colour coding systems may make it easier for consumers to identify unhealthy foods and beverages, even if they have limited health literacy.

The promotion of healthy diet is a key message in health promotion and disease prevention. The FAO and WHO encourage governments to create a positive food environment, ensuring high-quality food for all individuals. Coherent national policies aim to ensure that healthy dietary practices are implemented, are accessible to all and are sustainable for future generations. Unsurprisingly, the promotion of eating a healthy diet requires input from numerous stakeholders, ranging from central government to the food industry. Even when individuals have an adequate level of nutrition literacy and knowledge of healthy food choices, this cannot be acted upon if healthy foods are not readily available or affordable [32]. Evidence suggests that where access to healthy foods is limited, ultraprocessed foods become the key alternative. So, as well as improving the health literacy of the individual, it is vital to also ensure that there are incentives for producers and retailers to grow, use and sell fresh fruit, vegetables and other components of a balanced diet to allow health knowledge to be acted upon appropriately.

Interventions at the level of the healthcare practitioner

An adequate level of health literacy is critical to allow an individual to have greater control of their own health and the choices that they make. Health professionals and those involved in health promotion, disease prevention and policymaking must have a comprehensive understanding of the importance of health literacy and the role it plays as a social determinant of health. Health literacy ensures that individuals can confidently read nutrition labels, follow directions to take regular medication and ensure they make healthy lifestyle choices.

Health literacy awareness and training is relevant to a wide range of individuals, including traditional healthcare workers such as doctors, nurses, physiotherapists and occupational therapists. However, its scope extends even further, and it is also important for administrative staff, voluntary and community organisations, fire and rescue services and job centres. Health literacy training can be delivered by national agencies to support this [16,39−42]. This gives all professionals the knowledge and confidence to incorporate practical approaches and techniques to enhance service delivery to everyone [43].

In clinical settings, health literacy is not a parameter that is assessed or recorded. Therefore, it maybe unclear *which* individuals have poor literacy. There are, however, red flags that may indicate that an individual requires additional support. These include the following:

- Frequently missed appointments
- Incomplete registration forms
- Poor compliance with medication
- Inability to name medications, purpose or timing of the medication
- Identification of pills by looking at them or describing them, rather than reading the label
- Inability to provide a detailed history
- The individual asks fewer questions
- Patients who do not attend follow-up appointments or chase-up results
- A person with poor literacy, may use excuses such as 'I forgot my glasses' when asked to complete a form

To improve the quality and delivery of information to those with poor health literacy, simple steps can be adopted by all professionals in a clinical setting to create a positive health environment [44]. In particular, all front-of-life staff should adopt an attitude that is friendly, helpful and be able to signpost people to additional support if it is needed. Other steps include the following:

- Asking patients in advance to bring all necessary information to an appointment, for example all their medications. This could be delivered via a text or a phone call reminder by administrative support staff.
- Advise patients that they are welcome to bring someone with them to the appointment.
- Limit the number of administrative forms that need filling.
- Confidential assistance should be provided to those who may have difficulty filling in forms or making appointments.
- Written material should be written using simple words, short sentences and lots of white space. Only the essential information that is required should be included.
- The use of medical terminology should be avoided.

These simple strategies can improve the yield of the consultation and importantly can help patients to increase their understanding of their health and the healthcare system, thus empowering them to exert greater control of their health and well-being.

Communicating effectively

It is the responsibility of healthcare professionals to ensure that they communicate with patients in the manner which serves the individual patient's best interests. Communicating in a manner that it is open, without judgement or discrimination will engage patients, particularly those from disadvantaged groups. To enable empowerment of patients, there must be respect, patient choice, motivation, development of self-esteem and education. Encouraging an exchange of views and information will serve to develop a shared plan of self-care and management appropriate to the individual's social and cultural requirements.

Table 4.2 Empowering patients.

Issues to consider	Possible steps to improve communication
Does the patient have the cognitive skills to understand complex instructions, for example can they follow the directions on prescribed medication?	Simplify instructions. Check understanding. Consider using the simplest dosing regime, or the use of medication aides such as medication boxes. A support worker/advocate to champion the patient's individual needs.
Can the patient navigate their way around clinical settings or do they have physical impairments making this difficult?	Arrange transport services to allow patients to attend appointments. Encourage patients to attend appointments with carers or support workers. Try to arrange 'one-stop' appointments, where many issues are dealt with at one time.
Can the patient provide informed consent for procedures, initiation of treatment, etc.?	Ensure adequate time in appointments. Provide information ahead of time to allow patients enough time to review information. Arrange follow-up appointments. Information must be provided in a format that meets the individual needs of the patient.
Are there barriers for this patient embarking on a healthy lifestyle?	Engage the patient to develop shared care plans that fit into the patient's current lifestyle.
Are there communication difficulties?	Provide written materials, or resources in different formats, for example audio, visual, pictorial. Allow for extra time in clinical settings. Harness the use of digital technologies such as websites and podcasts.
Are there language barriers?	Use an advocate to translate for the patient.
Does the patient have adequate numeracy skills to interpret nutritional information, for example following recommended intake guidelines for fat and sugar?	Encourage the use of existing resources such as front-of-pack labelling which often uses 'traffic lights' to symbolise healthy and unhealthy food. Use of decision aids with icons and pictographs, clarifying risks and benefits of screening.

Efforts to improve clarity of information can improve comprehension in those with poor literacy. It is suggested that when only essential information is provided, and when such information is delivered at the start of the encounter, it can increase understanding by 0.7 and 0.6 on a 3-point scale, respectively [45]. Important communication issues to be considered are highlighted in Table 4.2. When conscious measures are taken to improve health literacy, health knowledge can be improved, the patient's resilience can be enhanced and patients are empowered to make positive lifestyle changes and to effectively manage long-term health conditions. This then reduces the burden on health- and social care services [8].

Patient-centred care

Placing the patient at the centre of any management plan is the key to health and behaviour change. Individuals are the products of the interaction between their genes and environment, so it is important to identify personal factors, environmental factors and specific behaviours which could be modified to promote good health.

Teach-back

Often a clinical contact or consultation will end with the professional asking a patient whether they understand the information that has been presented and if they have any questions. For some individuals, this may cause embarrassment, particularly if they are confused by what has been said or do not have the confidence to ask questions.

The 'teach-back' method is a way of confirming that a patient understands the information that has been delivered. It is effective in improving the quality of healthcare, patient safety, risk management and cost efficiency [40]. The method involves asking a patient to use their own words to summarise a discussion or management plan. Table 4.3 addresses issues to avoid and strategies that may support clinicians in a clinical setting.

Table 4.3 How to implement the 'teach-back method' effectively.
Avoid
Do not sound as though you are 'testing' the patient Do not be judgemental Avoid closed questioning such as, 'do you understand?' or 'is that ok?'. Avoid using complex medical terminology
Do
Allow enough time in the clinical encounter for this process, so that the patient is unhurried Use phrases such as follows: - Would you mind explaining your management plan back to me, so that I can make sure I have covered everything? - When you go home today, what will you tell your partner about what we have discussed? - We have covered a lot of different management options. What do you think you will choose? - We are starting some new medications. I just want to make sure I was clear about the side effects. Can you explain to me the kinds of things to look out for? - Can you tell me how you take your medication? - People often have trouble remembering how to follow this. Could you just run through how you will make this work for you? - We have covered a lot of information today. Why don't we list the main points to check we have covered everything? - Based on our discussion today, what changes do you think you might make tomorrow to help you with your diabetes? When the patient lists the key points, it can be helpful to provide a brief written summary they can keep and refer back to.

Chunk and check

In addition to the teach-back method, 'chunk and check' is an approach that can be adopted by professionals to deliver information in small amounts [40]. Professionals should take the time to offer 'chunks' of information, pause and then 'check' understanding. This prevents too much information being delivered in one go. It also makes processing this information easier and may allow the professional to assess understanding and allow the opportunity for plans to be modified to meet an individual's needs.

Goal setting

Many clinical encounters result in management plans which involve lifestyle change or the modification of risky behaviours. SMART is a simple, easy-to-follow framework that can be used to set and achieve a goal. It can be applied to many situations, for example introducing changes to a person's diet or embarking on a new exercise regime.

It stands for **specific, measurable, achievable, realistic and timely.**

Through open, shared discussion, identify one easily achievable goal that the individual would like to accomplish and work to develop a way to achieve this. It may be useful to identify possible barriers or limitations or establish whether there is a support network. Table 4.4 demonstrates initiating exercise as an example of this method.

Prescribing medication

A major concern with poor health literacy is patient safety in relation to medication adherence. In a study by Davis et al., 395 patients were shown five prescription labels and then asked if they were able

Table 4.4 An example of an SMART goal.

Specific	Describe exactly *what* will be accomplished. *Who* else will be involved? *Which* constraints are present?	I will exercise for 30 min by walking briskly around the block.
Measurable	Look for ways the progress can be monitored.	I can take measurements of my body, and check for changes every 2 weeks.
Achievable	This needs to be a reasonable goal, in the context of the patient.	I don't have access to a gym, but I can exercise outside. I already walk to the shops, so walking briskly is an achievable target for me.
Realistic	They need to be relevant to the patient and should meet personal interests, skills and resources. For someone with a sedentary lifestyle, getting them moving is more realistic than asking them to do a 10K run.	My neighbour walks her dog, so I could join her. It also means I have some company.
Timely	Identifying a time frame to complete the goal, including frequency and deadline, may support progress.	I will do this on Monday, Wednesday and Saturday, during the next month.

to understand what the instructions meant and whether they were able to demonstrate the dosage instructions. Participants had their literacy skills assessed using a 'Rapid Estimate of Adult Literacy in Medicine (REALM)' test, a reading recognition test, assessing understanding of health-related words. Correct understanding of the five labels ranged from 67.1% to 91.1%. In those with low levels of literacy, 70.7% were able to read the instructions, 'take two tablets by mouth twice daily', but concerningly, only 34.7% could demonstrate the number of pills to be taken daily, highlighting a disparity between the ability to read and comprehension. Professionals should therefore consider this when prescribing medication, particularly in groups at risk of poor literacy and those who are on multiple medications.

When prescribing, it can be helpful to explain to the patient:

- What the medication is for
- What the benefits are of taking the medication
- Potential side effects
- How the medication should be taken: be specific, for example 'two tablets in the morning before food'
- The duration of the treatment
- Whether the medication will be reviewed
- Whether there is a need for additional monitoring, for example blood testing

It is beneficial to use the simplest dosing regime, or use combination preparations to limit the number of pills needed. Enlisting the help of other allied professionals, such a specialist nurse or community pharmacist may also support individuals to take greater control of their medication, by providing regular follow-up to monitor and review adherence. In an intervention where adults with low literacy received verbal medicine dosage instructions from a pharmacy dispenser, participants reported greater understanding of their medication dosage regime than those who did not receive verbal instructions. In those who received the intervention, 88% were able to correctly describe their regime, compared with 70% of those who did not receive additional verbal counselling [46].

Brief advice

Making every contact count is an approach to behaviour change that encourages health- and social care professionals to engage in conversations that support individuals in making positive changes towards better health and well-being [47]. Evidence suggests that adopting healthy conversations as part of normal interactions can have significant impact. One such method is the ask, advise and assist framework which is demonstrated in Table 4.5. This framework enables all professionals in a clinical setting to raise awareness of health and lifestyle and support individuals to make change.

Table 4.5 The ask, advise, assist framework.	
Ask	Ask about a behavioural risk factor and record its status; for example do you smoke? Avoid asking how much or what they smoke or even if they want to stop.
Advise	Advise about the benefit of changing the behaviour in a personalised and appropriate way, for example the best way to stop smoking is with medication and support. Avoid mentioning risk, or asking the patient to stop.
Act	Signpost and promote local services.

There is good evidence that very brief advice which can take less than 30 s is a very effective way of promoting smoking cessation [48]. The National Centre for Smoking Cessation and Training in the United Kingdom estimates the number needed to treat (NNT) for brief advice to smokers to generate a long-term quit is 40 [49].

Promoting behavior change

One of the challenges that healthcare professionals face is supporting individuals to develop positive health behaviours. Many risky behaviours, such as smoking or alcohol excess, are habitual and result from a multitude of developmental, socioeconomic and cultural factors. The psychology of health-related behaviour change is explored in Chapter 3. Whatever approach is undertaken to help encourage a patient to adopt healthy behaviours, it is vital that the foundation of this is good communication. The individual should be placed at the centre of any management plan. By assessing the individual's thoughts, beliefs and attitudes, healthcare professionals can gain a greater understanding of how they can best support individuals to actively participate and engage in behaviour change. A key component of this process is appraising the individual's level of health literacy and ensuring that all information is presented appropriately.

Conclusion

Health literacy is a key determinant of health and is influenced by multiple, complex, interacting factors. To improve health literacy requires a collaborative approach involving government and policymakers, and front-line professionals responsible for health and education. For policymakers, an example of a system-based intervention is ensuring that the marketing and labelling of food is clear, transparent and accessible to consumers. This will support the public in making better health choices. Similarly, interventions through the education system, particularly in the early years, can embed the foundations of health literacy which is vital for lifelong health and well-being. For health practitioners, developing systems to support people with poor literacy through improved communication and training for all staff will help empower individuals to take control of their own health.

Summary

Health literacy is a key determinant of health, associated with significant economic implications.
Those with poor health literacy have more adverse health outcomes, including increased morbidity and increased risk of premature death.
For practitioners to support those with poor health literacy:
- There should be training in and awareness of health literacy in healthcare settings.
- Communication, both written and verbal, should be patient centred and appropriate to an individual's level of understanding.
- Patients should be empowered to take control of their health and well-being.

References

[1] World Health Organisation. Shanghai declaration on promoting health in the 2030 agenda for sustainable development. In: Shanghai: 9th Global conference on health promotion; 2016.

[2] World Health organisation, Europe. Health Literacy: the solid facts. 2013. Denmark.

[3] World Health Organisation. Health literacy toolkit for low- and middle-income countries. A series of information sheets to empower communities and strengthen health systems. 2015. Geneva.

[4] Parker R. Measuring health literacy: what? So what? Now what? Institute of medicine (US) roundtable on health literacy. Washington (DC) Washington DC: National Academies Press; 2009.

[5] Nutbeam D. The evolving concept of health literacy. Soc Sci Med 2008;67(12):2072−8.

[6] Berkman N. Health literacy interventions and outcomes: an updated systematic review. Agency for Healthcare Research and Quality; 2011.

[7] HLS_EU Consortium. Comparative report of health literacy in eight EU member states. The European Health Literacy Survey. Vienna: Ludwig Boltzman Institute Health Promotion Research; 2012.

[8] Public Health England. Local action on health inequalities Improving health literacy to reduce health inequalities. UCL Institute of Health Equity; 2015.

[9] Jayasinghe U. The impact of health literacy and life style risk factors on health-related quality of life of Australian patients. Health Quality Life Outcomes 2016;14(68).

[10] Johnson A. Health literacy, does it make a difference? Aust J Adv Nursing 2014;31(3):39−45.

[11] Martensson L. Health literacy – a heterogeneous phenomenon: a literature review. Scand J Caring Sci 2012; 26(1):151−60.

[12] Rothman R. Influence of patient literacy on the effectiveness of a primary care− based diabetes disease management program. J Am Med Assoc 2004;292(14):1711−6.

[13] Howard-Pitney B. The Stanford Nutrition Action Program: a dietary fat intervention for low-literacy adults. Am J Publ Health 1997;87(12):1971−6.

[14] BD W. Literacy education as treatment for depression in patients with limited literacy and depression: a randomized controlled trial. J Gen Intern Med 2006;21(8):823−8.

[15] Rowlands G. A mismatch between population health literacy and the complexity of health information: an observational study. Br J Gen Pract 2015;65(635):e379−86.

[16] Australian Institute of health and welfare. Australia's health. Health no.14 cat No. Aus 156. Canbera; 2012. 2012.

[17] Davis T. Low literacy impairs comprehension of prescription drug warning labels. J Gen Intern Med 2006; 21(8):847−51.

[18] Eichler K. The costs of limited health literacy: a systematic review. Int J Publ Health 2009;54(5):313−24.

[19] Sorensen K. Health literacy in Europe: comparative results of the European health literacy survey (HLS-EU). Eur J Pub Health 2015;25(6):1053−8.

[20] Consortium H-E. Comparative report on health literacy in eight EU member states. 2012.

[21] Neilsen-Bohlman L. Health literacy: a prescription to end confusion. Washington DC: National Academies Press (US); 2004.

[22] US department of education. A first look at the literacy of America's adults in the 21st century. 2005.

[23] Australian Commision on safety and quality in helth care. Health LIteracy: taking action to improve safety and quality. 2014. Sydney.

[24] Baker D. Health literacy and mortality among elderly persons. Arch Intern Med 2007;167(14):1503.

[25] Nutbeam D. Health literacy as a public health goal: a challenge for contemporary health education and communication strategies into the 21st century. Health Promot Int 2000;15(3):259−67.

[26] World health organisation and Food and agricultural organisation. Diet, nutrition, and the prevention of chronic diseases. Report of a WHO study group. WHO technical report series, No. 797. Geneva: World Health Organization; 1990.

[27] World health organisation. 1986. cited 2019 Nov 28. Available from: https://www.who.int/healthpromotion/conferences/previous/ottawa/en/index4.html, www.who.int.

[28] World helth organisation. 2018 [cited 2019 nov. Available from: www.who.int. https://www.who.int/newsroom/fact-sheets/detail/healthy-diet.

[29] World Health organisation. Health literacy the solid facts. 2013.

[30] World health organisation. International code of marketing of breast-milk substitutes. 1981. Geneva.

[31] Paakkari L. Does health literacy explain the link between structural stratifiers and adolescent health? Eur J Publ Health 2019;5:919—24.

[32] World health organisation. Report of the commission on Ending childhood obesity. 2016. Geneva.

[33] Australian Curriculum, assessment and reporting authorty. Revised australian curriuclm: health and physical education -foundation to year 10. 2013. Canberra.

[34] cited 2019 Dec, https://www.schoolsforhealth.org; 2019.

[35] Muesums Victoria. 2019. Available from: https://museumsvictoria.com.au/media-releases/museums-victoria-to-phase-out-all-sugary-drinks/.

[36] World Cancer research fund international. Curbing global sugar consumption: effective food policy actions to help promote healthy diets and tackle obesity. 2015. Geneva.

[37] World health organisation. Guiding principles and framework manual for front-of-pack labelling for promoting healthy diet. Geneva: Department of nutrition for health and development; 2019.

[38] World health organisation. Nutrition labels and health claims: the global regulatory environment/Corinna Hawkes. 2004.

[39] NHS edcuation for scotland. Health literacy place. 2015. cited 2019 Dec. Available from: http://www.healthliteracyplace.org.uk/media/1285/health-literacy-leaflet-final-7.pdf.

[40] Agency for healthcare research and quality. Agency for healthcare research and quality. 2017. cited 2019 Nov. Available from: https://www.ahrq.gov/health-literacy/quality-resources/tools/literacy-toolkit/tool3a/index.html.

[41] Australian commision on safety and quality in health care. [Online]. cited 2019 Dec. Available from: https://www.safetyandquality.gov.au/our-work/patient-and-consumer-centred-care/health-literacy/tools-and-resources-for-health-service-organisations#health-literacy-infographics.

[42] Canadian public health association. 2017. cited 2019 Dec. Available from: https://www.cpha.ca/vision-health-literate-canada-report-expert-panel-health-literacy.

[43] Health education England, NHS England. Public health England and the community health and learning foundation. Making the case for health literacy. Strategic report. East Midlands national demonstrator 2016-17. 2016.

[44] Graham S. Do patients understand. Perm J 2008;12(3):67—9.

[45] Peters E. Less is more in presenting quality information to consumers. Med Care Res Rev 2007;64(2):169—90.

[46] McKellar A. Using community medical auxiliary trainees to improve dose understanding among illiterate hospital outpatients in rural Nepa. Trop Doct 2005;35(1):17—8.

[47] Health education England. 2010. cited 2019 Dec, https://www.makingeverycontactcount.co.uk.

[48] Aveyard P. Brief opportunistic smoking cessation interventions: a systematic review and meta-analysis to compare advice to quit and offer of assistance. Addiction June 2012;107(6):1066—73.

[49] National centre for smoking cessation and training. 2014. cited 2019 Dec. Available from: https://www.ncsct.co.uk/publication_very-brief-advice.php.

The role of the healthcare system in social prescribing

Emma Ladds[1,2]

[1]*Academic Clinical Fellow, Nuffield Department of Primary Health Care Sciences, University of Oxford, Oxford, United Kingdom;* [2]*General Practitioner, National Health Service, Oxford, United Kingdom*

'*He who has a why to live for can bear almost any how.*' [1].
Friedrich Nietzsche

Introduction

The psychosocioeconomic determinants of health

There are few more impressive trends than that of life expectancy in the 20th century. As recently as 1900, the global average was still only 31 years, remaining under 50 years even in the wealthiest countries [2]. Over the next century, it increased to more than 40 years [3], to the point where, in 2012, the United Nations (UN) estimated there were almost 350,000 centenarians living worldwide [4].

This astounding increase is attributed, not to the medical discoveries and interventions that became hallmarks of the 20th century, but to the implementation of public health strategies and social measures [5]. Sanitation, vaccination, safer workplace regulation and other 'unglamorous' interventions remain the hidden heroes of global health.

The psychosocial contribution to health outcomes such as morbidity, mortality and quality of life is well recognised. This was initially highlighted in Michael Marmot's Whitehall Studies of English civil servants, which found that 70% of health outcomes were determined by societal factors, such as social status of employment grade or social support at home [6]. The significance of these remains evident today, through their clear correlation with gross health inequalities.

A famous example from 2010 describes a deprived Glaswegian boy with an average life expectancy 28 years below that of a child from an affluent area just 12 km away [7]. Sadly, similar patterns are encountered worldwide. Chicago has the greatest discrepancies of any large city within the United States (US). Those in affluent, 'white' Streeterville can expect to live to 90 years, while 9 miles away in the deprived, 'black' Englewood life expectancy is just 60 years of age [8].

Analyses of equally deprived areas have demonstrated that socioeconomic poverty alone does not account for all the health outcome discrepancies observed. Equally deprived populations in Manchester and Liverpool in the United Kingdom have higher life expectancies than comparable communities in Glasgow, while the wealthiest 10% of the Glaswegian population have lower life expectancies than the same group in other cities [9].

A Prescription for Healthy Living. https://doi.org/10.1016/B978-0-12-821573-9.00005-9

61

One explanation for these differences may be a disparity in psychological factors between populations, which can impact on health and well-being. Important factors may include a real or perceived lack of control over one's own life and fragmented societies with low levels of social capital and community support [10].

Societal and population trends

Universal Health Coverage, the provision of equitably accessible, high-quality healthcare without risking financial harm to the user, is the holy grail of global health [11–13]. However, it is a goal that is proving frustratingly difficult to achieve.

Demographic trends underlie some of the challenges. Population growth, aging societies, costly technological and pharmaceutical interventions and rising expectations make it increasingly expensive to provide 'appropriate' care for all.

Similarly, societal change has had a huge impact. Those living throughout the 20th century saw greater upheaval than had been previously encountered in human history. From horses to space shuttles, bloodletting to the Human Genome Project, individuals lived through two world wars, the fear of a nuclear holocaust and a general liberalisation of social opinion. Most significantly, they had to adapt to the homogenising influences of free movement and globalisation and the all-pervasive impact of a digital revolution that forever changed public and personal communication.

Despite undeniable improvements, the results have not been universally positive. Some individuals have embraced the pace and flexibility of the modern world; others are lost without the stability of traditional employment and may feel abandoned by the breakdown of well-defined social structures. For example, following the closure of the Redcar steel plants in Teeside, UK, employment rates, the local economy and a whole way of life all suffered significantly, leaving community members feeling lost and out of control [14].

To illustrate how societies are changing, more than 50% of individuals in the United Kingdom and just over a third of the US population now hold no religious beliefs [15,16], as opposed to 31% in the United Kingdom in 1983 and around 20% in the United States in 2008 — when the earliest comparable data are available [15,17]. The percentage of single-parent households in the United States has doubled in the past three decades [18], with such individuals more likely to feel socially isolated and to have higher rates of mental health problems such as anxiety and depression than those with more social support [19]. Similarly, more than 8 million individuals now live alone in the United Kingdom — a number that has more than doubled since the early 1970s [20]. As of January 2019, almost 80% of all Internet users actively used social media sites each month [20]. Of the 65% in 2016 who used Facebook [20], on average each had 155 virtual friends [21].

Despite improved connectivity through the digital revolution, studies suggest such alterations in communication also contribute to isolation and loneliness [22,23], mental health issues [24,25], uncertainty and a loss of meaning in life. Reductions in social media use have been associated with increased happiness, contentedness [26] and general well-being [27]. Similarly, real-life interpersonal relationships and spiritual or civic engagement can increase a sense of meaning for individuals. This is associated with improved social functioning [28], psychological well-being [29] and reduced morbidity and mortality [29,30].

Health inequalities undoubtedly result from social and economic factors, and the cost of these is largely met by health- and social care systems. In the United Kingdom, it is estimated that 20% of primary care consultations are primarily for social problems [31]. The financial cost to the UK's National Health Service (NHS) of nonhealth demand on GPs is at least £395m, equivalent to the salaries of 3750 full-time GPs [32].

Similarly, international unplanned hospital admission rates have increased steadily in recent decades, particularly among the elderly [33—36]. Social determinants, including loneliness, isolation and an inadequate home support system, frequently underpin many of these admissions [33].

Traditional models of healthcare are woefully inadequate in meeting such demand and frequently contribute to overmedicalisation of nonmedical problems. Over the past few decades, it has become apparent that more innovative approaches are required. This has led to the enthusiasm for social prescribing.

What is social prescribing?

Social prescribing, in its simplest sense, is a return to treating the 'patient' as a 'person'. It is an attempt to expand the options available to medical practitioners, patients and families when faced with problems rooted in psychosocioeconomic factors. Typically, these might include individuals with mild or long-term mental health problems, vulnerable groups such as the frail, elderly or socially isolated individuals and those who frequently attend healthcare services [37].

Although vast in scope, approaches often include referral to local voluntary, community and social enterprise services. These opportunities encourage individuals to develop meaning in their lives, generate relationships or take ownership of personal challenges and develop creative problem-solving strategies. Examples include voluntary work, education courses, group lunches, hobby classes, sports clubs, befriending or informal care organisations, self-help groups, nature conversation or even 'Men's Sheds' clubs [38,39].

While these initiatives are most relevant in primary care [40,41], similar approaches are increasingly being used at the primary—secondary care interface and are even under development in secondary care settings such as emergency departments and outpatient clinics [42]. Social prescribing has increased in popularity in recent years, particularly in the United Kingdom, where it is enshrined within national policies such as the NHS Long Term Plan and 2019 General Practice Contract [43,44].

However, the approach is also gaining momentum in North America [45], Australia [46] and Scandinavia [47].

Models of social prescribing

Social prescribing interventions are grounded in the values of 'personalisation', tailoring the management approach to the needs and values of the individual [48,49]. Therefore, schemes are diverse, highly variable and locally distinctive. They are determined by geographical factors, existing community assets and resources, perceived local need, relationships between stakeholders, available funding, infrastructure support and local leadership.

The diversity and breadth of schemes is both social prescribing's greatest asset and most significant challenge, facilitating a holistic approach to an individual and community at the expense of standardisation, generalisability and evaluative ease.

Approaches typically involve a signposting or connecting step from primary care to an external resource. A variety of transfer models have been proposed, ranging from simple information services, for example online directories, through to referrals to an outside agency that provides the social prescribing activities or resources [40,50].

Models frequently involve referral to a nonclinically trained individual, who acts in an enabling role to assess and establish the needs of an individual and codevelop and support management strategies that make best use of available community resources. Such individuals are variously described as 'link navigators', 'community connectors', or 'health advisors'. They exist within diverse organisational frameworks, including being employed by primary healthcare agencies, voluntary organisations or a partnership of cross-sector stakeholders.

Their involvement ranges from resource signposting through light, medium and intense interventions, depending on the required level of engagement [51]. For example, intense interventions might involve conducting a holistic needs assessment with an individual and helping codevelop and support management strategies to meet them — even attending or transporting the person to appointments or classes/clubs if required (Fig. 5.1).

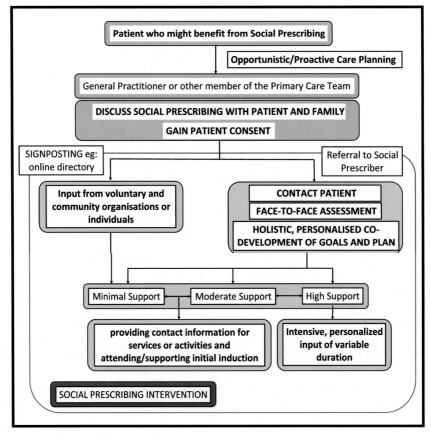

FIGURE 5.1

Social prescribing model.

Broader aspects of social prescribing may include motivating citizens to engage in local projects or community development. This could encompass neighbourhood watch schemes or litter collection days. With sufficient grassroots support, social prescribing schemes may rapidly become very popular [52]. It is essential to ensure sufficient support and resource within the community to facilitate such demand, promote service sustainability and fill service gaps for unmet needs. These elements depend on efficient communication and collaboration between sectors and appropriate funding streams. Therefore, two of the greatest challenges for many social prescribing initiatives are how best to develop intersectoral relationships and cross-sector working, while ensuring adequate funding for up-front investment in nontraditional interventions.

Why should social prescribing work?

A number of theoretical models have been proposed to underpin the theory of change behind social prescribing. These aim to provide mechanistic explanations for the link between societal factors and health outcomes. Importantly for health policy development, a better understanding of this relationship would facilitate a priori development of effective, efficient interventions, rather than relying on reactive adjustment to evaluation findings.

Identifying and promoting qualities that enhance health

Some theories as to why social prescribing can be successful explain it in terms of promoting individual qualities that enhance health. Antonovsky believes it is necessary to optimise an individual's resources and capabilities to allow them to create and maintain a state of health [53]. His *salutogenesis theory* states that this is possible by promoting an individual's understanding of their environment and enhancing their capacity to respond coherently to the stress of continual change [54]. Coherence represented an individual's ability to view life as structured, manageable and meaningful [54] through their thoughts, actions and general existence. More 'coherent' individuals are able to identify, use and reuse resources at their disposal to achieve better health outcomes [55]. These might be general or specific to situations [54] and operate at the level of the individual, family, community or wider society [56].

Practical applications of this theory are often seen within social prescribing, with broader models including communities. For example, *asset-based* approaches, such as peer support groups or group counselling sessions, encourage individuals to develop 'positive, empowering' attitudes, rather than 'negative, disempowering' attitudes in dealing with societal and health challenges. *Glass-half-full* approaches focus on supporting individuals to value their existing skills, knowledge, relationships, connections and potential within societies. For example, those with physical or learning disabilities may be supported and encouraged to participate in part-time, protected employment. Such approaches empower individuals and communities [57]. From 'isolates' disinterested in participation towards a collective aim, they are transformed into 'activists', eager, energetic and engaged participants [58].

Improving relationships

Other proposals support specific elements of these more general theories. Wakefield's 'Social Cure' [59], for example, focusses on the impact of relationships, particularly group membership. They argue that a sense of belonging and emotional support is offered within a group, as well as information and practical assistance, whereas social isolation heightens the risk of mental and physical illness [60]. In a study exploring two different social prescribing initiatives, they found consistent, statistically significant changes in well-being that were mediated by group membership, a sense of belonging and social support. These factors also positively impacted interactions with GPs and social prescribing staff, adherence to the suggested programmes and efficacy of treatment [61].

Enhancing control and autonomy

In contrast, others have focussed on how individuals build a sense of control and autonomy in their lives and the importance of this in improving health and well-being. Amartya Sen's *'Development as Freedom'* outlined the importance of human 'capabilities', i.e. their ability to use opportunity and freedom to generate valuable outcomes that would enable an individual to progress from poverty by retaking control of their life. He observed that throughout the world, there are relative disparities in political freedom, economic facilities, social opportunities, transparency and security. These could lead to feelings of a lack of control and powerlessness, which, Sen claims, are fundamental causes for the socioeconomic inequalities in health seen throughout different populations [62].

Whitehead et al. developed this concept, proposing a mechanistic link to explain how actual or perceived lack of control at the micro/personal, meso/community and macro/societal levels could result in societal and health inequalities. Lower control at the micro/personal level acts directly via greater exposure to health-damaging environments such as potentially dangerous working environments or having to live in cheaper areas with higher levels of pollution, and indirectly through chronic stress. At the meso/community level, the interaction between disadvantaged people and their disadvantageous environment results in a sense of collective threat and powerlessness that causes chronic stress and negative health outcomes. In contrast, community empowerment may encourage members to work together to challenge the unhealthy material conditions of their environment. They could work to improve or attract resources to their environment and enhance community spaces. Finally, at the macro/societal level, individuals are subject to cultural, social or political processes that can cause social exclusion or discrimination [63].

In reality, the different theories that could explain why social prescribing might be successful are not comprehensive or mutually exclusive. For example, the community and neighbourhood empowerment generated through the *asset-based* or *glass half-full* approach is thought to enhance an individual's experienced control over their life by promoting opportunities for direct participation in democracy and decision-making processes, increasing social contact through closer community networks and relationships and promoting confidence in an individual's — or community's — capacity to control their own circumstances [64].

Does social prescribing work?

Despite its theoretically rational basis, real-world evidence is required to demonstrate the impact of social prescribing. Providing a definitive answer is challenging, dependent on who is asked/asking, how the question is defined and the approach taken to answer it. A recent systematic review deemed the current evidence insufficient to provide definitive guidance on what, if anything, works [65].

Much of the problem lies in the challenge of evaluating complex interventions to 'wicked' problems, i.e. those that are firmly resistant to definitive solution [66]. Given that social prescribing aims to restore a personalised approach, there is necessarily great variation in the nature of interventions at both 'project' and 'individual participant' levels, confounded further by the wealth of additional influences that affect health and well-being.

Alongside these challenges, social prescribing interventions tend to be localised to a particular place or population; therefore, evaluations are frequently relatively small, making it difficult to generate meaningful quantitative results. Moreover, the expense of conducting thorough evaluations often results in poorly designed studies lacking standardised outcomes, which do not take account of the wider influences on health and well-being and with too short a duration to accurately determine any impact [65].

Arguably the greatest challenge lies in defining the aims and expected outcomes or outputs for social prescribing. These may be specific to each scheme − and indeed participants − and are subject to stakeholder agendas and local pressures. Therefore, an objective overview of the 'impact' of a particular scheme or social prescribing is perhaps an unobtainable ideal.

However, in the United Kingdom, commissioners, practitioners, providers, evaluators and other stakeholder groups have come together to create a consensus on what outcomes should be evaluated. This Common Outcomes Framework is designed to capture information relating to impacts on

- **the person**, for example physical or psychological health and well-being measures;
- **the community**, including experienced social capital or social determinants of health, such as sense of security, levels of homelessness, crime, or unemployment, and the make-up or dynamics of local community or voluntary sectors;
- **the health and care system**, for example the number of primary care or emergency department presentations.

Attempting to establish the cost-effectiveness and sustainability of any social prescribing scheme is also crucial to truly understand the value of an intervention [52].

Many evaluations of social prescribing schemes employ a mixed-methods approach, generating a combination of quantitative and qualitative data, which allows for quantitative demonstration of process elements and specific impacts.

A recent systematic review of 86 'nonclinical' interventions in primary care, such as befriending services and art therapy classes, revealed that less than half conducted any evaluation. However, within these, which included a number of small randomised control trials, improvements were seen in mental well-being, anxiety/and or depression scores or increases in self-esteem, confidence and empowerment. There were also positive changes in physical health and health behaviours, reductions in social isolation and loneliness scores and in measures of primary and secondary care usage [67].

A number of social prescribing evaluations have attempted to estimate the economic benefits resulting from interventions. Health Connections Mendip, in Frome, UK, demonstrated a 14% drop in unplanned hospital admissions between 2013 and 2017, with an associated 21% reduction in healthcare costs. This followed the introduction of a scheme which developed community support, connections and interventions that targeted individuals with chronic health problems or following hospital discharge [68]. Similarly, the Wellspring social prescribing project in Bristol, UK, which involved referral of a patients to a healthy living centre with activities, support and health services, demonstrated a return of £2.80 for every £1.00 invested [69], using a social return on investment calculation. However, critics have argued that such estimates will always be unreliable, given the large number of underlying assumptions and nonmarket proxies [70].

While the majority of available evidence currently arises from the United Kingdom, a systematic review of 36 interventions in the United States targeting patients' social and economic needs noted some improvements in various socioeconomic determinants of health, health outcomes and healthcare utilisation and cost measures — albeit noting mixed results and concerns about methodological flaws within many studies.

Despite the challenges in providing 'hard' quantitative evidence for efficacy, this does not mean social prescribing does not work or should not be promoted within health and social policies. Qualitative evidence reveals a high level of patient satisfaction with social prescribing schemes. They particularly value the trusting, supportive relationship frequently developed with a link worker and their knowledge of the community resources, alongside the space, time and legitimisation to focus on addressing social problems [71–74].

Moreover, there is extensive qualitative evidence to show improvements in managing a range of chronic physical and mental health conditions, impacts on a range of health-related behaviours such as healthy eating, physical activity and weight loss and subjectively reported increases in well-being, social isolation, resilience, confidence and problem-solving [71–74] (Fig. 5.2).

To aid in meaningful evidence development, social prescribing programmes need to be founded with a strong and transparent understanding of the intended impacts and proposed mechanisms to achieve these. It is essential to consider how each fits into and affects the communities and wider

One lunchtime in England, UK, a GP was asked to do a home visit by the district nursing team. They were concerned one of their elderly patients, Rose, might have a urine infection. Rose lived alone, had diabetes, poor vision, chronic hip pain and mild cognitive impairment. These problems made it difficult for her to leave the house. The GP knew her two daughters had successful, busy jobs in London and rarely visited and she suspected Rose was rather lonely. During the visit, the GP noticed Rose was very thin and there didn't seem to be much food, although there was a pile of unopened post.

When she returned to the practice, she referred Rose to Sally, the practice's social prescriber. Sally visited Rose and had a long chat. They discussed how hard it was for Rose to get to her many medical appointments and that she was confused by the post and all the bills she received. Rose disclosed that she was unable to get out to do any shopping and was dependent on another elderly neighbour to bring her some bread, milk and ready meals. Above all, Rose discussed how lonely she was and how angry and sad the loss of her health and independence made her.

Over the next few weeks, Sally worked with Rose to develop a plan. They agreed that Sally would contact a local voluntary transport service to help take Rose to her medical appointments. Together they referred Rose to a charity that provided help with the elderly in dealing with finances; Sally organised a local volunteer to help Rose do some online food shopping once a week and together they approached a local charity shop to see if they might have space for Rose to volunteer there. Soon Rose was working two mornings a week. Although her poor health and lack of full independence still upset her, she felt less isolated and scared and stopped dreading the future. She now had a purpose every day and could see some meaning to life again.

FIGURE 5.2

Case Study of a successful social prescribing referral.

health- and social care system in which they operate, as well as the impact each has on individuals and their families [74]. High-quality, mixed-methods evaluations need to become routine components of interventions and receive the necessary time, resources and expertise to be conducted effectively. Finally, policy debates need reframing such that the value of localised, high-quality, narrative evidence — traditionally considered 'low level' — is used appropriately to inform commissioning and policy development.

Challenges for social prescribing

The growing global popularity and enthusiasm for social prescribing-style approaches makes it essential to develop an appropriate awareness and scrutiny of what they can or cannot achieve and the challenges that must be overcome to do so effectively.

Arguably one of the greatest risks of social prescribing is that it is viewed as a panacea for all society's social ills, thus detracting attention and resources away from policies and approaches specifically targeting poverty, unemployment, homelessness, etc. To avoid this, interventions need clear aims, anticipated outcomes and boundaries. Funding streams must be transparent, and all should evaluate any detrimental external impact [75].

Moreover, social prescribing programmes do not include community resource mapping and development risk bypassing the local community. This sort of streamlined connecting/signposting transactional service is relatively simple to implement but does not necessarily bring investment, employment or economic growth to a deprived area. Nor does it develop social capital or community independence and resilience [76]. Similarly, such an approach may turn into a 'doing to' model, i.e. removing the incentive and opportunity for individuals and communities to help themselves; again, not sustainable in the longer term.

Alongside, these wider considerations, specific challenges and endless questions will arise when trying to design and commission an optimal service. What is the most suitable model to meet the local needs, stakeholder agendas and funding availability? Over what sort of population and geographic scale will this be most efficient and pragmatically possible? What sort of commissioning model would produce the best outcomes — payment for process or outcomes? How should healthcare professional 'buy-in' be encouraged, particularly in strained healthcare systems? How will outcomes and impacts be assessed and demonstrated?

One of the most fundamental elements in any 'link worker' model is the role of these individuals. Recent work has demonstrated great heterogeneity across the United Kingdom in how the role is interpreted and implemented [77]. Feedback from numerous projects emphasises the significance of these individuals in determining the efficacy of any 'connecting' intervention [78,79]; thus, it is vitally important to consider how they are selected, employed, trained, supported and protected [80].

There is a great risk that social prescribers becoming the 'dumping ground' for insoluble, complex problems that do not fit neatly within any other professional's remit. They may not be prepared or indeed appropriate to deal with such issues, and if we are to promote compassionate interventions within society, above all else, we must surely ensure that those providing them are treated with equal care.

How to get involved in social prescribing

Internationally, primary care teams tend to be the gatekeepers of and principal referrers to available social prescribing services — although self-referral is sometimes an option. Therefore, if patients, families or other community members are keen to explore possibilities, discussing approaching the local primary care providers is a sensible first step. Other social care, local authority or voluntary sector services; community notice boards and media outlets or online social media platforms and search engines may also be good sources of information for local activities, support groups or offer useful advice.

Given that social prescribing is thought to particularly benefit the isolated and vulnerable within communities, including the elderly or housebound, unemployed, or those with chronic physical and mental health conditions, solely relying on such proactive searching will only exacerbate any health inequalities. While all members of the health- and social care teams should use opportunistic or proactive care planning contacts to identify and refer any who may be able to benefit from social prescribing, the most successful programmes will likely include multiple, different outreach attempts.

Social prescribing is more often a hand held out to those in a dark place, rather than those in the dark searching for a hand. However, by whatever means they meet, there is no denying the impact and value of the partnership for some individuals and their communities.

Summary

- The majority of health outcomes are determined by psychosocioeconomic factors.
- Social prescribing aims to provide 'nonmedical' options for healthcare practitioners when confronted by complex problems with psychosocioeconomic roots.
- There are many diverse models, but all aim to provide personalised care. The majority include some kind of sign-posting and/or link worker.
- Some social prescribing schemes include an element of community development.
- Various theoretical models exist to explain how social prescribing might work, but there is insufficient evidence to definitely demonstrate effectiveness.
- Key challenges include defining the remit and limitations of interventions; developing and supporting link workers; ensuring adequate funding support; generating buy-in from, and developing relationships between, stakeholders and appropriately evaluating interventions.
- While social prescribing is currently most popular within the United Kingdom and countries with similar healthcare systems, for example Scandinavia and Australia, international enthusiasm is growing, and social prescribing is likely here to stay.

References

[1] Frankl V. Man's search for meaning. London: Random House; 2004.
[2] Prentice T. Health, history and hard choices: funding dilemmas in a fast-changing world. 2010.
[3] Life expectancy at birth, total (years) - data. 2012. data.worldbank.org.
[4] Gutman G. American Society on Aging. 2019 [cited 2019]. Available from: https://www.asaging.org/blog/global-look-oldest-old-and-centenarians-it-genes-diet-luck-or-all-combined.
[5] CDC. Ten great public health achievements—United States, 1900−1999. J Am Med Assoc 1999;281(16): 1481.

[6]   Marmot M. Fair society, healthy lives: the Marmot Review: strategic review of health inequalities in England post-2010. Institute of Health Equity; 2010.

[7]   CSDH. Final Report. Closing the gap in a generation: health equity through action on the social determinants of health. Geneva: World Health Organization; 2008.

[8]   Gourevitch MN, Athens JK, Levine SE, Kleiman N, Thorpe LE. City-level measures of health, health determinants, and equity to foster population health improvement: the city health dashboard. Am J Publ Health 2019;109(4):585–92.

[9]   Walsh D, Bendel N, Jones R, Hanlon P. It's not 'just deprivation': why do equally deprived UK cities experience different health outcomes? Publ Health 2010;124(9):487–95.

[10] Reid M. Behind the "Glasgow effect". Bull World Health Organ 2011;89(10):706–7.

[11] A/RES/217(III). UN; 1948.

[12] WHO. International conference on primary health care. Alma-Ata. USSR; 1978.

[13] United Nations General Assembly 2019 [Available from: https://www.un.org/en/ga/].

[14] Tighe C. Redcar struggles to recover one year after steel plant closure. Financial Times; 2016.

[15] Phillips D, Curtice J, Phillips M, Perry J. British social attitudes: the 35th report. London: The National Centre for Social Research; 2018.

[16] The American Family Survey. Brigham Young University. Center for the Study of Elections and Democracy; 2018.

[17] Kosmin B, Keysar A. American religious identification survey. Hertford, Connecticut: Trinity College; 2016.

[18] Benokraitis N. Marriages & families. 7th ed. PH; 2007.

[19] Brown GW, Moran PM. Single mothers, poverty and depression. Psychol Med 1997;27(1):21–33.

[20] Kemp S. We are social and hootesuite. 2019 [cited 2019].

[21] Dunbar RI. Do online social media cut through the constraints that limit the size of offline social networks? Roy Soc Open Sci 2016;3(1):150292.

[22] Primack BA, Karim SA, Shensa A, Bowman N, Knight J, Sidani JE. Positive and negative experiences on social media and perceived social isolation. Am J Health Promot 2019;33(6):859–68.

[23] Primack BA, Shensa A, Sidani JE, Whaite EO, Lin LY, Rosen D, et al. Social media use and perceived social isolation among young adults in the U.S. Am J Prev Med 2017;53(1):1–8.

[24] Shensa A, Sidani JE, Dew MA, Escobar-Viera CG, Primack BA. Social media use and depression and anxiety symptoms: a cluster Analysis. Am J Health Behav 2018;42(2):116–28.

[25] Twenge JM, Martin GN, Campbell WK. Decreases in psychological well-being among American adolescents after 2012 and links to screen time during the rise of smartphone technology. Emotion 2018;18(6):765–80.

[26] Tromholt M, Lundby M, Andsbjerg K, Wiking M. The Facebook Experiment. Does Social Media affect the quality of our lives? The Happiness Institute; 2015.

[27] Allcott H, Braghieri L, Eichmeyer S, Gentzkow M. The welfare effects of social media - NBER working paper No. 25514. National Bureau of Economic Research; 2019. https://www.nber.org/papers/w25514.

[28] Stavrova O, Luhmann M. Social connectedness as a source and consequence of meaning in life. J Posit Psychol 2015;11:470–9.

[29] Steptoe A, Fancourt D. Leading a meaningful life at older ages and its relationship with social engagement, prosperity, health, biology, and time use. Proc Natl Acad Sci USA 2019;116(4):1207–12.

[30] Cohen R, Bavishi C, Rozanski A. Purpose in life and its relationship to all-cause mortality and cardiovascular events: a meta-analysis. Psychosom Med 2016;78(2):122–33.

[31] Torjesen I. Social prescribing could help alleviate pressure on GPs. BMJ 2016;352:i1436.

[32] A very general practice. Citizens Advice; 2015.

[33] Garcia-Perez L, Linertova R, Lorenzo-Riera A, Vazquez-Diaz JR, Duque-Gonzalez B, Sarria-Santamera A. Risk factors for hospital readmissions in elderly patients: a systematic review. QJM 2011;104(8):639–51.

[34] Schuur J, Venkatesh A. The growing role of emergency departments in hospital admissions. N Engl J Med 2012;367:391—3.

[35] Tang N, Stein J, Hsia RY, Maselli JH, Gonzales R. Trends and characteristics of US emergency department visits, 1997-2007. J Am Med Assoc 2010;304(6):664—70.

[36] Wittenburg R, McCormick B, Hurst J. Understanding emergency hospital admissions of older people. Oxford: Centre for Health Service Economics & Organisation (CHSEO); 2014.

[37] Buck D. What is social prescribing? The King's Fund; 2017.

[38] Brandling J, House W. Social prescribing in general practice: adding meaning to medicine. Br J Gen Pract 2009;59(563):454—6.

[39] Polly M. Making sense of social prescribing. University of Westminster; 2017.

[40] Kimberlee R. Developing a social prescribing approach for Bristol. Bristol: Bristol CCG; 2013.

[41] Kimberlee R. Gloucestershire clinical commissioning group's social prescribing service: evaluation report. Gloucestershire CCG; 2015.

[42] Social A. Prescribing in secondary care pilot service evaluation report July 2018. Healthy London Partnership; 2018.

[43] NHSE. A five-year framework for GP contract reform to implement the NHS long term plan. 2019.

[44] NHSE. The NHS long term plan. 2019.

[45] Alderwick HAJ, Gottlieb LM, Fichtenberg CM, Adler NE. Social prescribing in the U.S. and England: emerging interventions to address patients' social needs. Am J Prev Med 2018;54(5):715—8.

[46] Hendrie D. Social prescribing: has the time come for this idea? Roy Aust Coll Gen Pract 2018. https://www1.racgp.org.au/newsgp/clinical/social-prescribing-has-the-time-come-for-this-ide.

[47] Jensen A, Stickley T, Torrissen W, Stigmar K. Arts on prescription in Scandinavia: a review of current practice and future possibilities. Perspect Public Health 2017;137(5):268—74.

[48] Transforming social care. Local authority circular (DH). London: Department for Health; 2007.

[49] UCL. Future of healthcare in Europe — meeting future challenges: key issues in context. London: UCL European Institute; 2012.

[50] Brandling JHW. Investigation into the feasibility of a social prescribing service in primary care: a pilot project. Bath, UK: University of Bath and Bath and North East Somerset NHS Primary Care Trust; 2007.

[51] Kimberlee R. What is social prescribing? Adv Soc Sci Res J 2015;2(1).

[52] Polley M, Fleming F, Wheatley J, et al. Making sense of social prescribing. University of Westminster: NHS England; 2017.

[53] Antonovsky A. Health, stress and coping. San Francisco: Jossey-Bass; 1979.

[54] Antonovsky A. Unraveling the Mystery of Health. How people manage stress and stay well. San Francisco: Jossey-Bass; 1987.

[55] Eriksson M, Lindstrom B. Antonovsky's sense of coherence scale and the relation with health: a systematic review. J Epidemiol Community Health 2006;60(5):376—81.

[56] Lindstrom B, Eriksson M. Salutogenesis. J Epidemiol Community Health 2005;59(6):440—2.

[57] Foot JH T. A glass half-full:how an asset approach can improve community health and well-being. Improvement and Development Agency; 2010.

[58] Kellerman B. Followership. Harvard. Harvard Business Press; 2008.

[59] Wakefield JRH, Bowe M, Setvenson C, et al. When groups help and when groups harm: origins, developments, and future directions of the 'Social Cure' perspective of group dynamics. Soc Pers Psychol Compass 2019;13(3).

[60] Haslam C, Jetten J, Cruwys T, Dingle GA, Haslam SA. The new psychology of health: unlocking the social cure. New York: Routledge; 2018.

[61] Halder MM, Wakefield JR, Bowe M, Kellezi B, Mair E, McNamara N, et al. Evaluation and exploration of a social prescribing initiative: study protocol. J Health Psychol 2018;1359105318814160.

[62] Sen A. Development as freedom. Oxford: Oxford University Press; 1999.

[63] Whitehead M, Pennington A, Orton L, Nayak S, Petticrew M, Sowden A, et al. How could differences in 'control over destiny' lead to socio-economic inequalities in health? a synthesis of theories and pathways in the living environment. Health Place 2016;39:51—61.

[64] The state of happiness: improvement and development agency. 2010. Available from: www.idea.gov.uk.

[65] Bickerdike L, Booth A, Wilson PM, Farley K, Wright K. Social prescribing: less rhetoric and more reality. A systematic review of the evidence. BMJ Open 2017;7(4). e013384.

[66] West Churchman C. Wicked problems. Manag Sci 1967;14(4):B141—6.

[67] Chatterjee H, Camic P, Lockyer B, Thomson L. Non-clinical community interventions: a systematised review of social prescribing schemes. Arts Health 2017;10(2).

[68] Abel J, Kingston H, Scally A, Hartnoll J, Hannam G, Thomson-Moore A, et al. Reducing emergency hospital admissions: a population health complex intervention of an enhanced model of primary care and compassionate communities. Br J Gen Pract 2018;68(676):e803—10.

[69] Kimberlee R. What is the value of social prescribing? 2016.

[70] Mook L, Mairoano J, Ryan S, Armstrong A, Quarter J. Turning social return on investment on its head. Nonprof Manag Leader 2015;26(2):229—46.

[71] Moffatt S, Steer M, Lawson S, Penn L, O'Brien N. Link Worker social prescribing to improve health and well-being for people with long-term conditions: qualitative study of service user perceptions. BMJ Open 2017;7(7). e015203.

[72] South J, Higgins T, Woodall J, White S. Can social prescribing provide the missing link? Prim Health Care Res Dev 2008;9(4):310—8.

[73] Wildman JM, Moffatt S, Steer M, Laing K, Penn L, O'Brien N. Service-users' perspectives of link worker social prescribing: a qualitative follow-up study. BMC Publ Health 2019;19(1):98.

[74] Faulkner M. Supporting the psychosocial needs of patients in general practice: the role of a voluntary referral service. Patient Educ Counsel 2004;52(1):41—6.

[75] Harrison K. Social Prescribing: let's not lead in without the evidence. BMJ Opin 2018;383:30.

[76] Graham B. Power to Change. 2019 [cited 2019]. Available from: https://www.powertochange.org.uk/blog/social-prescribing-two-good-two-bad/.

[77] Tierney S, Wong G, Mahtani KR. Current understanding and implementation of 'care navigation' across England: a cross-sectional study of NHS clinical commissioning groups. Br J Gen Pract 2019;69(687): e675—81.

[78] White J, Kinsella K. An evaluation of social prescribing health trainers in South and West Bradford. Leeds: Leeds Metropolitan University; 2010.

[79] Carnes D, Sohanpal R, Frostick C, Hull S, Mathur R, Netuveli G, et al. The impact of a social prescribing service on patients in primary care: a mixed methods evaluation. BMC Health Serv Res 2017;17(1):835.

[80] Frostick C, Bertotti M. The frontline of social prescribing - how do we ensure link workers can work safely and effectively within primary care? Chron Illness 2019;1742395319882068.

Mental health and wellbeing

The role of stress in health and disease

6

Athanasios Hassoulas

Programme Director MSc Psychiatry, School of Medicine, Cardiff University, Cardiff, United Kingdom

Introduction

The ability of the human population to manage stress has, through millions of years of natural selection, been integrally hardwired into a number of physiological systems. These enable an appropriate response to threats in the environment. The stress response is arguably as crucial to survival as are other basic functions of life such as eating and sleeping. Whether it be fleeing a predator's pursuit in the African Savannah or meeting an impending chapter deadline, the response to a stressor has remained consistent despite the nature of the threat changing over time and in different situations.

While the understanding of the stress response has improved significantly over the past century, we now find ourselves in a world where up to one quarter of people will, at some point in their lives, receive a clinical diagnosis of depression and anxiety. Burnout has become commonplace, and high levels of anxiety fuel risky health behaviours such as smoking, which often provide just a few minutes of relief from a constant onslaught of demands. This is the paradox of stress in the 21st century: despite a good understanding of the science underlying this primal response, we appear to be more vulnerable to crossing that diagnostic threshold where stress no longer serves an adaptive purpose.

Some may suggest that this comes down to a lack of resilience in recent generations, but this statement in itself lacks a basic understanding and appreciation of the neurobiology of stress that has taken eons to evolve. For some individuals, stress no longer serves an adaptive role. Society is on the verge of an anxiety epidemic in which the demands placed on individuals far outstrip the resources they have inherited to deal with these stressors. This raises important questions about the lifestyles that many individuals now lead.

As the stress response has been hardwired into interrelated biological systems that have several different functions, it is clear that dysregulation caused by persistent stress will inevitably disrupt other key biological systems. To understand a maladaptive stress response, it is important to initially consider how these systems perform as part of the normal, adaptive stress response.

Psychological theories of stress

Various approaches to stress have attempted to provide a universal definition of the emotional and physical responses that are perceived by each individual in a rather subjective manner. For instance, stress has been defined as *the body's nonspecific response to demands made on the organism* or as a *reaction to aversive events in the environment* [1,2].

Lazarus (1966) reported that the response to stress is influenced by the perception of the demands placed, and whether an individual feels they possess the necessary resources to meet such demands [3]. Some of the early definitions of stress therefore considered two crucial variables: the objective and universal physiological response, and the psychological appraisal of the stressor and the situation. Consequently, it is impossible to consider the physiological without considering the psychological, and vice versa.

Perhaps the most well-known and mainstream theory of stress is that of the **fight-or-flight response,** introduced by Cannon (1932) [4]. The premise of this particular theory is rather straight-forward: When faced with a threat, or an aversive stimulus that is perceived to be a threat, we are mobilised into action by either encountering the threat head-on (fight) or fleeing from the situation entirely (flight). Whether we stay to fight or take flight, survival is the desirable outcome.

This theory explains how behavioural responses have, over time through the power of natural selection, anatomically and physiologically crafted entire species as either supreme 'fighters' or swift and agile 'flee-ers'.

The physiological arousal that thrusts the organism into action in response to a threat is caused by the activation of the autonomic nervous system (ANS) and its two branches: the sympathetic division and the parasympathetic division. Furthermore, activation of key structures of the limbic system is fundamental in regulating the stress response. The amygdala is considered to be the 'threat detection' centre of the brain, whereas the hippocampus and hypothalamus are of vital importance in perceiving the danger and generating the fear that drives the stress response [5].

When faced with a stressor, corticotropin-releasing hormone (CRH) is produced by the hypothalamic paraventricular nucleus (PVN), following excitation of brainstem catecholamine-producing pathways that project to CRH-containing neurons of the hypothalamic PVN [6,7]. In turn, sympathetic and parasympathetic activity is mediated by projections to sympathetic and parasympathetic preganglionic neurons in the brainstem. Activation of sympathetic preganglionic α1-adrenoreceptors leads to an increase in sympathetic arousal, whereas activation of parasympathetic preganglionic α2-adrenoreceptors is found to decrease parasympathetic activity [8].

The sympathetic division is crucial in activating the stress response when confronted with an immediate threat, whereas the parasympathetic division hits the brakes on the stress response once the danger has passed. Both branches of ANS regulate normal function in relation to adaptive stress, by mobilising the organism in response to danger and subsequently dampening the response once the threat abated. This suggests that the stress response may be divided into various stages whereby the organism moves through experiences of alarm, sympathetic arousal and later parasympathetic compensation.

Selye (1956) proposed a three-stage approach to stress, in his general adaptation syndrome (GAS) model, that considers both sympathetic as well as parasympathetic involvement [9].

During the first stage, the organism confronts the stressor and is motivated into action as a result of sympathetic arousal. This alarm stage therefore corresponds with the fight-or-flight response described by Cannon [4]. Once the organism engages with the stressor, the stage of resistance ensues, with ongoing exposure to the aversive event, leading to parasympathetic compensation to restore homeostasis. This leads to the third and final stage of exhaustion, where physiological and psychological resources have been depleted and require replenishing (Fig. 6.1).

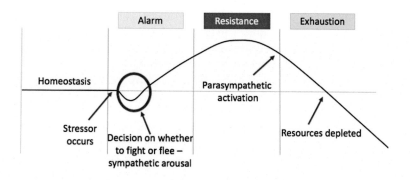

FIGURE 6.1

The three-stage general adaptation syndrome (Selye, 1956). Sympathetic arousal during the alarm stage leads to activation of the stress response, with parasympathetic activation restoring homeostatic balance [9].

Selye's GAS model has been pivotal to our understanding of the physiology underpinning the stress response, with an emphasis on the nonspecific and universal nature of this physiological reaction regardless of the specific nature of the threat or stressor [9]. However, while Selye perfectly described the physiology of stress by building on Cannon's fight-or-flight response, he gave little consideration to the significance that perception and other cognitive processes play in this process. Furthermore, the role of anticipatory anxiety, which describes the anxiety experienced in anticipation of an aversive stimulus, was largely overlooked.

It was Lazarus (1966) who considered how cognitive processes specifically influence the experience of stress. He performed a series of experimental and clinical studies [3] which culminated in a new theory of stress, the cognitive appraisal theory. This aimed to unify psychological processes with physiological responses [10]. Moreover, as anticipatory anxiety had been found to be regulated in part by the amygdala, this suggested that threat detection and the onset of the stress response occur prior to the actual presentation of an aversive event, i.e. the 'alarm stage' [11]. Dysregulation of neural circuits that include the amygdala as well as other important limbic structures may explain chronic negative expectation and anticipation observed in several psychopathologies.

Lazarus and Folkman proposed that cognitive appraisal was deemed to consist of two parts: primary and secondary appraisal [10]. The primary appraisal process is suggested to involve the individual's assessment of the situation within which the stressor, or threat, has been encountered. As such, the individual carries out an assessment of the level of risk and whether the situation is one that carries the potential to cause harm. Once this assessment has been made, secondary appraisal occurs whereby the individual evaluates whether the resources required to meet the demands of the situation are sufficient, and what coping strategies are required. Once appraisal of the threat and the stressor has been carried out, sympathetic activation follows if appropriate which enables the individual to react accordingly.

Cognitive appraisal has, as its starting point, the perception of the threat itself. This then influences whether the stress response is subsequently activated or not. The process of perception requires information from the environment, via the five senses, being processed and interpreted by key limbic structures that decipher the level of threat posed by a stimulus. Neural pathway transmission from sensory processing areas, such as the thalamus and sensory cortices, to the amygdala is vital in threat detection. In addition to sensory information, the amygdala also receives visceral input and is also reciprocally connected with key structures involved in memory and learning, such as the hippocampus [12].

The amygdala is therefore considered to be the threat detection centre of the brain, receiving information from the environment about the nature of the stressor as well as drawing from memory to decide what risk the stimulus possess. Furthermore, the amygdala regulates hypothalamic PVN activity. As such, where a stimulus is considered to pose a threat, activation of the PVN leads to the production of CRH and the physiological experience of stress. The cognitive appraisal of the environmental stimulus or event is therefore further supported by our neurobiological understanding of threat detection and fear conditioning (Fig. 6.2).

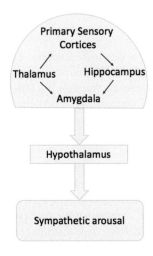

FIGURE 6.2

Sensory information is processed and relayed to the amygdala, where threat is appraised. Where threat is detected, a distress signal is sent to the hypothalamus, triggering the stress response.

Escape, avoidance and safety

The limbic system regulates the appraisal of an event and therefore raises the alarm in the face of danger. The amygdala signals to the hypothalamus that a threat has been detected, leading to autonomic arousal and activation of the stress response. The role of the amygdala in threat detection is well established [12–14]. The ability to effectively detect and respond with urgency to threats in the environment requires a type of associative learning whereby changes in synaptic strength occur in the amygdala itself.

This form of associative learning, or conditioning, underlies our experience of fear when a threat is detected. Fear conditioning is therefore a form of implicit, or nondeclarative, learning that produces automatic and unconscious responses to stimuli. Disruption of amygdala activity is implicated in certain neuropsychiatric diseases, where a deficit in fear learning is observed [15,16].

Fear conditioning is the driving force behind the escape response that is observed in all animal species when danger is detected. In humans, specifically, this escape response has been intimately linked to avoidance behaviour, whereby a stimulus that carries aversive qualities is avoided in future encounters. Avoidance is therefore a product of a type of negative reinforcement. This means that a desirable outcome, i.e. the avoidance of harm, is achieved when avoidance behaviour results in the aversive stimulus not being encountered.

As such, the acquisition and maintenance of avoidance behaviour becomes deeply engrained through a two-stage process that entails both classical and operant conditioning [17]. Specifically, the process involves the association of a previously neutral stimulus with an aversive event, with the previously neutral stimulus subsequently evoking a fear response. The second stage involves the reinforcement of the fear evoked by the previously neutral stimulus, which has effectively taken on the properties of the aversive event with which it has effectively been paired. This, in essence, provides an accurate behavioural model of avoidance behaviour as well as maladaptive fear conditioning that takes place in the development of specific and social phobias [18].

A well-known study demonstrating this model of avoidance behaviour in relation to the acquisition of a specific phobia is that of the Little Albert study [19]. Not much is known about the infant who, since the study was published, has been famously referred to by the pseudonym Little Albert. The experimental design involved a conditioning paradigm whereby the infant, the sole participant in this study, was presented initially with a live white rat. Baseline reactions to the white rat were recorded and showed no discomfort or agitation in the presence of the rat; instead the infant seemed quite curious by the furry creature's presence. The rat, therefore, was considered a neutral stimulus in this scenario.

The next phase of the experiment involved exposing the infant to an aversive auditory stimulus, a loud noise, which, as anticipated, produced a startle response in the infant. Over a series of trials that followed, the aversive auditory stimulus was paired with the rat. With time, the aversive stimulus was removed from the scenario, but the aversive qualities of the auditory stimulus had been successfully paired with the previously neutral stimulus. Presentation of the white rat alone then produced the conditioned startle response. The study effectively illustrated how a phobia could be acquired and maintained through the process of associative learning. Furthermore, the maintenance of rigid avoidance behaviour has been found to play a profound role in the pathophysiology of anxiety disorders such as obsessive-compulsive disorder (OCD) [20,21].

A recent psychological model of stress, however, suggests that previous accounts have over-emphasised the importance of aversive stimuli, or events, in triggering the stress response while ignoring the importance that the perception of safety, or the lack thereof, may play in this process. Brosschot et al. (2018) have introduced the generalised unsafety theory of stress, or GUTS, which emphasises the importance that perceived safety plays in regulating the stress response [22]. According to the model, the stress response is always active, or rather on constant standby, but is inhibited through perceived safety in the environment. The prefrontal cortex (PFC), crucial in decision-making and mediating behaviour, inhibits the stress response by keeping the 'brakes' on the amygdala during times of safety. In the absence of a safety cue or signal, the PFC is hypothesised to lift the brake on the amygdala, unleashing the stress response.

This account is consistent with early behavioural models on conditioned inhibition, where avoidance learning is not merely dependent upon the presence of stimuli that signal an aversive event but rather the presence of stimuli that signal a period of safety from the aversive event [23]. The GUTS model focuses on how safety signals are primary learned through the social context and how social environments may shape adaptive as well as maladaptive reactions to stress. Appraisal of the social context, therefore, is considered to be intricately linked to the experience of stress and pathological anxiety.

While the model retains the focus on appraisal, albeit of safety cues as opposed to aversive stimuli, it differs from earlier models of stress, suggesting that the appraisal of environmental cues lifts the brakes on a system that is quietly running in the background, as opposed to the parasympathetic division hitting the brakes on the system that was initially dormant and only activated once danger had been detected. Whether brakes are being lifted or applied, what appears to be regulating the entire process is the meaning that we ascribe to stimuli, events and cues in our environment. To better understand this rather complex relationship between the psychological and the physiological, a good starting point would be to explore further the basic physiology of stress.

Physiological and biological accounts of stress

The hypothalamus mediates a number of basic functions of life ranging from appetite control and sleep to regulating body temperature, reproductive behaviour and the stress response. Once information reaches the hypothalamus that the organism needs to respond to an immediate threat, autonomic arousal leads to the activation of two interrelated stress systems: the sympathetic–adrenal–medullary (SAM) and the hypothalamic–pituitary–adrenal (HPA) axis. Activation of the SAM and HPA axis leads to far-reaching changes in respiratory, gastrointestinal, cardiovascular, immune and central nervous systems [24]. When confronted with a stressor, the hypothalamic PVN is activated, with projections to both the sympathetic and the parasympathetic divisions of the ANS [25].

Activation of the hypothalamic PVN produces an immediate, short-acting stress response as well as a longer-lasting neurohormonal response, through the SAM and HPA axis systems, respectively. The short-lasting response occurs through the detection of a threat that leads to hypothalamic activation, signalling in the sympathetic nervous system and subsequent release of catecholamines from the adrenal medulla. The secretion of adrenaline and noradrenaline produces the observable effects and experiences of the initial stress response including **tachycardia, hyperventilation, dilated pupils and inhibited flow of saliva.** In addition, this immediate response also induces cognitive changes such as an **increase in concentration, alertness and vigilance.** As such, the initial physiological response corresponds with Cannon's fight-or-flight model, as well as the alarm stage in Selye's GAS (Fig. 6.3) [4,9].

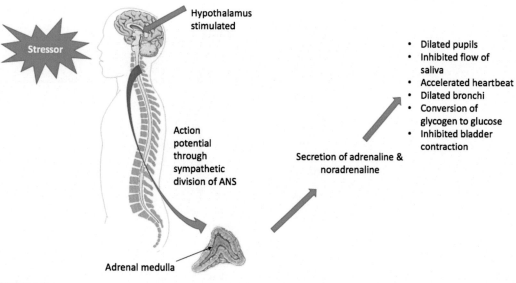

FIGURE 6.3

The sympathetic–adrenal–medullary (SAM) system produces the rapid stress response when a threat is detected in the environment.

Activation of the hypothalamic PVN furthermore produces CRH, which binds to receptors on the anterior pituitary gland and stimulates the release of adrenocorticotropic hormone (ACTH). The release of ACTH acts on the adrenal cortex, which in turn releases androgens and glucocorticoids, specifically cortisol, the 'stress hormone'. This chain of endocrine events describes the neurohormonal mechanisms of the HPA axis, which drives the longer-lasting stress response. Negative feedback is then exerted by cortisol on the hypothalamus and anterior pituitary, inhibiting further release of CRH and ACTH, restoring homeostasis (Fig. 6.4).

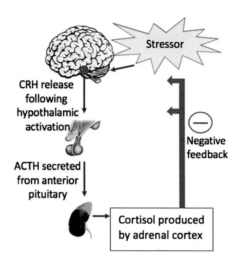

FIGURE 6.4

The hypothalamic–pituitary–adrenal (HPA) axis underpins the neurohormonal mechanisms that regulate the longer-lasting stress response through the production of cortisol.

Analogous to the definition of homeostasis, the concept of allostasis was introduced by Sterling and Eyer (1988) to describe how fluctuations in physiological systems enable organisms to meet stressful demands [26]. It is stated that to achieve and maintain equilibrium, physiological systems are in a state of constant flux as the demands that arise from a stressful situation differ greatly from nonstressful situations with fewer such demands.

It is thus suggested that neurohormonal mediators, i.e. catecholamines and glucocorticoids, produced by the interrelated stress systems play a crucial role in homeostatic maintenance and adaptation. Allostasis as a concept, therefore, goes beyond the definition of homeostasis by considering how such fluctuations influence long-term physiological and behavioural adaptation [27,28]. Adaptation is also believed to be influenced in part by the process of habituation, whereby the frequency and magnitude of a response to a specific stimulus decreases with repeated exposure to that particular stimulus. Failure to habituate, likewise, demonstrates impaired adaptation to stressful events.

Where a cumulative biological toll is observed due to chronic overstimulation of the stress system, this increase in allostatic load is considered to play a crucial role in the development of chronic pathologies, resulting in allostatic overload. This suggests that dysregulation of psychological and physiological processes that mediate the stress system may therefore subsequently lead to an impairment in physiological and behavioural adaptation.

There have been a number of questions raised, however, in relation to whether the concept of allostasis differs significantly from that of homeostasis. Specifically, it has been queried whether allostasis does in fact actually contribute a novel perspective to our understanding of the psychological and physiological processes that mediate the stress response [29,30]. Whether allostasis does enhance our understanding of homeostatic regulation or not, it is clear that stress-induced neurohormonal mediators are crucial in considering how cellular, genetic, epigenetic and physiological changes influence adaptation over time (Fig. 6.5).

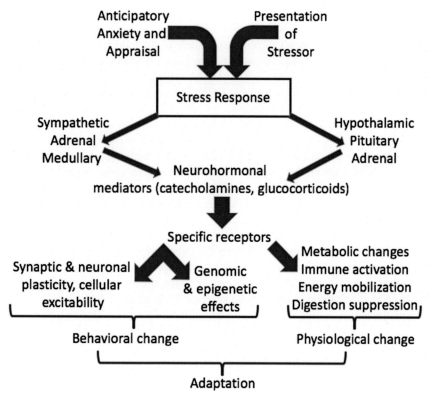

FIGURE 6.5

Behavioural and physiological changes, in response to stress, are influenced by neurohormonal mediators that act on specific receptors in maintaining homeostasis and adaptation to stressful events [24].

Adapted from: Godoy LD, Rossignoli MT, Delfino-Pereira P, et al. A comprehensive overview on stress neurobiology: basic concepts and clinical implications. Front Behav Neurosci 2018;12:127.

Stress and disease

The adaptive role that stress plays is crucial not only in terms of our activities of daily living but to our very survival. However, the stress response can become maladaptive. Vigilance becomes hypervigilance, and arousal transforms into crippling hyperarousal. Understanding when adaptive and beneficial stress, also referred to as *eustress*, crosses the diagnostic threshold is vital in relation to improving our knowledge of the stress system and the role that stress plays in disease. For instance, there is strong evidence to suggest that there is a genetic component to pathological stress. Genetic association studies have revealed that there are certain genes that could alter the stress response through their involvement with the HPA axis and sympathetic branch of the ANS.

A number of metaanalyses have hinted at the importance of serotonin transporter genes, such as SLC6A4, and the 5-HTTLPR genotype, among others, as being closely associated with pathological stress and low mood, potentially as a result of a disturbance in serotonergic signalling [31−33]. It is therefore not surprising that an association between certain genes and monoamine neurotransmitter systems has been identified in specific anxiety and mood disorders [34]. Furthermore, a relationship between candidate 'stress' genes and cardiovascular disease has also been suggested through these genes' association with inflammatory and immunological responses, as well as the renin—angiotensin—aldosterone system (RAAS).

Disruptions to key systems appears to occur as a result of the modifying effects of certain genes. Early life stressors and even maternal stress and depression are important environmental considerations in this regard [35,36].

If we consider one specific anxiety disorder, OCD, it is observed that candidate serotonin transporter genes may directly influence the functioning of neural circuits, such as the corticostriatal thalamocortical (CSTC) circuit, implicated in the pathophysiology of the condition [37]. Dysregulation of the CSTC circuit in OCD underpins the specific pathological signs of anxiety observed in this particular psychiatric condition, which include obsessional thinking and repetitively rigid rule following. Hyperactivity in the orbitofrontal cortex and ventromedial striatum appears, in part, to drive this process.

The nature of dysregulation throughout this neural circuit may also result in impaired decision-making and appraisal of stressors. In addition, the involvement of the amygdala, via disruption of the thalamoamygdala circuit, is also found to potential impact on the CSTC circuit, through a deficit in threat detection and subsequent appraisal. Where the stress response relies on early detection of a threat and the appropriate appraisal of the level of threat presented, dysregulation to circuits that include key cortical and subcortical structures involved in the stress response could ultimately lead to hyperarousal even in the absence of an aversive external stimulus (Fig. 6.6).

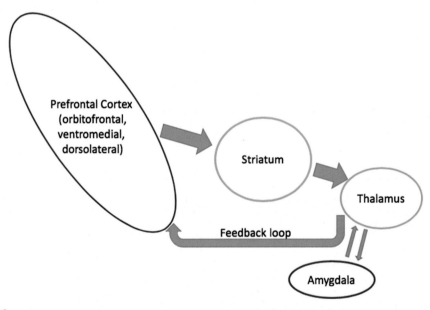

FIGURE 6.6

Schematic depiction of the corticostriatal thalamocortical (CSTC) circuit. Neurotransmitter systems within the circuit (including dopamine, serotonin, glutamate and GABA) are implicated in the pathophysiology of OCD and in the development of new treatments for the condition. *OCD*, obsessive-compulsive disorder.

While the gene—environment interaction is considered of paramount importance in the holistic understanding of the mechanisms that drive pathological anxiety, it is also important to consider a third variable in this equation: age. Lenroot and Giedd (2008) have studied adolescence as period of particular vulnerability given the neurohormonal changes that take place with the onset of puberty [38]. Surges in circulating levels of precursor androgens, such as dehydroepiandrosterone (DHEA), as well as sex steroid hormones, such as testosterone, not only exert a profound influence on the maturing adolescent body but also exert an influence on the structural organisation of the brain, in part eliciting new behaviours, for example social and reproductive.

Biological mechanisms that also contribute to grey matter changes in the brain, i.e. synaptic pruning, are also vital in preparing the adolescent brain for adulthood. Moreover, changes to HPA axis reactivity is a hallmark sign of the adolescent years, with elevated stress-induced responses influenced by key endocrine events [39]. As such, the widespread and far-reaching changes trigged by crucial neurohormonal changes that are unleashed during this stage of development provide more opportunity for suboptimal trajectories to occur where individuals may possess genetic as well as environmental susceptibilities.

For instance, it is suggested that in comparison with the adult brain, the adolescent brain may be more responsive to glucocorticoids, influencing to a larger degree gene expression specifically in the hippocampus. Changes in grey matter volume may place the individual at increased risk of depressive and anxiety disorders and once again appear to be associated with serotonin transporter genes, specifically the 5-HTTLPR genotype [40]. Adverse early life experiences have also been found to influence HPA axis reactivity, with children who have experienced significant stress and trauma not only presenting with poorer cognitive outcomes but also with delayed and stunted growth [41].

Dysregulation of the HPA axis, as a result of impaired negative feedback sensitivity, due to prolonged exposure to early life stressors, has also been found to have immunosuppressive effects as well as altering proinflammatory cytokines, such as TNF-α, interleukin 1β, 6 and 8 [42]. Elevated cytokines, which also play a part in activating the HPA system, have been implicated in the pathophysiology of depression and anxiety through the hypothesised gut—brain axis [43,44]. Disruption to these axes has additionally been found to provide new insight into gastrointestinal disorders such as irritable bowel syndrome (IBS). As such, the bidirectional relationship between the cellular components of the immune system and the systems that mediate the stress response illustrates how inextricably linked mental well-being is to physical health.

Beyond gastrointestinal disorders, the relationship between stress and disease has also been demonstrated in relation to the respiratory system. For instance, Cohen et al. (1999) illustrated that stress was found to be significantly associated with elevated IL-6 production as well as an increased risk of upper respiratory tract infections (URTIs), suggesting once again the intimate nature of the mechanisms that underlie the stress and immune systems [45]. Furthermore, the Finnish Twin Cohort Study investigated the association between chronic disease and stressful life events [46]. Specifically, the number of stressful life events experienced in the 5 years prior to the study was correlated with the risk of breast cancer in the 15 years that followed.

The findings revealed that three significant life events in particular, namely the death of a close relative or friend, death of a spouse, or divorce, were associated with an elevated risk of breast cancer in the 15-year follow-up period. The results of this particular cohort study imply that there is a clear link between the physiological changes induced, via the stress system, and the potential immunosuppressive effects of stress in breast carcinogenesis. As such, dysregulation of the stress system and metabolic pathways appear to influence the immune response in relation to tumour progression too. Moreover, Reiche et al. (2004) revealed that stress was not only associated with elevated cytokine production but also reduced natural killer (NK) cell activity, impacting on immune surveillance and tumour detection [47]. Once again, findings of this nature demonstrate the relationship between the stress system and chronic disease.

Health-compromising behaviours provide an additional pathway linking stress and disease; as such, behaviours may be acquired and maintained in response to the aversive nature of stressful life events. Such behaviours, which include alcohol consumption and tobacco smoking, might be effective in providing temporary relief from stressful events but, in the long-term, serve only to maintain and strengthen this form of avoidance behaviour and increase the risk of chronic diseases such as coronary heart disease, respiratory illness and cancer [48,49]. Prevention strategies employed to modify such health-compromising behaviours not only limit the impact of stress on chronic disease but have also been estimated to save the national health service, in Wales specifically, up to £450 million a year (Fig. 6.7) [50].

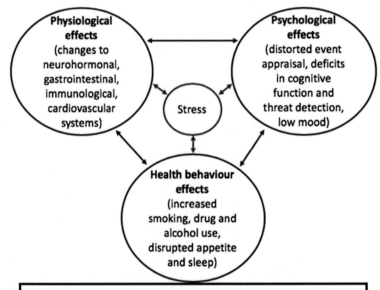

FIGURE 6.7

The interplay between pathways linking stress with disease.

Summary

The stress response is an automatic response to a perceived threat. We do not usually stop to think about what we are experiencing or what is actually even happening. The primal response launches a default protection mode to ensure that the necessary physiological resources are available to us in taking appropriate action and ensuring our ongoing survival.

For many of us, our physical environment today is rather different to that of our ancestors, but we have inherited all their primal response and urges. The brain has been hardwired to detect danger and respond by triggering a cascade of physiological events that produce the same behaviours in these situations that our ancestors would have exhibited too. Given that the response is somewhat beyond voluntary control, the nature of appraising an event may have altered over time in adjusting to our new physical environment, but the primal response led by the hypothalamus, activating both the SAM and HPA axis, remains unchanged.

It is also clear that the stress system does not exist in isolation. The experience of beneficial as well as pathological anxiety demonstrates that the stress response is closely linked to various other systems, all influencing each another. Stress is intricately interwoven into the immunological, cardiovascular, gastrointestinal and neurohormonal systems, and vice versa.

Prolonged HPA axis activation has been proven to influence cytokine production, and similarly, immunological events have been found to exert an influence over the regulation of the HPA axis through the production of various inflammatory mediators. Furthermore, prolonged periods of stress following adverse life experiences appear to influence a range of health outcomes, from urinary tract infections to tumour surveillance and coronary heart disease [51].

Disruption to key neural circuits has also been found to trigger the stress response even in the absence of an external stimulus. Such disruptions to specific circuits may either lead to deficits in threat appraisal or even trigger the stress response in the absence of any external stimulus, with a mere thought possessing the power to initiate the experience of stress. Individuals diagnosed with post-traumatic stress disorder (PTSD), for instance, are plagued by flashbacks of distressing memories, with these internal events prompting a severe stress response characterised by hyperarousal, hypervigilance and excessive avoidance behaviour. To alleviate some of the distress experienced by these intrusive and very aversive flashbacks, patients may acquire certain health-compromising behaviours that provide just some temporary relief from the stress experienced by these memories. As such, cooccurring substance abuse may be observed in several anxiety and mood disorders.

First-line treatment for pathological anxiety, on the basis of the biopsychosocial model, includes a combination of pharmacological intervention and evidence-based psychological therapeutic modalities. For instance, cognitive-behavioural therapy (CBT) is considered the first-line, evidence-based therapeutic modality for generalised anxiety disorder (GAD), panic disorder and many other anxiety disorders [52]. CBT targets distorted cognitions through a process of reappraisal and revaluation. Furthermore, the approach aims to evaluate emotional responses to events and stimuli, thereby modifying behavioural responses through a process of cognitive change. Given the pathophysiology of most anxiety disorders, selective serotonin reuptake inhibitors (SSRIs) have proven to be an effective component of first-line management of these conditions. The SSRI pathway acts on numerous processes that influence cognitive function, hormone production, emotional regulation, reward-seeking behaviour, autonomic arousal and motor function [53]. Similarly, monoamine reuptake inhibitors such as serotonin—noradrenaline reuptake inhibitors may be used in the treatment of pathological anxiety and low mood.

With advances in our understanding of the interrelated mechanisms that drive stress, new and perhaps rather controversial treatments have recently been investigated following a period of little pharmacological development in the management of anxiety disorders. For instance, methylenedioxymethamphetamine (MDMA) and 4-phosphoryloxy-*N*,*N*-dimethyltryptamine (psilocybin) have recently been proposed to offer exciting new possibilities in the treatment of PTSD [54]. Advances in our understanding of the physiological, psychological and behavioural underpinnings of stress may usher in a new era of managing not only complex mental illnesses but also physical disease. These advances may also equip us with the knowledge we need to perhaps cope a little better in a fast-paced world filled with stressors that our species has never before had to encounter.

Summary

- Stress is a psychological and physiological response to threat that plays a crucial and adaptive role in our everyday lives.
- The stress response has far-reaching effects, as it is intimately interwoven in immunological, gastrointestinal, cardiovascular and neurohormonal systems.
- The hypothalamus plays a crucial role in activating two interrelated stress systems, the SAM and HPA axis, in response to threat.
- Appraisal of an aversive event or stimulus is crucial in activating the stress response, thereby leading to sympathetic arousal.
- Secretion of catecholamines and stress hormones, such as cortisol, is crucial in mobilising the organism in response to danger.
- Genetic and/or environment factors can be associated with dysregulation of crucial stress systems.
- Abnormal activation of the SAM and HPA axis can impact on mental and physical health.
- Severe and prolonged experiences of stress have been associated with psychiatric disorders including mood, anxiety and eating disorders, and psychosis.
- Stressful life events have furthermore been implicated in immunosuppression, gastrointestinal disturbances and cardiovascular disease.
- Effective management of stress can improve physical and mental well-being.

Acknowledgements

I would like to thank my wife, Eliana, my mother, Nicky, my sister, Antonia, and my late father, Vasili, for all their love and support in my own struggles with obsessive-compulsive disorder. Thank you also to Dr Emma Short for the opportunity to contribute to this book.

References

[1] Selye H. Stress without distress. 1974.
[2] Rosenhan DL, Seligman MEP. Abnormal psychology. 1989.
[3] Lazarus RS. Psychological stress and the coping process. 1966.
[4] Cannon WB. The wisdom of the body. 1932.
[5] Jankord R, Herman JP. Limbic regulation of hypothalamo-pituitary-adrenocortical function during acute and chronic stress. Ann N Y Acad Sci 2008;1148:64—73.

[6] Herman JP, Cullinan WE. Neurocircuitry of stress: central control of the hypothalamo-pituitary-adrenocortical axis. Trends Neurosci 1997;20(2):78—84.

[7] Aguilera G, Liu Y. The molecular physiology of CRH neurons. Front Neuroendocrinol 2012;33(1):67—84.

[8] Won E, Kim YK. Stress, the autonomic nervous system, and the immune-kynurenine pathway in the etiology of depression. Curr Neuropharmacol 2016;14(7):665—73.

[9] Selye H. The stress of life. New York: Mc Graw-Hill; 1956.

[10] Lazarus RS, Folkman S, Dawsonera. Stress, appraisal, and coping. 1984.

[11] Schulkin J, McEwen BS, Gold PW. Allostasis, amygdala, and anticipatory angst. Neurosci Biobehav Rev 1994;18(3):385—96.

[12] Janak PH, Tye KM. From circuits to behaviour in the amygdala. Nature 2015;517(7534):284—92.

[13] Ohman A. The role of the amygdala in human fear: automatic detection of threat. Psychoneuroendocrinology 2005;30(10):953—8.

[14] Peck CJ, Salzman CD. Amygdala neural activity reflects spatial attention towards stimuli promising reward or threatening punishment. Elife 2014;3.

[15] Hamann S, Monarch ES, Goldstein FC. Impaired fear conditioning in Alzheimer's disease. Neuropsychologia 2002;40(8):1187—95.

[16] Zhang X, Ge TT, Yin G, et al. Stress-induced functional alterations in amygdala: implications for neuropsychiatric diseases. Front Neurosci 2018;12:367.

[17] MOWRER OH. Two-factor learning theory: summary and comment. Psychol Rev 1951;58(5):350—4.

[18] Krypotos AM, Effting M, Kindt M, et al. Avoidance learning: a review of theoretical models and recent developments. Front Behav Neurosci 2015;9:189.

[19] Watson JB, Rayner R. Conditioned emotional reactions. Am Psychol 1920;55(3):313—7. 2000.

[20] Hassoulas A, McHugh L, Reed P. Avoidance and behavioural flexibility in obsessive compulsive disorder. J Anxiety Disord 2014;28(2):148—53.

[21] Hassoulas A, McHugh L, Morris H, et al. Rule-following and instructional control in obsessive-compulsive behavior. Eu J Behav Anal 2017;18(2):276—90.

[22] Brosschot JF, Verkuil B, Thayer JF. Generalized unsafety theory of stress: unsafe environments and conditions, and the default stress response. Int J Environ Res Publ Health 2018;15(3).

[23] Rescorla RA, Lolordo VM. Inhibition of avoidance behavior. J Comp Physiol Psychol 1965;59:406—12.

[24] Godoy LD, Rossignoli MT, Delfino-Pereira P, et al. A comprehensive overview on stress neurobiology: basic concepts and clinical implications. Front Behav Neurosci 2018;12:127.

[25] Geerling JC, Shin JW, Chimenti PC, et al. Paraventricular hypothalamic nucleus: axonal projections to the brainstem. J Comp Neurol 2010;518(9):1460—99.

[26] Fisher S, Fisher S, Reason JT. Handbook of life stress, cognition, and health. 1988.

[27] McEwen BS. Allostasis and allostatic load: implications for neuropsychopharmacology. Neuropsychopharmacology 2000;22(2):108—24.

[28] Ramsay DS, Woods SC. Clarifying the roles of homeostasis and allostasis in physiological regulation. Psychol Rev 2014;121(2):225—47.

[29] Day TA. Defining stress as a prelude to mapping its neurocircuitry: no help from allostasis. Prog Neuropsychopharmacol Biol Psychiatry 2005;29(8):1195—200.

[30] Davies KJ. Adaptive homeostasis. Mol Aspect Med 2016;49:1—7.

[31] Ising M, Holsboer F. Genetics of stress response and stress-related disorders. Dialogues Clin Neurosci 2006; 8(4):433—44.

[32] Lam D, Ancelin ML, Ritchie K, et al. Genotype-dependent associations between serotonin transporter gene (SLC6A4) DNA methylation and late-life depression. BMC Psychiatry 2018;18(1):282.

[33] McGuffin P, Rivera M. The interaction between stress and genetic factors in the etiopathogenesis of depression. World Psychiatry 2015;14(2):161—3.

[34] Barnett JH, Xu K, Heron J, et al. Cognitive effects of genetic variation in monoamine neurotransmitter systems: a population-based study of COMT, MAOA, and 5HTTLPR. Am J Med Genet B Neuropsychiatr Genet 2011;156(2):158—67.

[35] Araya R, Hu X, Heron J, et al. Effects of stressful life events, maternal depression and 5-HTTLPR genotype on emotional symptoms in pre-adolescent children. Am J Med Genet B Neuropsychiatr Genet 2009;150B(5): 670−82.

[36] Capron LE, Ramchandani PG, Glover V. Maternal prenatal stress and placental gene expression of NR3C1 and HSD11B2: the effects of maternal ethnicity. Psychoneuroendocrinology 2018;87:166−72.

[37] Nestadt G, Grados M, Samuels JF. Genetics of obsessive-compulsive disorder. Psychiatr Clin 2010;33(1): 141−58.

[38] Lenroot RK, Giedd JN. The changing impact of genes and environment on brain development during childhood and adolescence: initial findings from a neuroimaging study of pediatric twins. Dev Psychopathol 2008;20(4):1161−75.

[39] Romeo RD. The teenage brain: the stress response and the adolescent brain. Curr Dir Psychol Sci 2013; 22(2):140−5.

[40] Little K, Olsson CA, Whittle S, et al. Association between serotonin transporter genotype, brain structure and adolescent-onset major depressive disorder: a longitudinal prospective study. Transl Psychiatry 2014;4: e445.

[41] Loman MM, Wiik KL, Frenn KA, et al. Postinstitutionalized children's development: growth, cognitive, and language outcomes. J Dev Behav Pediatr 2009;30(5):426−34.

[42] Silverman MN, Pearce BD, Biron CA, et al. Immune modulation of the hypothalamic-pituitary-adrenal (HPA) axis during viral infection. Viral Immunol 2005;18(1):41−78.

[43] Carabotti M, Scirocco A, Maselli MA, et al. The gut-brain axis: interactions between enteric microbiota, central and enteric nervous systems. Ann Gastroenterol 2015;28(2):203−9.

[44] Farooq RK, Asghar K, Kanwal S, et al. Role of inflammatory cytokines in depression: focus on interleukin-1β. Biomed Rep 2017;6(1):15−20.

[45] Cohen S, Doyle WJ, Skoner DP. Psychological stress, cytokine production, and severity of upper respiratory illness. Psychosom Med 1999;61(2):175−80.

[46] Lillberg K, Verkasalo PK, Kaprio J, et al. Stressful life events and risk of breast cancer in 10,808 women: a cohort study. Am J Epidemiol 2003;157(5):415−23.

[47] Reiche EM, Nunes SO, Morimoto HK. Stress, depression, the immune system, and cancer. Lancet Oncol 2004;5(10):617−25.

[48] Friedman M, Rosenman RH. Association of specific overt behavior pattern with blood and cardiovascular findings; blood cholesterol level, blood clotting time, incidence of arcus senilis, and clinical coronary artery disease. J Am Med Assoc 1959;169(12):1286−96.

[49] Rosenman RH, Friedman M, Straus R, et al. Coronary heart disease in the Western Collaborative Group Study. A follow-up experience of 4 and one-half years. J Chronic Dis 1970;23(3):173−90.

[50] Hale J, Phillips CJ, Jewell T. Making the economic case for prevention−a view from Wales. BMC Public Health 2012;12:460.

[51] Johnston DW. The current status of the coronary prone behaviour pattern. J R Soc Med 1993;86(7):406−9.

[52] Excellence NIfHaC. Generalised anxiety disorder and panic disorder in adults: management (update). [CG1113]. https://www.nice.org.uk/guidance/cg113.

[53] Zhou M, Engel K, Wang J. Evidence for significant contribution of a newly identified monoamine transporter (PMAT) to serotonin uptake in the human brain. Biochem Pharmacol 2007;73(1):147−54.

[54] Mithoefer MC, Grob CS, Brewerton TD. Novel psychopharmacological therapies for psychiatric disorders: psilocybin and MDMA. Lancet Psychiatry 2016;3(5):481−8.

The importance of a good night's sleep

7

Devina Leopold and Thom Phillips

General Practitioner, National Health Service, Cwmbran Village Surgery, Cwmbran, United Kingdom

What is sleep?

Sleep is a state of rest, accompanied by changes in the brain and body. Without exception, every creature in the animal kingdom requires sleep [1].

Humans spend approximately a third of their lives sleeping. This essential biological activity provides the energy needed to perform even the simplest of tasks, such as walking, talking and concentrating. Sleep also affects mood, appetite and libido. Sleep is a biological necessity and is essential for health and daily functioning [2].

Sleep is a state of unconsciousness. The heart rate falls, blood pressure is lowered and respiratory and metabolic rates reduce. Body temperature drops by $1-2°$, and the production of urine is reduced. Sleep is promoted by the release of the hormone melatonin from the pineal gland. It is produced with a circadian rhythm and reaches peak levels as night falls. When the sun rises, melatonin production ebbs, and the body begins to wake up.

Other hormonal changes also occur during sleep. Growth hormone surges, and there are modulations in T3 and T4, impacting metabolic rate. Levels of cortisol and prolactin decrease as a person first falls asleep, and then rise before waking [3,4].

How does sleep affect the mind and body?

There are several explanations as to why humans and other animals spend such a large proportion of their lives sleeping. Sleep plays a significant role in brain development [4,5] and is necessary for a number of bodily functions. At all stages of life, the brain is active during sleep, consolidating memories, processing emotions and clearing out waste materials such as Tau and β-amyloid that can otherwise accumulate, eventually impairing neural function [5].

Without adequate sleep, cognitive skills such as speech, memory, innovative and flexible thinking are impaired [6]. Some studies show that sleep-deprived individuals have impaired functioning of the immune system and that processes such as muscle growth, tissue repair and protein synthesis occur mainly, or only, during sleep. There is a clear diurnal variation in the activity of T cells likely influenced by the release of proinflammatory cytokines such as IL-1, IL-6, IL-17 and TNF-α [7]. For example, sleep loss delays the night-time peak of IL-6, and sleep deprivation alters IL-6 rhythm across

A Prescription for Healthy Living. https://doi.org/10.1016/B978-0-12-821573-9.00007-2

the day, including daytime over secretion and night-time under secretion leaving the body in a persistently proinflammatory state [8]. It has been suggested that routinely sleeping less than 6—7 hours a night may double an individual's cancer risk [9]. There are several possible mechanisms which could explain this, including disruption of neuroendocrine function, metabolic dysfunction and increased release of reactive oxygen species [3,8].

Inadequate sleep is also associated with an increased risk of metabolic disorders, including obesity, type 2 diabetes and cardiovascular disease in addition to psychiatric disorders such as Alzheimer's disease [10].

The sleep cycle

During a typical night, sleep occurs in cycles of 90—110 min. Following the monitoring of brain waves, muscle tension and observation, the cycles have been split into two key categories: rapid eye movement (REM) sleep and non—rapid eye movement (NREM) (Fig. 7.1).

- **Non-REM sleep** accounts for most of the time during sleep and has four stages. During the deepest stages, three and four, the brain waves are slow, and respiratory and heart rates are at their lowest [3]. NREM sleep dominates early in the night and is important for cementing new memories. Deep NREM sleep duration declines with age, which is associated with a decline in memory [11].
- **REM sleep** occurs when the brain is very active, but the body is limp or intermittently paralysed, apart from the eyes, which move rapidly behind closed eyelids. Most dreaming occurs during REM sleep, which tends to be in the last part of the sleep period. It appears that during this time, the brain melds past and present knowledge, there is healing of emotional wounds reducing anxiety and depression and there is strengthening of memories and the inspiration of creativity [12]. REM sleep helps with social interactions and communication.

Each night, around four to five periods of quiet sleep alternate with four to five periods of REM sleep. In addition, several short periods of waking for 1—2 min occur approximately every 2 h or so, and more frequently towards the end of the night [13].

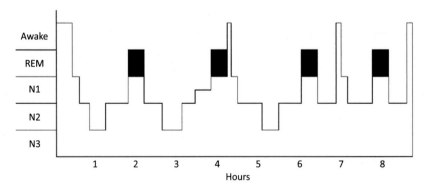

FIGURE 7.1

Normal hypnogram showing sleep stages.

Usually, individuals do not remember the times that they wake if the episodes last less than 2 min. However, if someone is distracted during the wakeful times, for example, by a partner snoring or the traffic noise, the wakeful times tend to last longer and they are more likely to be remembered [14].

How much sleep is necessary?

Normal sleep requirements vary widely, and there is no standard definition of what is normal. The National Sleep Foundation recommends 8 h per night [15]. Most adults on average sleep for 6—9 h. Infants and children generally sleep for longer durations, and the amount of sleep required decreases with age [16].

In 2016, the American Association of Paediatrics endorsed the following sleeping guidelines for children of different ages [17]:

- 4 months—1 year: approximately 12—16 h of sleep a day
- 1—2 years: 11—14 h of sleep a day
- 3—5 years: 10—13 h of sleep a day
- 6—12 years: 9—12 h of sleep a day
- 13—18 years: 8—10 h of sleep a day

Teenagers and sleep

Research has shown that the biological sleep rhythms of teenagers are different to adults. It is proposed that the hormonal upheaval of puberty pushes the release of melatonin back from around 10 p.m.—1 a.m., and therefore the onset of sleep in teenagers is later than in adults [18]. The delay in melatonin production may further be compounded by the behaviour of teenagers, who spend a lot of time accessing social media [19]. Spending time on computer devices that emit blue light has been shown to impede sleep initiation [20]. Sleep is crucial for teenagers, because while they are sleeping, they release growth hormone which is essential for the growth which occurs during adolescence. They need more sleep than both children and adults, but they tend to have less than either [21,22].

In adolescence, the brain is still developing, and sleep is essential for healthy brain development. The brain's prefrontal cortex (PFC), responsible for complex thinking and decision-making, as well as emotional regulation, is among the last areas of the brain to develop and undergoes significant maturation during teenage years. This part of the brain is especially sensitive to the effects of sleep deprivation. The PFC is essentially a 'filter'. It is responsible for coordinating complex cognitive behaviours, planning, decision-making and regulating behaviour. Imaging studies have shown that during periods of sleep deprivation, the blood flow to the PFC is reduced [23], significantly impairing its function [24].

Teenagers may also be dealing with stresses associated with schooling and social changes, and they might be unaware of the importance of sleep and the effects of sleep deprivation. Multiple studies have shown a link between lack of sleep and a decrease in academic performance [25]. Just 1 h of sleep lost per night can result in performing two academic years inferior to an individual's ability and could mean the difference between a grade A and a grade C [26].

Getting older and sleep

It has been reported that over 40% of adults aged over 65 years have difficulties initiating or maintaining sleep [27]. Sleep problems in the elderly are complex and often multifactorial. With age, there is increased prevalence of primary sleep disorders such as circadian rhythm disturbances, restless leg syndrome and sleep disordered breathing. Older adults are also more likely to suffer from medical conditions such as depression or to be taking medications such as diuretics [28], β-blockers [29] or ACE inhibitors [30] that disrupt sleep quality or quantity.

Sleep disorders

About one-third of adults do not get as much sleep as they would like. 'Poor sleep' can describe difficulties in falling asleep, trouble staying asleep, early wakening or unrefreshing sleep [31]. The American Academy of Sleep Medicine characterises sleep problems into six broad categories:

- Insomnia
- Sleep-related breathing disorders, for example obstructive sleep apnoea (OSA), sleep-related hypoventilation
- Central disorders of hypersomnolence
- Circadian rhythm sleep−wake disorders, for example delayed sleep−wake phase disorder
- Parasomnias, for example night terrors and sleep paralysis
- Sleep-related movement disorders, for example restless leg syndrome

Insomnia is a sleep disorder characterised by difficulty in falling asleep and/or remaining asleep [32]. It is often associated with fatigue, mood disturbances, problems with interpersonal relationships, occupational difficulties and a reduced quality of life [33].

Psychological factors, including depression, anxiety or stress, are the cause in 50% of cases of insomnia. Emotional turmoil can cause the sympathetic nervous system to become overactive, causing heart rate, metabolic rate, production of cortisol, adrenaline and temperature to all go up, all of which may block the initiation of sleep [34]. Physical health problems such as restless legs, heartburn, hot flushes, pain or sleep apnoea contribute to 40% of cases of insomnia [35]. Other causes are poor sleep hygiene and external stimulants, such as loud noises or bright lights.

Treatment for poor sleep

A significant number of consultations, especially in primary care, relate to poor sleep. In the first instance, it is important to try and establish whether there is an underlying cause. Take a full sleep history including the length of time it has been a problem and whether the issue is initiation or maintenance. Explore behaviours around sleep, and establish how sleep fits into the patient's psychosocial situation. Shift workers and individuals who travel through time zones often are particularly prone to poor sleep health. There is a specific clinical entity known as shift work disorder; it is considered a 'circadian rhythm sleep disorder' by the International Classifications of Sleep Disorders [36].

A full medication review is necessary so that consideration can be given to any drugs which could be causing the sleep problem. There are certain tools available, such as the Epworth Sleepiness Score or Philadelphia Sleep Quality Index, to detect cases of sleep apnoea.

The first approach should be nonpharmacological. Strategies include physical activity, sleep restriction, stimulus control, cognitive behavioural therapy for insomnia (CBT-I) or light therapy.

Sleep restriction is an evidence-based approach to improving insomnia [37]. Patients should be advised to stay awake while avoiding stimulants for as long as possible and then go to bed and sleep, making sure to wake up at their normal waking time. The premise being that slowly the body's sleep—wake cycle is reset, and eventually the person goes to bed earlier and earlier. Sleep restriction involves maximising sleep efficiency, which describes the relationship between the number of hours spent in bed and the number of hours actually asleep. This is often difficult to communicate to patients as, on the surface, it seems an extreme approach, but can be very effective [38].

Stimulus control describes a set of instructions designed to help patients to deal with the stimuli associated with sleep habits. For example, advising patients to leave their bedroom if they have been in bed, but not asleep for more than 20 min, or immediately going to bed when they start to feel sleepy. The bedroom should be used for sleep only; external stimuli such as televisions, phones and computers should be removed.

CBT-I is a talking therapy, which can help to change the way people think and behave. The National Institutes for Health (NIH) has recently concluded that CBT-I is a safe and effective means of managing chronic insomnia [39]. CBT-I incorporates many of the techniques and strategies described, including stimulus control, with elements of quantifying the problem via sleep diaries and a strong emphasis on relapse prevention.

Physical activity has been shown to improve both the quality and quantity of sleep [40]. As well as being physically tiring, activity reduces stress levels, which helps to improve sleep. Activity should be preferably be undertaken earlier in the day during daylight hours to ensure early daylight exposure, but exercising at any time of the day has been shown to improve sleep quality [41]. Furthermore, exercising outside can also help expose the rods and cones on the retina of the eyes to natural sunlight, which is important in maintaining the circadian rhythm.

Sunlight is an important signal for triggering the sleep—wake cycle. Commercially available lights, which emit a minimum 2500Lx, can be used in sleep therapy and can help shift workers adjust to their irregular schedules. Light has two primary effects on the sleep—wake cycle. It has an alerting effect, and it also has the ability to gradually shift sleeping patterns earlier or later, depending on the timing of light exposure. During light therapy, the user sits near a light box for a prescribed amount of time at the correct time of day to modulate melatonin release and reset the sleep—wake cycle. Exposure to bright light early in a person's wake period, and dim light at the end of the wake period, has the effect of moving the internal clock to an earlier set point. Conversely, exposure to bright light later in the day delays the timing of the internal clock. The timing of exposure to light can help establish a better sleep/wake pattern (Fig. 7.2).

There is some evidence that some dietary supplements can aid sleep quality and quantity. Both zinc and magnesium aspartate have been shown to improve sleep—wake cycles in elderly populations suffering from primary sleep disorders [42]. A study in Italy compared giving elderly individuals 225 mg magnesium and 11.25 mg zinc, mixed with 100 g of pear pulp, an hour before bed every night

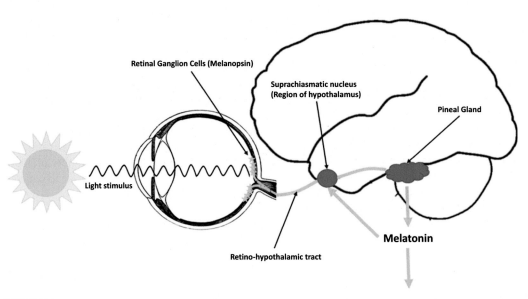

FIGURE 7.2

Melatonin production: Melatonin release is regulated by a region of the hypothalamus called the suprachiasmatic nucleus (SCN). Light hitting the retinal ganglion cells send an inhibitory signal to the SCN which causes melatonin to be released by the pineal gland. Increases in circulating melatonin work on a negative feedback loop with the SCN to regulate levels.

for 8 weeks, to a control group who had pear pulp only. The group receiving magnesium and zinc reported significantly improved perception of sleep quality and quantity, as measured across five sleep evaluation tools, and there were improved objective markers of sleep measured using an actiogram. Furthermore, taking small doses of melatonin is thought to 'reset' the internal sleep—wake cycle and improve sleep initiation [43].

Pharmacotherapy guidelines vary around the world, but generally, sedative drugs such as benzodiazepines and selective serotonin reuptake inhibitors (SSRIs) should be avoided as far as possible as the sleep they provide is suboptimal. Sedation lacks the non-REM component of the sleep cycle, which has been implicated in reduced declarative memory and procedural learning capacity [44]. Studies have shown that there is also an increase of falls, accidents, cognitive impairment and dementia in patients who use these drugs either acutely in the case of falls and cognitive impairment or long term in the case of dementia [45]. These risks are further compounded by the highly addictive potential of some of the drug classes.

A list of simple strategies that can be shared with patients to improve their sleep patterns is outlined in Box 7.1.

Box 7.1 Simple lifestyle modifications to improve sleep patterns
Tips for improving sleep
- **Reduce any anxiety-provoking thoughts and worries** by learning to mentally decelerate before bed [1]. Many individuals find the practise of mindfulness or meditation to be helpful.
- **Talk through any problems or about anything you are worried about.** Some people find it useful to keep a notebook by their bed. This can be used to jot down worries or 'to do' jobs so that they do not keep playing on your mind at night-time.
- **Limit screens** in the bedroom. Do not have a mobile phone, tablet, television or computer in the bedroom at night and have at least 1 h of screen-free time before going to bed.
- Create a **sleep-friendly bedroom,** ideally a room that is dark, cool, quiet and comfortable. It might be worth investing in thicker curtains or a blackout blind to help block out early summer mornings and light evenings. Try a warm yellow or red light bulb as a night light if needed instead of blue LED-powered lights.
- Aim to do some **exercise** every day. Try to exercise outside during daylight.
- **Cut out or drink less** caffeine, particularly in the 4 hours before bed.
- **Avoid smoking.** The nicotine in cigarettes increases your heart rate and alertness, so impairs your ability to fall asleep. Sleep may also be interrupted as there may be withdrawal symptoms during sleep.
- **Avoid alcohol** especially in the hours before bedtime as it disrupts your sleep cycle. It is a sedative, initially causing drowsiness; however it affects your sleep patterns so you spend less time in REM sleep throughout the night and you wake up feeling tired. Alcohol can also make you need to pass urine overnight.
- **Don't fight it.** If you feel tired, it is a cue to head to bed. Try and work with your natural sleep–wake cycle to improve sleep quality.
- **Go to bed and get up at the same time each day.** Avoid long weekend lie-ins. Late nights and lie-ins can disrupt your body clock and leave you with weekend 'jet lag' on Monday morning. Set an alarm for bedtime, and try to avoid naps after 3 p.m.

Conclusion

Sleep is vital to human survival. Insufficient sleep can increase the all-cause mortality risk by 13%. This includes all causes of death, including fatal car accidents, strokes, cancer and cardiovascular disease [46]. Assessing a patient's sleep health is an often-overlooked aspect of a patient's presentation and has implications for a number of associated health problems. Getting a good amount of high-quality sleep will help to enhance daily functioning, mood, health and well-being.

Summary
- Sleep is an essential biological activity necessary for health and daily functioning.
- During sleep, the heart rate falls, blood pressure is lowered and respiratory and metabolic rates reduce. Body temperature drops by 1–2°, and there is decreased urine production.
- Sleep plays a significant role in brain development and is important for the consolidation of memories and processing emotions.
- Inadequate sleep can impair cognitive skills such as speech, memory and innovative and flexible thinking.
- Sleep-deprived individuals may have impaired immune function, increased cancer risk, metabolic dysfunction and an increased risk of developing psychiatric disorders such as Alzheimer's disease.
- Sleep requirements are age dependent. Most adults need around 8 h.
- Assessing a patient's sleep status is an important component of a holistic approach to healthcare.

References

[1] Walker MP. Why we sleep. The new science of sleep and dreams. Penguin Randomhouse, UK, 2018.

[2] Williamson AM, Feyer A-M. Moderate sleep deprivation produces impairments in cognitive and motor performance equivalent to legally prescribed levels of alcohol intoxication. Occup Environ Med October 2000;57(10):649–55.

[3] Foster RG, Kreitzman L. The rhythms of life: what your body clock means to you! Exp Physiol 2014;99(4): 599–606.

[4] Heraghty JL, et al. The physiology of sleep in infants. Arch Dis Child 2008;93(11):982–5.

[5] Winer JR, et al. Sleep as a potential biomarker of tau and β-amyloid burden in the human brain. J Neurosci 2019;39(32):6315–24.

[6] Miyata S, et al. Poor sleep quality impairs cognitive performance in older adults. J Sleep Res 2013;22(5): 535–41.

[7] Grandner MA, Sands-Lincoln MR, Pak VM, Garland SN. Sleep duration, cardiovascular disease, and proinflammatory biomarkers. Nat Sci Sleep 2013;5:93.

[8] Wright Jr, Kenneth P, et al. Influence of sleep deprivation and circadian misalignment on cortisol, inflammatory markers, and cytokine balance. Brain Behav Immun 2015;47:24–34.

[9] Haus EL, Smolensky MH. Shift work and cancer risk: potential mechanistic roles of circadian disruption, light at night, and sleep deprivation. Sleep Med Rev 2013;17(4):273–84.

[10] John U, Meyer C, Rumpf HJ, Hapke U. Relationships of psychiatric disorders with sleep duration in an adult general population sample. J Psychiatr Res 2005;39:577–83.

[11] Fogel S, Martin N, Lafortune M, Barakat M, Debas K, Laventure S, Latreille V, Gagnon JF, Doyon J, Carrier J. NREM sleep oscillations and brain plasticity in aging. Front Neurol 2012;3:176.

[12] Miller CB, Espie CA, Bartlett DJ, Marshall NS, Gordon CJ, Grunstein RR. Acceptability, tolerability, and potential efficacy of cognitive behavioural therapy for Insomnia Disorder subtypes defined by polysomnography: a retrospective cohort study. Sci Rep 2018;8(1):6664.

[13] DeLeon CW, Karraker KH. Intrinsic and extrinsic factors associated with night waking in 9-month-old infants. Infant Behav Dev 2007;30(4):596–605.

[14] Hirshkowitz M, et al. National Sleep Foundation's sleep time duration recommendations: methodology and results summary. Sleep Health 2015;1(1):40–3.

[15] American Academy of Pediatrics. Recommended amount of sleep for pediatric populations. Pediatrics 2016; 138(2):e20161601.

[16] Barnes M, et al. Setting adolescents up for success: promoting a policy to delay high school start times. J Sch Health 2016;86(7):552–7.

[17] Woods HC, Scott H. # Sleepyteens: social media use in adolescence is associated with poor sleep quality, anxiety, depression and low self-esteem. J Adolesc 2016;51:41–9.

[18] Cho YM, et al. Effects of artificial light at night on human health: a literature review of observational and experimental studies applied to exposure assessment. Chronobiol Int 2015;32(9):1294–310.

[19] Dorofaeff TF, Simon D. Sleep and adolescence. Do New Zealand teenagers get enough? J Paediatr Child Health 2006;42(9):515–20.

[20] Wheaton AG, et al. Short sleep duration among middle school and high school students—United States, 2015. MMWR (Morb Mortal Wkly Rep) 2018;67(3):85.

[21] Drummond SP, Brown GG. The effects of total sleep deprivation on cerebral responses to cognitive performance. Neuropsychopharmacology 2001;25(Suppl. 5):S68–73.

[22] Kamphuis J, et al. Sleep restriction in rats leads to changes in operant behaviour indicative of reduced prefrontal cortex function. J Sleep Res 2017;26(1):5–13.

[23] Dimitriou D, Le Cornu Knight F, Milton P. The role of environmental factors on sleep patterns and school performance in adolescents. Front Psychol 2015;6:1717.

[24] https://www.sleepfoundation.org/excessivesleepiness/content/improve-your-childs-school-performance-good-nights-sleep.

[25] Foley DJ, Monjan AA, Brown SL, Simonsick EM, Wallace RB, Blazer DG. Sleep complaints among elderly persons: an epidemiologic study of three communities. Sleep 1995;18(6):425–32.

[26] Rosen RC, Kostis JB. Biobehavioral sequelae associated with adrenergic-inhibiting antihypertensive agents: a critical review. Health Psychol 1985;4:579.

[27] Cicolin A, Mangiardi L, Mutani R, Bucca C. Angiotensin-converting enzyme inhibitors and obstructive sleep apnea. Mayo Clin Proc 2006;81:53–5.

[28] Erickson VS, et al. Sleep disturbance symptoms in patients with heart failure. AACN Clin Issues 2003;14(4): 477–87.

[29] Kraus SS, Rabin LA. Sleep America: managing the crisis of adult chronic insomnia and associated conditions. J Affect Disord 2012;138(3):192–212.

[30] World Health Organization. The ICD-10 classification of mental and behavioural disorders: clinical descriptions and diagnostic guidelines. Geneva: World Health Organization; 1992.

[31] Kostis JB, Rosen RC, Holzer BC, et al. CNS side effects of centrally-active antihypertensive agents: a prospective, placebo-controlled study of sleep, mood state, and cognitive and sexual function in hypertensive males. Psychopharmacology (Berl) 1990;102:163–70.

[32] Kräuchi K. The thermophysiological cascade leading to sleep initiation in relation to phase of entrainment. Sleep Med Rev 2007;11(6):439–51.

[33] Parish JM. Sleep-related problems in common medical conditions. Chest 2009;135(2):563–72.

[34] Sateia MJ. International classification of sleep disorders. Chest 2014;146(5):1387–94.

[35] Friedman L, Benson K, Noda A, Zarcone V, Wicks DA, O'Connell K, et al. An actigraphic comparison of sleep restriction and sleep hygiene treatments for insomnia in older adults. J Geriatr Psychiatr Neurol 2000; 13(1):17–27.

[36] Taylor DJ, et al. Sleep restriction therapy and hypnotic withdrawal versus sleep hygiene education in hypnotic using patients with insomnia. J Clin Sleep Med 2010;6(02):169–75.

[37] Espie CA, et al. Randomized clinical effectiveness trial of nurse-administered small-group cognitive behavior therapy for persistent insomnia in general practice. Sleep 2007;30(5):574–84.

[38] Reid KJ, et al. Aerobic exercise improves self-reported sleep and quality of life in older adults with insomnia. Sleep Med 2010;11(9):934–40.

[39] Lang C, Brand S, Feldmeth AK, Holsboer-Trachsler E, Pühse U, Gerber M. Increased self-reported and objectively assessed physical activity predict sleep quality among adolescents. Physiol Behav 2013;120: 46–53.

[40] Rondanelli M, Opizzi A, Monteferrario F, Antoniello N, Manni R, Klersy C. The effect of melatonin, magnesium, and zinc on primary insomnia in long-term care facility residents in Italy: a double-blind, placebo-controlled clinical trial. J Am Geriatr Soc 2011;59(1):82–90.

[41] Dijk DJ, Cajochen C. Melatonin and the circadian regulation of sleep initiation, consolidation, structure, and the sleep EEG. J Biol Rhythm 1997;12(6):627–35.

[42] Bottiggi KA, et al. Long-term cognitive impact of anticholinergic medications in older adults. Am J Geriatr Psychiatr 2006;14(11):980–4.

[43] Diem SJ, et al. Use of non-benzodiazepine sedative hypnotics and risk of falls in older men. J Gerontol Geriatr Res 2014;3(3):158.

[44] Hafner M, et al. Why sleep matters—the economic costs of insufficient sleep: a cross-country comparative analysis. Rand Health Q 2017;6:4.

[45] Vartanian O, et al. The effects of a single night of sleep deprivation on fluency and prefrontal cortex function during divergent thinking. Front Hum Neurosci 2014;8:214.

[46] Hirshkowitz M, Whiton K, Albert SM, Alessi C, Bruni O, DonCarlos L, Hazen N, Herman J, Hillard PJA, Katz ES, Kheirandish-Gozal L. National Sleep Foundation's updated sleep duration recommendations. Sleep Health 2015;1(4):233—43.

Gratitude: being thankful is proven to be good for you

Sonal Shah

General Practitioner, National Health Service, London, United Kingdom

'Gratitude turns what we have into enough'
Melodie Beattie

Introduction

The term 'gratitude' means different things to different people. The Oxford Dictionary defines it as *'The quality of being thankful; readiness to show appreciation for and to return kindness'*. Gratitude has also been described as an emotion, an attitude, a moral virtue, a habit, a personality trait, or a coping response [1]. The practice of gratitude is embedded in our history and modern-day culture; Cicero from Roman times stated that gratitude was the 'mother' of all human feelings. In the present day, Americans celebrate Thanksgiving every year, and many individuals will say thanks as part of their daily religious prayers. In the past half century, gratitude has also become an area of interest for the medical community and has been shown to have clear health and well-being benefits. The mind and body are intrinsically linked, and improving one will no doubt help the other.

Navigating life's ups and downs is difficult. Everyone experiences challenges and hardships, and things often do not go according to plan. Sources of stress can include work, family life, relationships, finances, health problems and a lack of time for oneself. This can lead to feeling weighed down with negative thoughts and emotions, which are often ignored or pushed aside. The result can be low mood, which can affect sleep and concentration, and can contribute to anxiety, depression or worsening of physical health problems.

The practice of gratitude is a simple yet extremely effective way of combating these negative emotions by reminding us of what is good in our lives. We are able to put things in perspective by learning to train the mind to focus on the things for which we feel thankful.

The benefits of gratitude

Over the past 20 years, researchers in the field of positive psychology have been reviewing the practice of gratitude and have found benefits in five key areas: emotional, social, personality, career and health. The evidence overwhelmingly shows that practicing gratitude is associated with a significant improvement in overall well-being.

Professor Robert Emmons, a leading scientist in the field of positive psychology, reports that gratitude is integral to our well-being. He explains that processing a life experience through a grateful lens allows a person to realise the power they have to transform an obstacle into an opportunity, a loss into a potential gain and recasting negativity into positive channels for gratitude [2].

Emmons and his team have conducted studies to measure the effect of an 'attitude of gratitude'. In one well-quoted study, participants were asked to list in a journal five things they were grateful for, each week, 10 weeks [1]. This was compared with a group that were asked to recall five hassles or burdens from the previous week and a neutral group who were asked to keep a record of five events from the previous week, good or bad. Furthermore, participants were also asked to record their mood, self-report any physical health problems and give an overall judgement concerning how their lives were going. These were quantified into a well-being score. It was reported that participants who kept a daily gratitude journal were 25% happier than other participants; they reported fewer health complaints and were found to spend significantly more time exercising. A similar study conducted over a 3-week period found that participants who practiced gratitude reported higher levels of alertness, enthusiasm, determination, attentiveness and energy [1]. In general, it appears that more grateful people are happier, more satisfied with their lives, less materialistic and less likely to suffer from burnout. Consistently grateful people are emotionally intelligent, forgiving and less likely to be depressed, anxious or lonely [3].

How are the positive effects of gratitude mediated?

The feeling of gratitude helps us to realise and appreciate what we have in the present moment. Gratitude exercises force the brain to look for that which is good. The regular practice of gratitude encourages the brain to become skilled at identifying opportunities for personal development and growth. It cultivates cognitive flexibility, creativity and social connections [1]. Improved personal resources are associated with positive emotions. Psychologist Barbara Fredicksons states that positive emotions can broaden mindsets and build enduring personal resources over time [4], resulting in a positive self-reinforcing feedback cycle.

Martin Sieglman and Christopher Paterson, researchers in the field of positive psychology, developed a comprehensive list of 24-character strengths and virtues, including gratitude, love and judgement [5]. When these character strengths were correlated with well-being, gratitude was the single best independent predictor of well-being [6]. Gratitude is one of the top five character strengths consistently and robustly associated with life satisfaction [7].

In his other work, Sieglman has described 'positive relationships' as one of the five key components in his theory of well-being [8], the others being positive emotions, engagement, meaning and accomplishment. Research suggests that gratitude cultivates connections between people by inspiring them to be more generous, kind and helpful or 'prosocial' [9]. Sara Algoe and her team postulated the theory that gratitude serves to 'find, remind and bind' [10]. We 'find' people who we can build relationships with, we are 'reminded' of their goodness when we seek reasons to be grateful for them and we 'bind' together by making others feel appreciated and encourage them to engage in behaviors that strengthen those relationships.

Gratitude in the workplace

In the workplace, gratitude can help employees to feel acknowledged and valued, and it can help to prevent burnout. Around the world, many employers use gratitude as a tool to help support the well-being of their workers. In Japan, the term 'Otsukaresama desu' loosely translates to 'thank you for your hard work'. It is used frequently throughout the working day, and it epitomises the Japanese work culture. This is of particular interest in healthcare, where physician burnout can affect patient care, productivity, turnover, and financial performance [11]. In 2017, the American Institute for Healthcare Improvement (IHI) released a white paper that identified nine critical components of a system for ensuring a joyful, engaged workforce [12]. 'Rewards and Recognition' was listed as one of the facets. Developing programmes that encourage rewards, recognition and validation are simple ways that departments can support healthcare professionals [13]. In 2018, the American College of Emergency Medicine launched the gratification in emergency medicine (GEM) project [14]. When emergency physicians were asked 'what matters most?' in terms of satisfaction in the workplace, 90% reported that external recognition was crucial and 43% appreciated words of affirmation. 35% of respondents ranked being recognised for clinical service as most important, but only 8% felt such recognition.

Developing a departmental culture of gratitude, formally or informally, may help with staff motivation and in making clinicians feel valued. In a double-blinded, randomised controlled trial, practitioners from five public hospitals wrote down what they were grateful for, twice a week for 4 weeks [15]. The group were compared with a control group who did not practice gratitude and a 'hassle' group who documented their 'hassles'. It was reported that the gratitude group experienced a decline in stress and depressive symptoms and that, relative to the control group, the gratitude group reported lower depressive symptoms (-1.50 points; 95% CI [-2.98, -0.01]; d $= -0.49$) and perceived stress (-2.65 points; 95% CI [-4.00, -1.30]; d $= -0.95$) at follow-up. The simplicity and impact of gratitude has seen the development of tools to help support patients give thanks [16], as well as simple departmental changes that invite a culture of gratitude, for example the hanging of gratitude trees around healthcare settings [17].

Gratitude and health

Gratitude and its effect on physical and mental health is rapidly gaining interest by those in the medical arena. This area is, however, in its infancy, and many of the studies are small and need replication. Gratitude, unlike other measurable entities such as heart rate or blood pressure, is difficult to quantify, and there are few validated assessment tools. This makes it difficult to determine the reliability of gratitude studies. Furthermore, many of the studies rely on self-reporting and therefore are very subjective.

While there is emerging interest in the health benefits of gratitude, this field of research does not receive significant amounts of funding compared with research involving diagnostic or therapeutic interventions, limiting aspects such as study size and duration. When studies are carried out, it is difficult to exclude other confounding factors, limiting the conclusions that can be drawn from the results. Despite this, studies have shown that gratitude is a promising therapeutic adjunct in conditions such as heart failure, depression and chronic pain. A growing number of institutions, for example the Greater Good Science Centre in association with the University of California and the Special and Affective Neuroscience Laboratory at the University of California, are investing in gratitude research.

In 2009, a large American metaanalysis assessing the role of positive psychology interventions, which included gratitude, showed that such interventions may have a role in the treatment of depression and anxiety [18]. A component of the metaanalysis involved the review of 25 studies specifically looking at depression, with gratitude as a therapeutic intervention. Results showed that the practice of gratitude was correlated with a reduction in depressive symptoms.

The practice of gratitude may help to confer resilience to depression and suicide, by increasing the sense of meaning in life [19]. In a study of 814 undergraduate students, with a mean age of 20.13, the relationship between gratitude, depression, suicidal ideation and self-esteem was examined [20]. Gratitude was found to be inversely related to suicidal ideation, and this association was mediated through enhanced self-esteem and reduced depressive symptoms. Students who were grateful tended to have a higher level of self-esteem. It was proposed that higher levels of self-esteem were related to feelings of being loved, cared for and valued and that self-esteem could act as a buffer in preventing suicide.

Gratitude has also been used as a direct intervention in suicide prevention. One small study of 52 patients considered the effects of nine positive psychology exercises on suicidal inpatients. The exercises included writing a gratitude letter, counting blessings or acts of kindness. Participants completed a daily exercise, and levels of optimism and hopelessness were measured on a 5-point Likert scale. Those who completed the exercise alongside standard treatment showed a moderate increase in levels of optimism and reduced feelings of hopelessness compared with those who received standard treatment only. Importantly, 72.3% of patients commented on the high ease of completion and the generation of positive emotions related to the exercise [21].

Gratitude may help to build resilience and can play a role in helping individuals to overcome trauma and posttraumatic stress disorder (PTSD). A number of studies have been published assessing symptoms of PTSD in those who have experienced traumatic life events, including Vietnam war veterans [22], survivors of destructive earthquakes [23] or those who have witnessed college shootings [24]. It has been shown that higher levels of gratitude are associated with lower levels of PTSD. The findings indicate that health professionals could consider a multifaceted approach to buffer the effects of trauma by cultivating resilience and enhancing gratitude in posttrauma interventions as a means of decreasing PTSD [24].

The practice of gratitude can contribute to positive emotional well-being, which is highly beneficial for recovery and long-term prognosis in people with physical illness [25]. There may also be a role in pain management [26]. One study, carried out over a 6-month period, involved monitoring patients over the age of 50 years with known osteoarthritis [27]. The participants were asked to complete a program of positive psychology interventions for 6 weeks, alongside their usual treatment, which included pharmacotherapy such as paracetamol, nonsteroidal antiinflammatory drugs and opioids. The interventions included writing down three good things that had happened that day or expressing thanks to someone. Symptoms were quantified using the Western Ontario Macmaster Osteoarthritis Index, which rates pain, stiffness and difficulty with physical functioning, resulting in a WOMAC score. Patients who completed the programme showed a 12.6-point improvement in their score compared with baseline, indicating an improvement in their symptoms. The change in the score for the neutral group was less than 2 points. The largest effect was seen for physical function, as opposed to pain or stiffness, suggesting a functional improved in some patients despite the lack of an associated improvement in pain or stiffness. It was suggested that it was the way in which patients *perceived* pain that improved, which indicates that, with training, the brain can learn to focus on positive experiences rather than minor symptoms. Previous studies have shown that when the practice of gratitude is combined with mindfulness, there is a resultant improvement in symptoms of pain, anxiety, pain interference and self-efficacy for patients with osteoarthritis [28]. Similar effects have been observed in patients with spinal cord injury, multiple sclerosis and postpolio syndrome [29].

There is also evidence that gratitude can improve sleep. Perhaps one of the best known and frequently quoted studies, counting blessings versus burdens, considered the effect of gratitude on patients with neuromuscular disease [1]. After completing a daily 21-day gratitude exercise, participants had improved sleep metrics, sleeping nearly 50 min more than a control group ($P < .05$) and feeling more refreshed upon waking ($P < .05$). Although the results were not statistically significant in other domains, participants practicing gratitude also reported less physical pain, greater well-being and self-satisfaction.

Having positive affect, such as that generated by practising gratitude, may influence the immune and cardiovascular systems directly. In a similar manner to mindfulness and meditation, it is hypothesised that gratitude can activate the parasympathetic nervous system, which may reduce the production of cortisol and other inflammatory mediators via the hypothalamic–pituitary–adrenal axis (HPA), thus buffering the impact of stress [25]. A recent review by Cousin et al. concluded that gratitude may positively impact cardiovascular parameters such as endothelial dysfunction and prognostic inflammatory markers and may improve adherence to health behaviours. However, such research is in its infancy, and methods are not robust enough to draw clinically significant conclusions [30]. Different studies have addressed serum levels of biomarkers commonly related to inflammation, such as high sensitivity C-reactive protein (CRP), interleukin 6 (IL-6), tumour necrosis factor-alpha (TNF-α) and soluble intracellular adhesion molecule-1 (sICAM, a marker of endothelial function). While some studies have identified that gratitude may positively influence levels of the biomarkers, results have not been consistent. Redwine et al. investigated 70 patients with stage B heart failure who kept a gratitude diary for 8 weeks. It was found that this group had reduced levels of CRP, TNF-α, sICAM-1 and IL-6 compared with a 'treatment as usual' control group [31]. It was also reported that the gratitude group experienced a greater increase in heart rate variability compared with the control group. A randomised controlled trial of 119 healthy females found that a gratitude intervention led to a reduction in diastolic blood pressure by 2.00 mmHg (CI 0.05–3.88, $P = .041$). Potential confounding factors such as age, BMI and baseline diastolic values had all been accounted for [32].

For patients who have recently suffered from cardiac events, those with higher levels of optimism and gratitude 2 weeks after the event report greater improvements in emotional well-being 6 months later [33] in addition to greater levels of physical activity and better adherence to medical therapies [34].

How to incorporate gratitude into your personal life and clinical practice

Gratitude motivates people to make positive changes in their lives and in the world around them. Unlike other interventions, gratitude is simple and easy to practise and does not have any adverse side effects. As a healthcare provider, you can support your patients and colleagues to cultivate this practice by the use of gratitude exercises, which are also tools that you can use yourself. Examples of such exercises are detailed in the following.

Five good things

This is the simplest, easily practised exercise and can be tailored to suit individual needs. Individuals should sit quietly for 5 min and bring to mind five things they are most grateful for. Doing this at the same time each day, for example during the daily commute or at bed time, may encourage this practice to become a daily habit. A digital prompt maybe useful. Involving others may help with adherence, for example inviting the whole family to state what they are thankful for, at dinner time. A key piece of advice would be to encourage people to be as specific as possible, for example being grateful for getting a seat on a train.

Gratitude journal

Writing a short daily journal entry about positive experiences encourages reflection and provides clarity to thinking. This practice may also help break down problems and create solutions, as it improves the way the brain focuses and engages both sides of the brain.

Writing a letter of thanks

Write a letter of thanks to someone you are thankful to, it could be long or short, but it should be specific about the reason you are grateful. It could be a loved one, family member, friend, coworker or a teacher/mentor. This letter could then be kept aside, posted or even given to the recipient. This practice also helps to cultivate and strengthen relationships.

Gratitude jar

Starting with an empty jar or box, decorate it with ribbons or glitter, so that it appears bright and inviting. Encourage individuals to record or make a note of positive experiences on a piece of paper that is then placed in a jar. The container is a physical reminder of things that the individual is grateful for, and it can also be used during adverse times. In the latter scenario, a piece of paper can be taken and read at random, which may support the cultivation of gratitude and help strengthen resilience.

Gratitude walk

This exercise involves going for a walk with the goal of being grateful for all the things in the local environment. During this time, encourage observing sights, smells and sounds that may otherwise be missed. The practice is particularly useful to ground people and to develop gratitude for things in the present moment. Use the opportunity to take in your surroundings, look for the things you might otherwise miss—sights, sounds and smells. This practice helps to remind you that you are present in the very moment.

Conclusion

Cultivating an attitude of gratitude develops a positive sense of well-being. It can lead to better physical and mental health, as well as improved personal resilience and stronger social connections. Unlike other medical interventions, it is inexpensive, has no side effects, requires no equipment and is easy to incorporate into daily life.

Summary

- The practice of gratitude has benefits in five key areas: emotional, social, personality, career and health.
- Gratitude is associated with higher levels of alertness, enthusiasm, determination, attentiveness and energy.
- Grateful individuals are happier, more satisfied with their lives, less materialistic and less likely to suffer from burnout.
- Consistently grateful people are emotionally intelligent, forgiving and less likely to be depressed, anxious or lonely.
- The practice of gratitude may help to confer resilience to depression and suicide.
- Gratitude may be beneficial in conditions such as PTSD, osteoarthritis, sleep disorders and in recovering from cardiovascular events.
- Unlike other medical interventions, gratitude is inexpensive, has no side effects, requires no equipment and is easy to incorporate into daily life

References

[1] Emmons R, McCullough M. Counting blessings versus burdens: an experimental investigation of gratitude and subjective well-being in daily life. J Pers Soc Psychol 2003;84:377—89.

[2] Emmons R. How gratitude can help you through hard times. 2013 [Online], https://greatergood.berkeley.edu/article/item/how_gratitude_can_help_you_through_hard_times.

[3] Achor S. The happiness advantage. s.l. New York: The Crown Publishing Group; 2010. p. 98.

[4] Fredrickson B. The broaden-and-build theory of positive emotions. Philos Trans R Soc Lond B Biol Sci 2004;359(1449).

[5] Peterson C, Seligman M. Character strengths and virtues. s.l. American Psyhcological Association; 2004.

[6] Kaufman S. Which character strengths are most predictive of well-being? American Scientific. 2015 [Online], https://blogs.scientificamerican.com/beautiful-minds/which-character-strengths-are-most-predictive-of-well-being/.

[7] Grenville-Cleave B. Positive psychology. UK: Icon Books; 2012.

[8] Seligman M. Authentic happiness. s.l. Simon & Schuster Ltd; 2004.

[9] Allen S. The science of gratitude. Greater good science center at UC Berkeley. US: John Templeton Foundation; 2018.

[10] Algoe S. Find, remind, and bind: the functions of everyday relationships. Soc Pers Psychol Compass 2012;6:455—69.

[11] Dewa C. How does burnout affect physician productivity? A systematic literature review 2014;14:325.

[12] Perlo J. IHI framework for improving joy in work. Cambridge, Massachusetts: IHI White Paper. Institute for Healthcare Improvement; 2017. s.n.

[13] Shah M. American College of Emergency Physicians. ACEP Now. 17 May 2019 [Online], https://www.acepnow.com/article/emergency-physicians-can-combat-burnout-with-gratitude/2/?singlepage=1.

[14] American College of Emergency physicians Wellbeing committee. The GEM (gratification in emergency medicine) Project Facet 1: rewards and recognition. 2018.

[15] Cheng. Improving mental health in health care practitioners: randomized controlled trial of a gratitude intervention. J Consult Clin Psychol February 2015;83(1):177—86.

[16] https://www.thnx4.org/about-thnx4. [Online].

[17] Connolly E. Medical bag. 30 November 2018 [Online], https://www.medicalbag.com/home/lifestyle/give-thanks-the-role-of-gratitude-in-combating-burnout/.

[18] Sin N, Lyubomirsky S. Enhancing well-being and alleviating depressive symptoms with positive psychology interventions: a practice-friendly meta-analysis. J Clin Psychol 2009;65(5):467–87. Wiley Interscience.

[19] Kleiman E. Gratitude and grit indirectly reduce risk of suicidal ideations by enhancing meaning in life: evidence for a mediated moderation model. J Res Pers October 2013;47(5):539–46.

[20] Lin CC. The relationships among gratitude, self-esteem, depression, and suicidal ideation among undergraduate students. Scand J Psychol December 2015;56(6):700–7.

[21] Huffman J. Feasibility and utility of positive psychology exercises for suicidal inpatients. Gen Hosp Pyschiatry 2014;36:88–94.

[22] Kashdan TB. Gratitude and hedonic and eudaimonic well-being in Vietnam war veterans. Behav Res Ther February 2006;44(2):177–99.

[23] Lies J. Gratitude and personal functioning among earthquake survivors in Indonesia. J Posit Psychol 2014; 9(4):295–305.

[24] Vieselmeyer J. The role of resilience and gratitude in posttraumatic stress and growth following a campus shooting 2017;9(1):62–9.

[25] Lamers S, Bolier L. The impact of emotional well being on long term recovery and survival in physical illness: a meta-analysis. J Behav Med 2012;35(5):538–47.

[26] Emmons R, Stern R. Gratitude as pscyotherapuetic intervention for osteoarthritis. Pain Med 2013;69:846–55.

[27] Hausmann L, Youk A. Testing a positive psychological intervention for oesteoarthritis. Pain Med 2017; 18(10):1908–20.

[28] Swain N. Gratitude Enhanced Mindfulness (GEM): a pilot study of an internet-delivered programme for self-management of pain and disability in people with arthritis. J Posit Psychol 2019:420–6.

[29] Muller R, Gertz K. Effects of a tailored positive psychology intervention on well-being and pain in individuals with chronic pain and a physical disability. A feasibility trial. Clin J Pain 2016;321:32–44.

[30] Cousin L. Effect of gratitude on cardiovascular health outcomes: a state-of-the-science review. J Posit Psychol 2020. https://doi.org/10.1080/17439760.2020.1716054.

[31] Redwine L. Pilot randomized study of a gratitude journaling intervention on heart rate variability and inflammatory biomarkers in patients with stage B heart failure. Pscyhosom Med 6 August 2016;78:667–76.

[32] Jackowska M. The impact of a brief gratitude intervention on subjective well-being, biology and sleep. J Health Psychol 10, 2016;21:2207–17.

[33] Millstein R. The effects of optimism and gratitude on adherence, functioning and mental health following an acute coronary syndrome. Gen Hosp Psychiatr 2016;43:17–22.

[34] Legler. State gratitude for one's life and health after an acute coronary syndrome: prospective associations with physical activity, medical adherence and re-hospitalizations. J Posit Psychol 2019;14(3):283–91.

Loneliness as a risk factor for chronic disease

9

Sonal Shah

General Practitioner, National Health Service, London, United Kingdom

Hunger takes care of your physical body. Loneliness takes care of your social body, which you also need to survive and prosper.
John Cacioppo

Introduction

It is well established that behavioural factors such as smoking and excessive alcohol consumption increase the risk of morbidity and premature mortality. Adverse health outcomes are also associated with environmental factors such as air and water pollution. There is now developing evidence that social factors, including loneliness and social isolation, have a role to play in poor health.

Loneliness and social isolation are increasingly being recognised as important determinants of health, with evidence suggesting that persistent loneliness can be as damaging as smoking. Social disconnection not only impacts psychological and emotional well-being but also has a negative effect on physical health.

What is loneliness?

Loneliness is common, with up to 50% of the adult population in Australia reporting that they feel lonely [1]. Loneliness is a complex human emotion, describing a state of solitude or a lack of meaningful relationships. Loneliness is different to being 'alone'. Many people are able to live alone but not feel lonely, conversely being surrounded by people does not stop one experiencing loneliness. Loneliness is a *perceived social isolation* and discontent with the *quality* of their relationships rather than the *quantity* of relationships they have [2]. It may be argued that loneliness is the social equivalent of physical pain, hunger or thirst.

For many individuals, relationships are perceived as meaningful when they feel supported, acknowledged and valued by others. Those who are lonely describe themselves as missing a 'connection' with people. A study of 20,000 American adults revealed that 40% felt they lacked companionship and that their relationships were not meaningful. Of these individuals, 54% felt that no one knew them well [3].

The human race is a social species who depend on a safe, secure and shared environment. Historically, populations lived as tribes, with each individual relying on their tribe for survival. Those who were isolated were more susceptible to predatory attack and were at increased risk of starvation. The survival of the fittest phenomenon therefore meant that to ensure survival, the human brain evolved to seek out companionship and community belonging. The perception of loneliness increases an individual's vigilance for threats along with feelings of vulnerability, which in turn creates a psychological desire to 'reconnect' with others. In the same way that hunger makes the brain seek out food, loneliness makes humans seek out the protection of a group.

At a subconscious level, lonely individuals develop a tendency to be hypervigilant, expecting negative interactions and remembering negative events. This can result in a reduction in social interactions and further social isolation. The self-reinforcing loneliness loop is accompanied by feelings of hostility, stress, pessimism, anxiety and low self-esteem [4].

While the evolutionary role of loneliness is beneficial in the short term, persistent hypervigilance for social threats causes continued activation of the sympathetic nervous system, so that the body remains in a 'fight or flight' mode. Altered physiological functioning creates a chronic inflammatory state, which impacts on sleep and increases morbidity and mortality [5].

The prevalence of loneliness

In 2016, the European Quality of Life survey asked adults living across Europe to rate how lonely they had felt in the past 2 weeks. Overall, 6% of respondents reported feeling lonely, with some countries, such as Albania and Greece, having rates of up to 11%. In the United Kingdom (UK), 10% of the adult population report feeling lonely [6], and in Japan, there are more than half a million people under 40 years of age who have not left their house or interacted with anyone for at least 6 months [7]. One in four Australian adults describe themselves as lonely, with one in two reporting feeling lonely for at least 1 day in the past week [1].

Who does loneliness affect?

Increasing numbers of young people report loneliness and social isolation. In the United Kingdom, 40% of individuals aged 16−24 years say they feel lonely often or very often [6]. This is mirrored in the United States (US), where the 18−22 age group is the loneliest group of all [3]. These findings have significant implications for future healthcare planning, as the effects of loneliness accrue over time to accelerate physiological aging. A longitudinal study of over 1000 children in New Zealand followed socially isolated children into adulthood. The study reported that there was a dose-dependent relationship between social isolation and poor adult health (risk ratio of 2.58%, confidence interval 1.46−4.56) [8]. This effect was present that even when other adverse health risk factors and health damaging behaviours had been taken into consideration. Compared with those who felt connected, individuals who were isolated had a higher incidence of cardiovascular risk factors including hypertension, hypercholesterolaemia and higher glycated haemoglobin levels.

Life in the 21st century has significantly different working and living patterns compared with those in previous times. There are fewer geographical limitations, and individuals frequently move away from the family home to study or work. Traditional models of employment have become more flexible, and there are more opportunities for individuals to work from home. Therefore, sources of social contact and daily interaction are changing.

Furthermore, there is the parody of technology and the Internet: Social media provides a platform to connect individuals all around the world, and it can allow families to stay in contact and can permit people with shared interests to form bonds. However, it can also be a cause of loneliness. A study of 1787 American adults aged between 19 and 32 years, with high social media use, found higher levels of perceived social isolation compared with those with lower use [9]. Social media use was defined by the amount of time spent on social media platforms such as Facebook, Twitter and Instagram, and social isolation was determined from self-reporting tools. The adults with the highest social media use time were twice as likely to have perceived social isolation compared with those with the lowest use. When social media use is restricted, loneliness can improve [10]. However, there is a degree of controversy surrounding the role of social media and loneliness: A large US survey of over 20,000 adults aged over 18 years found that social media use alone is not a predictor of loneliness and that those defined as very heavy users of social media had a loneliness score similar to those who never use social media [3].

It has been shown that positive relationships are one of the most important factors in shaping personal well-being [11]. National statistics in the United Kingdom suggest that those who are married or in a civil partnership have higher levels of life satisfaction, a greater feeling that life is worthwhile, and higher levels of happiness than those who are cohabiting, single, divorced or widowed. Married Australians are the least lonely compared with those who are single, separated or divorced, and those living with a partner are less lonely than those who are single or divorced [1].

In first half of the 20th century, it was rare for individuals to live in single occupancy households. In the 1970s, only 17% of households in the United Kingdom contained one person [11]. Over time, there have been increasing rates of separation and divorce, and one of the effects of the aging population is that the prevalence of widows and widowers is rising. In both the United Kingdom and Canada, the most common type of household is single occupancy, representing 28.3% and 28.2% of all households, respectively [12,13].

Loneliness appears to improve in middle age and then becomes more prevalent in the elderly. In England, it has been found that more than 2 million people over the age of 75 years live alone, and more than 1 million older people say they frequently go for over a month without speaking to a friend, neighbour or family member [14]. Loss of social contacts due to retirement, bereavements and worsening health can all play a causative role in loneliness, with increased frailty and disability also contributing to age-related social isolation [15,16].

Loneliness is a major social issue which impacts on community well-being, health and welfare services and the economy. In the United States, it has been reported that insurance companies spend an additional $134 per month or $1608 per year on the healthcare needs of older individuals aged over 65 years who are socially isolated [15]. In the United Kingdom, these economic costs have been projected to be even higher, with one research study suggesting that disconnected communities could be costing the UK economy £32 billion every year, partly due to increased spending on health services, policing and reduced productivity [17].

Loneliness can be the result of a complex interplay between multiple factors, and it is often challenging to understand how these interact. The key individual factors, social factors and life events believed to be involved in the aetiology of loneliness are outlined in Table 9.1.

Tackling the root causes of loneliness requires a united, multiagency approach, with input from the government, community groups, health- and social care providers, charities and individuals themselves. In 2018, significant concerns relating to the effects of chronic loneliness in the United

Table 9.1 Factors potentially leading to loneliness.

Individual factors	Life events	Social factors
Poor physical and mental health	Becoming a new parent	Lack of community services
Confidence and emotional state	Relocation/migration	Transport and accessibility
Disability	Separation/divorce	
Income	Retirement	
Mobility issues	Bereavement	
Language barriers		
Employment, working patterns and work—life balance		

Kingdom resulted in the appointment of the first ever Minister for Loneliness. Their role is to work in partnership with the Commission on Loneliness, a cross-party organisation that brings together representatives from relevant trusts, businesses and charities to develop a national multilevel strategy to combat loneliness [18].

Adverse health effects of loneliness

Social relationships have a significant impact on health, morbidity and mortality. It has been reported that a lack of relationships is a risk factor for all-cause mortality that has a similar magnitude as the well-established risk factors such as smoking and alcohol and exceeds the risks of physical inactivity and obesity [19]. Loneliness increases the risk of premature death due to all-cause mortality by 26% and is claimed to be as damaging to health as smoking 15 cigarettes a day [19]. Data from 308,849 individuals have shown that individuals who are socially isolated at baseline have a 50% higher risk of all-cause mortality compared with those with strong social connections, even when factors such as age, sex, initial health status and cause of death are accounted for [20].

The pathogenic effects of loneliness are mediated through both direct and indirect mechanisms. Individuals who are lonely are more likely to display adverse health behaviours including smoking and physical inactivity [21,22]. Poor sleep patterns are also notable: Individuals who are lonely have an increased number of 'microawakenings' during sleep, affecting its quality [23]. A lack of good quality sleep is detrimental to mood and can increase the risk of an individual developing depression [24]. It has been suggested that microawakenings occur when individuals do not feel secure in their social environment. They are prevented from achieving deep restorative sleep as they are continually scanning for threats. Studies of the Hutterites, a closed, highly religious farming community in South Dakota, have shown very low levels of loneliness, which corresponded with rare microawakenings [25].

The hypothalamic—pituitary—adrenal (HPA) axis works alongside the sympathetic adrenome-dullary axis to respond to threats and stressors. Activation of an acute stressor will lead to the stimulation of sympathetic nerve fibres to innervate most organ systems and lead to the immediate release of noradrenaline and adrenaline. The HPA system responds to stress by stimulating the release of hormones (CRH and ACTH) to activate the secretion of the glucocorticoid hormones (cortisol) into the blood circulation. These glucocorticoids are essential to the regulation of the body and can cross the blood—brain barrier. However, chronic stress, such as that associated with perceived loneliness, can adversely modulate the HPA axis, for example altering the circadian rhythm or increasing cortisol production [26]. The impact of loneliness on the neuroendocrine system has been confirmed in an extensive number of studies [27,28]. Lonely individuals have been found to have significantly higher concentrations of cortisol in urine, saliva and plasma than controls [29] and a 21% higher cortisol awakening response compared with nonlonely individuals [5].

In a population study of 229 adults aged 50—68 years, it was found that lonely individuals had increased systolic blood pressure compared with those who are not lonely [29]. Furthermore, the lonely cohort demonstrated greater overnight urinary concentrations of epinephrine. Therefore, it was been proposed that excess epinephrine causes vasoconstriction, leading to hypertension.

Loneliness may impact the body on a cellular level, causing upregulation of inflammatory gene expression and down regulation of antiviral gene expression [30]. In one small study, 14 high versus low lonely individuals had DNA microarray analysis performed. It was identified that 209 genes were differentially expressed in circulating leucocytes [31], including upregulation of genes involved in immune activation, transcription control and cell proliferation and downregulation of antiinflammatory genes, including genes supporting mature B lymphocyte function, type I interferon response and impaired transcription of glucocorticoid response genes. This change in pro- and antiinflammatory signalling was not attributable to differences in cortisol levels or other demographic factors [32]. Other studies have demonstrated how this inflammation can impact the immune response. Lonely individuals exposed to the common cold virus show increased levels of glucocorticoid receptor resistance and have a three times increased risk of developing a cold compared with those who are not lonely [33].

There is a large body of evidence which shows that loneliness can trigger the brain to alter behaviour. A variety of mechanisms are involved, including activation of afferent vagal nerves and cytokine activity, resulting in behaviours known as 'sickness behaviours'. These include loss of appetite, sleepiness, social withdrawal, fatigue, increased perception of pain and anhedonia. The evolutionary purpose behind sickness behaviour is thought to provide a motivational response to facilitate recuperation and recovery from illness and disease [34]. This may partially explain how loneliness can lead to symptoms of depression: It is estimated that loneliness increases the risk of experiencing depression by 15.2%, and being depressed increases the likelihood of being lonely by 10.6% [1]. Furthermore, there is some emerging evidence that experiences of social pain from rejection or isolation activate the same neural regions that are typically implicated in physical pain processing [35]. Social exclusion or relationship loss may be as emotionally distressing as experiences of physical pain. Conversely, there is evidence to suggest those who experience unexplained levels of pain may also have early experiences of social pain. It is postulated that this experience of 'pain' is again part of an evolutionary adaptation mechanism to promote social bonding and ultimately survival.

The key health consequences of loneliness are summarised in Table 9.2.

Tackling loneliness is a health and social priority

Loneliness is a clear risk factor for chronic disease and premature death and should be a priority for all policymakers and public health officials. Emphasis on population-based surveillance and the development of evidence-based interventions may prolong life and reduce the burden of illness and health

Table 9.2 Health consequences of loneliness.

Health consequences of loneliness	
Lower intake of fruit andvegetables	Odds ratio of 2.52 [21]
Obesity (*note also a cause for loneliness too* [36])	Odds ratio of 1.51 [22]
Increased sedentary behaviour	Odds ratio of 1.21 [22]−2.40 [21]
Increased alcohol intake	Odds ratio of 1.36 [21]
Smoking	Odds ratio of 1.55 [22]−1.60 [21]
Depression	Odds ratio of 1.91 [37]
Social anxiety	Odds ratio of 1.21 [37]
Higher rates of suicide	Odds ratio of 1.35 [37]
Increased incidence of cardiovascular risk factors, including hypertension	Odds ratio of 1.27 [38]
Increased mortality postcoronary artery bypass grafting (both at 30 days and 5 years)	Relative risk of 2.61 at 30 days Relative risk of 1.78 at 5 years [39]
Increased incidence of stroke risk	Relative risk of 1.32 [40]
Loneliness (not social isolation) increase in risk of dementia	Odds ratio of 1.64 [41]
Increased risk of all-cause mortality after a diagnosis of breast cancer Increased risk of breast cancer mortality	Hazard ratio of 1.66 [42] Hazard ration of 2.14 [42]

spending [15]. It is important that methods are developed to identify those at risk of loneliness, so that appropriate support can be offered. Many countries have taken steps through routine general practice or family-based medicine to assess social networks and identify those at risk of isolation, on a regular basis. These data can then be used not only to help the individual but also to build a picture of the needs of a population and thus develop targeted services taking into account affordability, language barriers and accessibility.

As the causes of loneliness are multifaceted, so must be the approach in tackling loneliness. Many health professionals have moved away from the traditional practice of prescribing medication to manage disease. Instead, there is now the development of the role of 'social prescribing'. As described in Chapter 5, this involves health professionals being able to 'prescribe' interventions such as singing, gardening or walking to individuals to manage a range of conditions. These ventures provide individuals with support, the opportunity to learn new skills and most importantly to connect with other people and develop relationships.

In the clinical setting, it is important to identify if someone is lonely. While direct questioning is possible, many individuals may be embarrassed by their situation and may not volunteer relevant information. Others may not even recognise it as a problem, and instead experience nonspecific symptoms such as insomnia, fatigue and pain or present frequently with medically unexplained problems. It should also be noted that 'being alone' does not always mean that someone will be lonely. There should be a low threshold to consider the impact of loneliness on health, particularly in at risk groups, for example those who may have experienced bereavement, suffer with mental health issues or even those with chronic health issues. Identifying these individuals provides the opportunity to offer tailored support. Simple strategies that can be considered include the following:

- Encouraging positive connections by exploring new ways of meeting people, for example by attending a community group such as a gardening group.
- Advise that the individual aims to leave the house every day. They could visit areas such as parks or a café, which may increase the opportunity to be around people and make new relationships.

Countless organisations both formal and informal exist internationally, nationally and locally to support people. These range from face-to-face meetups to those that use the latest digital technology. There are also many that cater for people with specific needs, for example:

- coffee mornings for carers, caring for people with dementia,
- support groups for parents of children with autism,
- telephone-based befriending services for housebound patients,
- social media platforms to connect people with similar interests, backgrounds and ideas,
- phone apps designed for people in busy cities to connect.

Another way for individuals to connect with others and to share their time and skills is through volunteering. For many people, helping others can give them a sense of purpose, and many charities welcome help from individuals from all backgrounds. Individuals who are lonely may feel as though they have endless amounts of time to fill, so it can be helpful to start a new hobby or study something of interest. This may also serve to build confidence and self-belief. Increasingly, there are also organisations setup to support older people embrace digital intelligence. By teaching them to use social media platforms and applications such as Facebook, Instagram and WhatsApp, geographical limitations are removed, and people may feel more connected.

For many people, changing old habits can be hard. Many have been isolated for a long time, some may feel as though they have been repeatedly rejected, and as discussed will be cautious and anxious about social interactions. So, encouraging them to take their time, be persistent and not compare [14] themselves to others may help increase their confidence and self-esteem. Patients should be given the opportunity to express their thoughts and feelings in an open and nonjudgemental way. Active listening and supporting patient to set small, realistic and achievable targets may be the best way to make long-lasting changes. Using techniques such as SMART goal setting (Smart, measurable, achievable, realistic, timebound) and motivational interviewing may be helpful in achieving this. Many causes of loneliness cannot be changed, but encouraging patients to learn ways of focusing on the things that they can control may be useful. In addition to the steps taken to combat loneliness, support patients in developing additional methods of self-care such as exercise, mindfulness, meditation or even gratitude, which all have huge clinical benefits.

Conclusions

Loneliness is a serious health issue which has implications for individuals, societies and health- and social care providers. Loneliness directly impacts a person's physical and mental health. To be connected with other is fundamental to the human existence. Humans are designed to coexist in groups, providing comfort, support and meaning to one another. Without the 'safety of the herd', there is threat to an individual's survival, via a variety of neuroendocrine and social mechanisms. It is therefore the responsibility of every member of the 'herd' to find those who have been left behind and bring them back. In a time where demands on health services are outstripping resources, health professionals are uniquely positioned to play a key role to address health and well-being in a holistic and humane manner. Not only is there a clinical necessity to identify people who may be lonely but also a deep-rooted moral one too.

Summary

- Loneliness is a complex human emotion, describing a state of solitude or a lack of meaningful relationships.
- Loneliness can be the result of a complex interplay between multiple factors including issues affecting an individual, life events and the social environment.
- Loneliness has neurobiological effects which directly impact an individual's health and wellbeing, increasing their risk of developing chronic disease and of premature mortality.
- Health consequences of loneliness include obesity, increased sedentary behavior, increased incidence of cardiovascular risk and stroke, and increased risk of depression and suicide.
- Loneliness should be a priority for all policy makers and public health officials due to its impact on community wellbeing, health and welfare services and the economy.
- Many countries have taken steps through routine general practice or family-based medicine to assess social networks and identify those at risk of isolation.
- Tackling the root causes of loneliness requires a united, multi-agency approach, with input from the government, community groups, health and social care providers, charities and individuals themselves.

References

[1] Australian psychological society and Swinburne University. Australian loneliness report. 2018.
[2] Peplau LP, Loneliness D. A sourcebook of current theory, research, and therapy. New York: Wiley; 1982.
[3] Cigna. New cigna study reveals loneliness at epidemic levels in America. May 1, 2018 [Cited: 8 Jan 2020], www.cigna.com. https://www.cigna.com/newsroom/news-releases/2018/new-cigna-study-reveals-loneliness-at-epidemic-levels-in-america.
[4] Newall N. Causal beliefs, social participation, and loneliness among older adults: a longitudinal study, vol. 26; 2009. p. 2—3.
[5] Hawkley L, Cacioppo J. Loneliness matters: a theoretical and empirical review of consequences and mechanisms. Ann Behav med 2010;40(2).
[6] Griffith J. The lonely society? London mental health foundation. 2010.
[7] Japan home to 541,000 young recluses, survey finds. The Japan Times; September 7, 2016. https://www.japantimes.co.jp/news/2016/09/07/national/japan-home-541000-young-recluses-survey-finds/#.XfYgcC2caJY.
[8] Caspi A. Socially isolated children 20 years later: risk of cardiovascular disease. Arch Paedatric adolescent med 2006;160(8):805—11.
[9] Primack B. Social media use and perceived social isolation among young adults in the U.S. Am J prevent med 2017;53(1):1—8.
[10] Hunt M. No more fomo: limiting social media decreases loneliness and depression. J Soc Clin Psychol 2018; 37(10).
[11] Office of national statistics. Measuring national well-being in the UK: international comparisons, 2019. March 6, 2019. https://www.ons.gov.uk/peoplepopulationandcommunity/wellbeing/articles/measuringnationalwellbeing/internationalcomparisons2019#personal-well-being.
[12] Families and Households in the UK: 2017. s.l. Office for national statistics; 2017.

[13] Grenier E. More Canadians living alone and without children, census figures show. CBC news. August 2, 2017. https://www.cbc.ca/news/politics/census-2016-marriage-children-families-1.4231163.

[14] Age UK. Evidence review: loneliness in later life. 2015.

[15] AARP public policy institute. Medicare spends more on socially isolated older adults. 2017.

[16] Age UK. Loneliness: the state we're in: a report of evidence compiled for the Campaign to End Loneliness. 2012. Oxfordshire: s.n.

[17] Eden Project. The cost of disconnected communities, an exective summary. 2017.

[18] UK government. 2018. www.gov.uk. https://www.gov.uk/government/collections/governments-work-on-tackling-loneliness.

[19] Loneliness and social isolation as risk factors for mortality: a meta-analytic review. Holt-Lunstad, J. 2015; 10:227—37.

[20] Holt-Lunstad J. Social relationships and mortality risk: a meta-analytic review. Pub lib sci 2010;7(7).

[21] Algern M. Social isolation, loneliness, socioeconomic status, and health-risk behaviour in deprived neighbourhoods in Denmark: a cross-sectional study. Pop Health 2020;10.

[22] Lauder W. A comparison of health behaviours in lonely and non-lonely populations. Psychol Health Med 2006;11(2):233—45.

[23] Cacioppo JT. Do lonely days invade the nights? Potential social modulation of sleep efficiency. Psychol sci 2002;13:385—8.

[24] Finan P. The effects of sleep continuity disruption on positive mood and sleep architecture in healthy adults. Sleep 2015;38(11).

[25] Kurina L, Kurina L. Loneliness is associated with sleep fragmentation in a communal society. Sleep 2011; 34(11).

[26] Cacioppo J. The neuroendocrinology of social isolation. Annu Rev Psychol 2015;66(9):9.1—9.35.

[27] Adam E. Day-to-day dynamics of experience—cortisol associations in a population-based sample of older adults. Proc Natl Acad Sci U S A 2006;103(45):17058—63.

[28] Brown E. Loneliness and acute stress reactivity: a systematic review of psychophysiological studies. Psychophysiology 2018;55(5).

[29] Hawkley L. Loneliness is a unique predictor of age-related differences in systolic blood pressure. Psychol Aging 2006;21(1).

[30] Cole S. Myeloid differentiation architecture of leukocyte transcriptome dynamics in perceived social isolation. Proc Natl Acad Sci Unit States Am 2015;49:15142—7.

[31] Powell N. Social stress up-regulates inflammatory gene expression in the leukocyte transcriptome via β-adrenergic induction of myelopoiesis. Proc Natl Acad Sci Unit States Am 2013;110(41).

[32] Cole S. Social regulation of gene expression in human leukocytes. Genome Biol 2007;8(9).

[33] Cohen S. Chronic stress, glucocorticoid receptor resistance, inflammation, and disease risk. Proc Natl Acad Sci Unit States Am 2012;109(16).

[34] Eisenberger N. Sickness and in health: the Co-regulation of inflammation and social behavior. Neuropsychopharmacology 2017;42:242—53.

[35] Eisenberger N. The neural bases of social pain: evidence for shared representations with physical pain. Psychosom Med February, 2012;74(2):126—35.

[36] JUng F. Overweight and lonely? A representative study on loneliness in obese people and its determinants 2019;12:440—7.

[37] Beutel M. Loneliness in the general population: prevalence, determinants and relations to mental health. BMC Psychiatr 2017;17(97).

[38] Valtora N. Loneliness, social isolation and risk of cardiovascular disease in the English longitudinal study of ageing. Eur J prevant cardiol 2018;25(13):1387—96.

[39] Herlitz J. The feeling of loneliness prior to coronary artery bypass grafting might be a predictor of short-and long-term postoperative mortality. Eur J Vasc Endovasc Surg 1998;16(2):120−5.

[40] Increased incidence of cardiovascular risk. Heart 2016;102(13):1009−16.

[41] Holwerda TJ. Feelings of loneliness, but not social isolation, predict dementia onset: results from the Amsterdam Study of the Elderly (AMSTEL). J Neurol Neurosurg Psychiatr February, 2014;85(2):135−42.

[42] Korenke, C. Social networks, social support, and survival after breast cancer diagnosis. J Clin Oncol 2006; 24(7):1105−11.

The health implications of achieving a satisfactory work—life balance

10

Emma Short[1] and Laura Sheldrake[2]

[1]*Department of Cellular Pathology, Division of Cancer and Genetics, Cardiff University, University Hospital of Wales, Cardiff, United Kingdom;* [2]*General Practitioner, National Health Service, Southampton, United Kingdom*

What is 'work—life balance'?

There are many different definitions of work—life balance, and they vary in different countries, industries, and communities. One definition is:

> The balance between your lifestyle and your work which you find to be satisfactory.

It is important to note that this balance is a personal choice. Some individuals gain their sense of self-worth and self-esteem from their job and will happily work a 70-hour week. At the other end of the spectrum, some people work purely for the monetary benefits and choose to work the minimum number of hours that is financially viable. It is up to an individual to decide their own priorities.

Another important point is that an ideal work—life balance is not a static concept. It varies over time as individuals go through different life phases and experience different life events.

Why is it important?

It is vital to achieve a desirable work—life balance as it impacts on mental health and well-being, physical health, relationships and family dynamics [1—4]. A survey carried out by the Mental Health Foundation, a charity based in the United Kingdom (UK), found that a third of employees were unhappy or very unhappy about the amount of time they devoted to work and that more than 40% of workers were neglecting other aspects of their life because of work, which could increase their susceptibility to mental health problems [5]. The survey reported that long working hours were associated with depression, anxiety and irritability and that nearly two-thirds of employees had experienced negative effects of work on their personal life, including a lack of personal development, physical and mental health problems and poor relationships.

A Prescription for Healthy Living. https://doi.org/10.1016/B978-0-12-821573-9.00010-2

Burnout

Individuals who do not have a satisfactory work—life balance are at risk of developing burnout. Burnout is included in the 11th Revision of the International Classification of Diseases (ICD-11). It is referred to as an *occupational phenomenon* rather than a medical condition, yet it is a reason why patients may seek help from Health Services.

'Burnout' is a syndrome thought to result from chronic workplace stress which has not been successfully managed [6,7]. It encompasses three dimensions [6,7]:

1. Feelings of energy depletion or exhaustion
2. Increased mental distance from one's job, or feelings of negativism or cynicism related to one's job
3. Reduced professional efficacy

In practise, this translates to patients reporting '*I'm exhausted, I don't care about my work and I'm no good at it*'.

Burnout has been recognised by the World Health Organization (WHO) as an area which needs addressing. The WHO is currently developing evidence-based guidelines about mental well-being in the workplace [6].

There are several reasons why individuals may be at an increased risk of developing burnout. In the workplace, the volume of work may be excessive, the environment could be highly pressurised, employers may not show appreciation for their employees and workers can feel unrecognised, undervalued and as though they have no control over their workload. Furthermore, working hours can be long, and it may be difficult to have adequate breaks throughout the working day. The stresses of work can be compounded if individuals have high achieving type A personalities with perfectionist tendencies.

If workers do not have a support system outside of work, the impact of burnout can be magnified, and it can be harder to see personal situations in their true perspective. Other lifestyle factors which can increase the risk of burnout include poor sleep, taking on too many responsibilities and not engaging in pleasurable activities.

Work—life balance and burnout in physicians

Burnout is well recognised in the medical profession [8] with a prevalence exceeding 50% across medical students, doctors in training programmes and consultants [9].

Working as a doctor is an important and worthy career, yet it comes with many pressures, including high levels of responsibility, long hours and the emotional impact of managing illness and death. A recent General Medical Council Training Survey in the United Kingdom (2018) found that nearly a quarter of doctors in training were 'burnt out' because of their work and over half of trainees reported that they often or always feel worn out at the end of the working day [10]. A 2016 paper [9] noted that postgraduate medical training in the United Kingdom was characterised by a poor work—life balance, with trainees in certain specialties feeling that they had to prioritise work over home life, that frequent transitions at work and home could disrupt personal relationships, that there was a lack of time to cope with personal pressures outside of work and there was low morale. Work—life imbalance was especially severe for those with children and especially women, who faced a lack of less-than-full-time positions and discriminatory attitudes [9].

Burnout not only impacts the doctor's health and well-being, but it also affects the quality of patient care: There is evidence that burnout has a negative impact on the quality of doctor's performance and on patient's perceptions of outcomes [8]. Factors which appear to increase the risk of burnout include younger age, female sex, negative marital status, long working hours and low reported job satisfaction [8].

Although there seems to be an increasing awareness of burnout in the medical profession and of the importance of achieving a healthy work—life balance to reduce burnout, this is not reflected in doctors' experiences. A study performed in America reported that burnout and satisfaction with work—life balance in US physicians worsened from 2011 to 2014, with 54.4% (n = 3680) of physicians reporting at least one symptom of burnout [11].

Identifying burnout

It is important that doctors and healthcare professionals are able to identify symptoms of burnout in themselves as well as in their patients. There are not rigorous diagnostic criteria, but some useful signs of burnout are listed in Box 10.1. It is important that an underlying adjustment disorder or a disorder associated with stress, anxiety, fear or mood has been considered and ruled out [7].

Box 10.1 Signs of burnout

- Tiredness and lack of energy
- Physical, mental and emotional fatigue
- Irritability
- Feeling that you do not care about your work
- Feeling that your work is pointless
- Feeling that you are no good at your job
- Reduced productivity and efficiency at work
- Reduced standards at work
- No sense of fulfilment/pride about work
- Feeling overwhelmed
- Neglecting aspects of your life outside of work
- Being preoccupied about your work and being unable to 'switch off' even when not at work
- Unsatisfactory relationships with family, friends and colleagues
- Unhealthy behaviours, for example smoking, excessive alcohol consumption, poor eating habits, lack of exercise
- Feelings of low mood/stress
- Difficulty concentrating
- Altered sleep habits
- Poor physical health

Achieving a desirable work—life balance and preventing burnout

Societies are likely to be happier and to work more efficiently and productively if individuals have a work—life balance which works for them. While it may not be possible to 'have it all' at any one point in time, it is possible to achieve an 'all' that is satisfactory.

How to achieve a work—life balance?

There are several steps that can be taken to try to achieve a desirable work—life balance. These strategies can not only can be discussed with patients, but they can also be used by physicians and healthcare professionals.

- *Take time to reflect.* Life can be so busy that it can be hard to take time to assess the reality of life. It is very important to pause and to take stock. Ask patients whether they are happy with their life and, if not, ask why? It can be helpful to prompt patients to think about where they want to be in 1, 5 and 10 years-time and to consider whether they are on the correct pathway to achieve their goals.
- *Take responsibility.* Nobody is going to tell someone how to achieve their optimum work—life balance. It is something for which individuals must take responsibility. Encourage patients to decide what their priorities are and then act accordingly. A well-known saying states that: '*You will never find time for anything. If you want time, you must make it* (Charles Buxton)'.
- *Prioritise.* To be good at anything in life, time needs to be devoted to it. Time is finite, and it is not possible increase the number of hours in the day. This means people need to choose carefully how to use their time. Support patients in considering what their priorities are and what is really important to them. Time is precious, so spend it well.
- *Learn to say no, do not take on too much and set your own boundaries.* This goes hand in hand with the fact that time is limited. If individuals take on too much, they can end up feeling overwhelmed and stressed, and not doing anything well. Before agreeing to do anything, it can be useful to encourage patients to ask themselves three questions:
  a. Will I enjoy this/will it make me happy?
  b. Will it be beneficial?
  c. Can I do it without it causing unnecessary stress?

There are some things in life that have to be done, and there is no choice in the matter. However, there are other activities, tasks, projects or events where there is a choice. Sometimes these can be fun and interesting, but if doing them means that people put themselves under excess pressure, advise them to think twice before they agree.

It is important that individuals learn to say 'no' without feeling the need to justify their reasons. Some people find it helpful to practise using phrases such as '*thank you for the opportunity but I won't get involved at this point in time*'. This then leaves the door open if they change their mind in the future. Once people understand the boundaries that have been set, they are less likely to try and push someone into doing things they do not want to do.

- *Work efficiently.* It can be tempting to believe that long hours need to be put in at work for efforts to be rewarded. While this may be true at times, it is not always the case. It is more important to think about the quality rather than the quantity of work. If work is performed efficiently and productively, tasks can often be completed in a shorter time than if the stage is reached when focus is lost.

- *Leave work at work.* In the digital era, it is all too easy to bring work home. It is now common to check work emails outside working hours and to carry on working on mobile devices. This results in the mind being continually active and never getting the opportunity to 'switch off'. As well as having an adverse impact on an individual's sense of well-being, it can be harmful to relationships due to continual distraction.

One reason for preoccupation with work is because there are concerns about the jobs which will need to be completed the following day. If that is the case, it can be beneficial to write a 'to do' list and leave it at work. Once the list is written, tasks are out of the mind and on a piece of paper, which can reduce anxiety about forgetting anything.

For individuals who work from home, it can be helpful to have designated work-free areas and work-free times of the day.

It is extremely worthwhile to draw a mental line between work and non-work time. This often requires a considerable conscious effort initially but will have a positive impact on health and relationships.

- *Forget about perfection.* Throughout life, it is usual to be encouraged to do your best and to try your hardest. While this may be necessary at certain times and for some roles, in other instances, it is absolutely fine to just get by. 'Good' can change to 'good enough'. Forgetting about perfection means different things to different people, for example it might mean stopping doing any ironing, getting the minimum mark needed to pass an exam or just completing a race rather than achieving a personal best. Whatever realm of life individuals choose to 'get by' in rather than excel in, support them in being kind to themselves and to accept that this is important for their health and well-being.
- *Take a break.* Many individuals have jobs which require self-discipline and self-direction. It is common to work solidly for hours without a break, which is not good for anyone. As well as the physical health risks associated with sedentary behaviour, when people work continually, they lose concentration and productivity falls. It is much better to take regular breaks, so as not to get overwhelmed. Similarly, if someone is starting to experience the signs of burnout, they may need to take a break from work completely to allow them to rest and recover before they reach crisis point.
- *Get help where you can and when you need it to make life easier.* This is relevant on two levels. First, if someone is feeling stressed, anxious or that they are losing control, advise them to seek help before the situation becomes untenable. This might mean talking to a family doctor, a counsellor, an occupational health advisor, a friend or a mental health charity. Talking through difficulties can be helpful, and strategies can be explored to ease the burden. Secondly, encourage individuals to do what they can to make life easier for themselves. This will vary according to an individual's circumstances but could include options such as online shopping or letting children have school dinners rather than packed lunches.
- *Eat well, exercise, sleep well.* As described throughout this book, diet, activity levels and sleep all have a profound impact on well-being. If individuals concentrate on devoting a small amount of time and effort to these, they often find that they are more resilient to the challenges of life and are better able to cope with any difficulties they encounter.

- *Make small changes.* Sometimes people realise that their lives are not how they had hoped or planned and they become aware they need to make changes. These changes may be so great that they can't be achieved in a short time frame. Individuals may want to change career or move elsewhere. If this is the case, support patients in focusing on small changes which can be made immediately. This will allow them to start to take control of their situation, and the impact of small changes can be huge. Suggestions could include making a concerted effort to leave work on time or starting a new hobby. It is beneficial if changes are made which will have a positive impact on quality of life.
- *Do what you enjoy and make time for yourself.* It is all too common for people to put themselves at the bottom of their priority list. However, when people neglect to look after themselves, they will not have the emotional reserves to care for those around them or to work to the best of their abilities. Advise patients to make the time to do what makes them happy. This can often be simple things such as going for a walk, baking a cake or taking a bath. It is also important to make time to nurture significant relationships.
- *Get organised.* A few simple actions can have an enormous effect on helping life to run to run smoothly. A practical example includes batch cooking. Some people find it helpful to have a clear-out of their home. By reducing items to those which are loved or needed, there is less clutter, the house is easier to clean and items are easier to find.
- *Don't feel guilty.* This is one of the key aspects in achieving a work—life balance. Support your patients to not feel guilty for the decisions they make. Advise them to choose the path which is best for them and to have faith in their selection. It is common to hear that individuals feel they should be working more or spending more time with family, but it is important that such individuals have the confidence to stick by their decisions.
- *Be careful with technology and social media.* Digital devices offer major pros and cons in the attainment of work—life balance. While they can allow more flexibility in terms of time and geography, for example they can enable individuals to work from home, they can also blur the line between work-time and downtime. It can be useful to address how much screen time individuals succumb to and whether this is a valuable and meaningful way to spend time.
- *Be prepared to be flexible.* There are times when individuals may need to leave work early or arrive late, for example to attend an important event in their personal lives. Employers and colleagues are more likely to be supportive if the individual concerned has been a good team player in the past. Good communication and flexibility are key. If someone has worked flexibility to help others previously, others are likely to reciprocate when it is needed.

A stark reminder of the importance of achieving a good work—life balance comes when the regrets of people who are dying are considered. The top five regrets are as follows [12]:

1. I wish I'd had the courage to live a life true to myself not the life others expected of me.
2. I wish I hadn't worked so hard.
3. I wish I'd had the courage to express my feelings.
4. I wish I had stayed in touch with my friends.
5. I wish I had let myself be happier.

It is important that everyone, patients and healthcare practitioners alike, choose their priorities well so, when the time comes, they will be able to say, in the words of Edith Piaf, *Non, Je Ne Regrette Rien.*

Summary

- A useful definition of work—life balance is *'The balance between your lifestyle and your work which you find to be satisfactory'*.
- This balance is unique to each individual and varies over time.
- It is important to achieve a desirable work—life balance as it impacts on mental health and well-being, physical health, relationships and family dynamics.
- A poor work—life balance can increase the risk of an individual developing burnout.
- Burnout is characterised by feelings of energy depletion or exhaustion, mental distance from one's job and reduced professional efficacy.
- There is a high prevalence of burnout amongst medical professionals.
- There are many signs of burnout which include tiredness, irritability, apathy towards work, feelings of being overwhelmed, poor physical and mental health.
- There are many strategies which can be explored by healthcare professionals and patients which aim to promote a positive work—life balance. For example, taking time to reflect, setting boundaries, prioritising, accepting help and spending time doing activities which bring enjoyment.

References

[1] Haar JM, Russo M, Suñe A, Ollier-Malaterre A. Outcomes of work—life balance on job satisfaction, life satisfaction and mental health: a study across seven cultures. J Vocat Behav 2014;85(3):361—73.

[2] Whose life is it anyway? Available from: www.mentalhealth.org.uk.

[3] Lunau T, Bambra C, Eikemo TA, van der Wel KA, Dragano N. A balancing act? Work—life balance, health and well-being in European welfare states. Eur J Publ Health 2014;24(3):422—7.

[4] Työterveyslaitos K, Arbejdsmiljøinstituttet GL, Statens arbeidsmiljøinstitutt A, Arbetslivsinstitutet K, Kauppinen K. SJWEH supplements. Finnish Institute of Occupational Health; 2005. p. 14—21. Available from: http://www.sjweh.fi/show_abstract.php?abstract_id=1235.

[5] https://www.mentalhealth.org.uk/a-to-z/w/work-life-balance.

[6] www.who.int/mental_health/evidence/burn-out/en/.

[7] https://icd.who.int/browse11/l-m/en#/http://id.who.int/icd/entity/129180281.

[8] Amaofa E, Hanbali N, Patel A, et al. What are the significant factors associated with burnout in doctors? Occup Med 2015;65:117—21.

[9] Rich A, Viney R, Needleman S, et al. 'You can't be a person and a doctor': the work-life balance of doctors in training. A Qualitative Survey BMJ Open 2016;6:e013897. https://doi.org/10.1136/bmjopen-2016-013897.

[10] General medical Council National training survey initial findings report. 2018. https://www.gmc-uk.org/about/what-we-do-and-why/data-and-research/national-training-surveys-reports.

[11] Shanafelt TD, Hasan O, Dyrbye L, et al. Changes in burnout and satisfaction with work-life balance in physicians and the general US working population between 2011 and 2014 Mayo. Clin Proc 2015;90(12):1600—13.

[12] https://www.mindful.org/no-regrets/.

Happiness and health

11

Alka Patel

Lifestyle Medicine Physician, General Practitioner and Health/Lifestyle Coach, Lifestyle First, London, United Kingdom

> *Happiness is a way of travel, not a destination.*
> **Roy L. Goodman**

Introduction

The concept of happiness is described in literature dating back to over 2500 years ago. Confucius wrote that happiness could be achieved through the practice of *jen,* which describes performing actions to enhance the welfare of others.

In traditional Buddhist teachings, reaching a state of nirvana — the ultimate spiritual goal — focuses on displaying compassion to all [1].

The Greek philosopher, Aristotle, stated that happiness encompasses a range of emotions. He believed that achieving happiness was the purpose of human life. He devoted a significant amount of time to studying the concept of happiness and was convinced that a genuinely happy life resulted from both physical and mental well-being.

Epicureanism is a system of philosophy based on the teachings of Epicurus, founded around 307 BCE. It teaches that happiness is the result of hedonism and sensory pleasures, resulting in a state of tranquility, freedom from fear and the absence of bodily pain.

By the late 18th century, the President of the United States (US), Thomas Jefferson, spoke of the pursuit of happiness as a fundamental human right in the Declaration of Independence. He championed the notion that people should be free to pursue joy and to live their life in a way that made them happy, providing they do not do anything illegal or violate the rights of others.

At around the same time in the United Kingdom (UK), the concept of Utilitarianism was born, when John Stuart Mill and Jeremy Bentham developed ethical and consequentialist theories that promoted actions that would bring about the most happiness for the greatest number of people. Such concepts underlie the humanitarian ethos that suffering is inherently wrong and that everyone, everywhere, should have the opportunity and the right to be happy.

It is significant that many of the historical perspectives on happiness highlight the important interrelationship between happiness and altruism.

A Prescription for Healthy Living. https://doi.org/10.1016/B978-0-12-821573-9.00011-4

Defining happiness

Happiness is an emotion with many different definitions that vary across different countries, communities, cultures and between individuals. Although happiness is a subjective concept, it is important that attempts are made to provide an adequate definition to permit research into its cause and effects.

The World Database of Happiness [2] is a collection of research findings regarding happiness, and it defines happiness as:

the subjective enjoyment of one's life as a whole

Some authorities regard happiness as having two components: pleasure and contentment. In her 2007 book, *The How of Happiness*, Sonja Lyubomirsky describes happiness as [3]:

the experience of joy, contentment or positive well-being, combined with a sense that one's life is good, meaningful and worthwhile

This definition captures the notion that happiness is associated with a deeper sense of meaning and purpose in life, alongside momentary positive emotions.

Ladislav Kováč, in an article addressing the biology of happiness, states that:

Happiness is experienced both as fleeting sensations and emotions, and consciously appreciated as a permanent disposition of the mind. It encompasses two inseparable aspects: hedonia (pleasure of the senses) and eudaimonia (pleasure of reason: living well and doing well).

Hedonic happiness refers to increased pleasure and decreased pain, whereas eudaimonic happiness refers to the notion that people feel happy if they experience life purpose, challenges and growth ([4.5]]).

Measuring happiness

Happiness is a subjective concept, which is therefore difficult to quantify. It is largely measured using self-assessment scales and questionnaires, for example:

(i) The Oxford Happiness Questionnaire [6] comprises 29 statements, to which individuals must give their reaction on a scale from 'strongly agree' to 'strongly disagree'. The statements included are phrased in a positive or negative manner, for example *'I feel able to take anything on'* or *'I feel that I am not especially in control of my life'*.

(ii) The Subjective Happiness Scale [7] is a very simple four-item tool to assess an individual's overall happiness. Individuals must rank their response to each item on a scale of 1−7. For example:

Some people are generally very happy. They enjoy life regardless of what is going on and getting the most out of everything. To what extent does this characterisation describe you?

(iii) The Satisfaction with Life Scale [8] is a 5-item scale used as a measure of life satisfaction. In a similar manner to other tools, respondents rate their responses to statements on a scale of 1−7, for example:

In most ways my life is close to ideal

The conditions of my life are excellent.

(iv) The Positive and Negative Affect Scale (PANAS) [9] is a questionnaire that consists of two 10-item scales which measure positive and negative affect. Positive affect is measured using the words interested, excited, enthusiastic, proud, alert, inspired, attentive and active. Negative affect is measured using the words distressed, upset, guilty, scared, hostile, irritable, ashamed, nervous, jittery and afraid. Respondents indicate on a scale of 1−5, the extent to which they feel each of these emotions in the given moment.

The World Database of Happiness contains both an overview of scientific publications on happiness and a record of all research findings. The database contains information on how happy people are across a wide range of circumstances in 165 different nations. It holds the results of over 1000 self-assessment scales, which are gathered in the 'Measures of Happiness Collection'.

Happiness around the world

The World Happiness Report [10] is an annual publication produced by the United Nations Sustainable Development Solutions Network. It is a survey of the state of global happiness that ranks 156 countries by how happy their citizens perceive themselves to be. Since 2012, the Gallup World Poll questionnaire has been used to measure satisfaction in 14 areas of life: business and economy; citizen engagement; communications and technology; diversity/social issues; education and families; emotions/well-being; environment and energy; food and shelter; government and politics; safety/law and order; health; religion and ethics; transportation and work.

The questionnaire uses the Cantril Self-Anchoring Striving Scale [11], which asks respondents to visualise a ladder with 10 at the top at 0 at the bottom. A score of 10 represents the best possible life and 0 the worst possible life. Participants are asked to rank where they place their current situation on the ladder for each of the 14 domains being addressed. The results for the period 2016−18 are illustrated in Fig. 11.1.

Nordic countries account for four out of the top five happiest nations, with Finland leading the pack. Finland's happiness scores have been rising since 2014. Some of the major factors contributing to Finish happiness levels include the residents feeling safe and feeling positive about their environment, community, public services and education system [10].

Happiness is increasingly being recognised on political agendas, and nations are beginning to plan policy according to statistics of well-being. New Zealand has a specific well-being budget, and the United Arab Emirates (UAE) has a Minister of State for Happiness alongside a national programme for happiness and positivity based on three pillars. These encompass:

• including consideration of happiness in all policies, programmes and services of government bodies and at work,
• promotion of positivity and happiness as a lifestyle modification,
• development of benchmarks and tools to measure happiness.

Bhutan measures Gross National Happiness over nine domains: psychological well-being, health, education, time use, cultural diversity, good governance, community vitality, ecological diversity and living standards. It is used to support policymaking and track the effectiveness of policies over time.

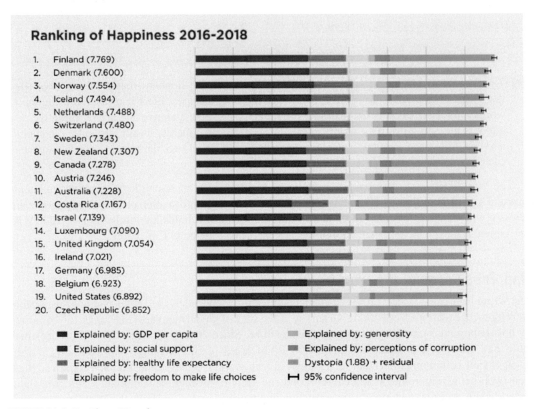

FIGURE 11.1 Ranking of happiness.

World Happiness Report 2019.

Health and social impact of happiness

The sensations of joy and happiness are mediated through the neurotransmitters dopamine and serotonin, and hormones such as oxytocin and endorphins. However, happiness is not merely a psychological or emotional state. It has a physiological impact and plays a role in health and disease.

Happy people are healthier and live longer

Happy people may have a greater life expectancy. An analysis of the relationship between happiness and the risk of death was undertaken in the United States using a General Social Survey — National Death Index Dataset [12]. Between 1978 and 2002, in excess of 31,000 individuals, more than 18 waves completed the survey, and the results were correlated with death certificate data from 1979 to 2008. Participants were asked:

Taken all together, how would you say things are these days — would you say that you are very happy, pretty happy, or not too happy?

It was reported that compared to 'very happy' people, the risk of death over the follow-up period was 6% higher among those who were 'pretty happy', and 14% higher among those who were 'not happy', even after controlling for lifestyle factors. Happiness is related to social determinants of health such as social relationships, socioeconomic status and religious attendance. Its relationship to life expectancy may be mediated through these factors, for example mortality risk is lower in married individuals compared with those unmarried, and this might be linked to increased social support.

In a UK-based study, 228 volunteers aged 45−59 years were asked to rate their levels of happiness on a workday and on a day off. Blood pressure and heart rate were monitored every 20 min, and happiness was ranked at the same time the physical measurements were taken. Volunteers also provided regular saliva samples [13]. The study reported that participants with higher happiness ratings had significantly lower levels of salivary cortisol, with a difference of 32.1% between the lowest and highest happiness quartiles ($P = .0009$). Cortisol is a primary stress hormone, and persistently raised levels are associated with an increased risk of a variety of chronic diseases, including cardiovascular disease and gastrointestinal disease, as detailed in Chapter 6. This study suggests that happiness and stress levels are inversely correlated, but it is not clear from this study whether this is cause or effect.

Further studies have shown that positive psychological well-being protects against cardiovascular disease, including myocardial infarction and heart failure, independently of traditional risk factors [14]. One study followed 6808 participants for 4 years [15]. Optimism was assessed using a 6-item Life Orientation Test-Revised. Statistical analysis using multiple regression models showed that optimism was independently inversely associated with heart failure after adjusting for sociodemographic and behavioural, biological and other psychological factors. A dose−response relationship was evident − as optimism increased, the risk of developing heart failure decreased, with a 48% reduced odds among people with the highest versus lowest optimism. The mechanisms that underlie this effect are not fully understood. While some hypothesise that optimism can have a direct impact on biological parameters such as serum levels of interleukin-6, C-reactive protein, fibrinogen, carotid intima-medial thickness, blood lipid profile and serum antioxidants, others propose that the protective effects of happiness relate to associated high levels of social support.

Furthermore, it has been shown that optimists and happy people with positive psychological well-being are more likely to engage in health-promoting behaviours such as eating healthier foods, exercising more, sleeping better and abstaining from smoking. It is likely that such positive health behaviours interact with biological factors, for example higher parasympathetic control and reduced inflammation, in a bidirectional relationship, although cause and effect has not been fully elucidated [14].

Happy people are more productive

There is a positive correlation between happiness and employee productivity [16]. Happy workers tend to perform more effectively as leaders, generate more sales, take fewer days off due to sickness and receive higher pay with better performance ratings. The effect of increasing happiness levels on productivity was examined in a trial of 270 subjects [17]. They were randomly assigned to watch, or not watch, a 10-min comedy clip before taking a numerical test in which a payment was given for each correct answer. Participants were asked to rate their happiness on a linear scale of 1−7 in response to the question: How happy are you at the moment? Productivity was assessed by determining the number of correct responses in a given time period. Results showed that a rise in happiness, even if superficial, was associated with greater productivity (correlation coefficient = 0.118).

Happiness and socioeconomic factors

In 2015, more than 30,000 people took part in the Happiness and Longevity study [12]. 31.4% of participants described themselves as very happy, 56.9% as pretty happy and 11.6% as not happy. The results showed that demographic and socioeconomic factors were related to the perception of happiness: Individuals who reported being very happy were generally older, and married. They had higher levels of education, higher incomes and regularly attended religious services. Yet, even in disadvantaged groups, only small proportions reported being not happy, for example among those in poverty, 21.2% reported not being happy. Individuals with poor living conditions and low incomes may encounter difficulties in meeting their basic needs, so are more susceptible to physical, mental and emotional problems [18].

It is proposed that deep connecting relationships with family, friends and colleagues are significant factors in life satisfaction and human happiness. Happy people are believed to be more socially connected. A German study of 15,000 individuals found that happier people tend to trust others more, have a higher desire to vote, perform more voluntary work and participate in public activities more frequently. They also have a higher respect for law and order, are more attached to their neighbourhood and extend help to others wherever they can [19].

Happiness in the healthcare profession

Working in the healthcare service can be demanding and stressful. However, a 2019 US Physician Happiness Survey of over 5000 doctors found that 71% of practicing doctors described themselves as happy although 19% felt very or somewhat unhappy. 57% of the cohort reported enjoying life 'a lot' or 'a great deal' with 1% saying 'not at all' [20]. These results are illustrated in Figs. 11.2 and 11.3.

Many doctors choose a career in medicine due to an enjoyment of science and a desire to help others. 74% of doctors report that friends and family make them happiest, with 20% stating that the most gratifying aspect of their lives is helping patients [20].

In the United Kingdom, the 2018 Medscape UK Doctors' Professional Satisfaction Survey of 800 doctors found that 83% were happy with their chosen career [21]. However, the UK Doctors' Burnout and Lifestyle Survey revealed that too many bureaucratic tasks and long working hours were major causes of stress [22].

Although the majority of practicing healthcare professionals may be happy with their vocation, it is of great significance that there is a higher suicide rate among doctors, especially female doctors, compared with the general population and other professional groups. The reasons for this often include depression, bipolar disorder or substance misuse, as for the rest of the population. However, doctors may face additional pressures such as long working hours, a lack of sleep, the emotional impact of dealing with suffering, illness and death, high levels of responsibility, postgraduate examinations, a lack of resources in the health service and potential complaints or referral to regulatory bodies. Furthermore, doctors may not seek help for mental health problems [23].

In 2019, the UK General Medical Council published its report, 'Caring for doctors, caring for patients' [24], confirming that doctors have an ABC of core needs to stay happy and motivated in work. The ABC refers to:

A: Autonomy — the need to have control over work.

B: Belonging — the need to be connected, cared for, valued, respected and supported.

C: Competence — the need to experience effectiveness.

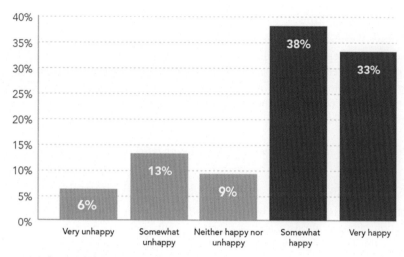

FIGURE 11.2 Self-ranking happiness of US physicians.

US Physician Happiness Survey 2019, CompHealth/AAFP.

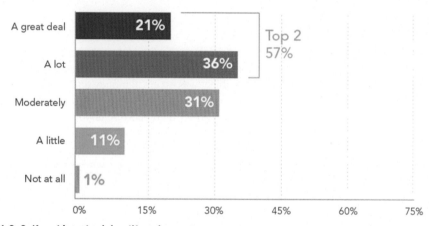

FIGURE 11.3 Self-ranking physician life enjoyment.

US Physician Happiness Survey 2019, CompHealth/AAFP.

Interventions to improve happiness

Happiness is a subjective emotion, and an individual's emotional state is the result of an incredibly complex interaction between many psychological, physical, social and environmental factors. However, there are several simple strategies that can be implemented to improve happiness and positive mental well-being. These are outlined in the following. They can be utilised by healthcare practitioners for their own well-being, as well as being discussed with patients as a component of a holistic approach to healthcare.

The power of relationships

There is a positive correlation between happiness and social connectivity. A study looking at 222 undergraduate students reported that the top 10% of consistently very happy people were 'highly social', did not spend much time alone and had stronger romantic and other social relationships than the less happy groups. However, it is not possible to infer cause and effect from these data. The happiest group experienced positive feelings most of the time but occasionally reported negative moods, suggesting that they have a functioning emotional system that reacts appropriately to life events [25].

Action: Focus on a relationship: invest time and energy into cultivating, affirming and enjoying it.

The power of altruism

Caring for other people is associated with increased levels of happiness, with research showing a strong correlation between altruism and well-being, health and longevity. It is important to note, however, that health benefits are only observed when 'helping behaviours' are not experienced as overwhelming [26].

Action: Consider volunteering for an organisation focused on an area of interest to you, but only take on what you can comfortably fit into your life.

The power of compassion and kindness

Compassion encompasses kindness and helping others. Individuals who experience or receive compassion often 'pay it forward', creating a cascade of cooperation. In one experiment, subjects were placed into groups of four individuals, and each subject was given 20 money units. Each individual had to decide how many money units to keep for themselves or contribute to a group project. It was observed that individuals were influenced by other group members' behaviours, demonstrating that individuals often mimic the behaviours they encounter [27]. Acts of generosity increase neuronal activity in the ventral striatum, which is a key component of the brain's reward system. Being generous therefore not only benefits the recipient but also rewards the individual who performs the behaviour [28].

Action: Make a conscious effort to show compassion, kindness and generosity to others and to support those in need.

The power of working towards a goal

The experience of engaging in an activity which moves someone towards a goal is associated with increased happiness. An example of this is the composition of music: The process of writing the music creates happiness in addition to the satisfaction the musician feels once the composition is complete. Some authorities describe a 'flow state' in which an individual is fully immersed in a feeling of energised focus and full involvement in an activity that requires complete concentration. However, merely being 'in the zone' does not guarantee increased positive effect. The activity undertaken must be suitably complex, with a potential for growth and the development of new skills [29].

Action: Choose one significant, meaningful goal and commit time and effort to working towards it.

The power of optimism

An optimistic attitude has been linked to improvements in health and life expectancy. A review of data from the Panel Study of Income Dynamics, from 1968 to 2015, found that males were more likely to be alive in 2015 if they were more optimistic [30]. Optimism was assessed using the Life Work Out question: *Have you usually felt pretty sure your life would work out the way you want it to or have there been more times when you have not been very sure about it?* Optimism is associated with reduced levels of cortisol [31], and it promotes the production of dopamine, which improves mood, increases motivation and is known to be responsible for giving entrepreneurs the courage to take risks [32].

Action: Visualise your best possible future and how you feel when you achieve it. Journaling can be helpful — a useful exercise is 'Three Good Things': recall and write down three positive things that happen during the day, however big or small.

The power of anticipation

It is interesting to note that the greatest peak in happiness may result from the anticipation of an event, rather than the event itself [33].

Action: Make a conscious effort to plan future activities to allow the experience of anticipatory pleasure.

The power of gratitude and mindfulness

The practices of gratitude and mindfulness have a profound effect on happiness and satisfaction, as described in Chapter 8 and in Concept boxes on page 297.

Action: Elicit the body's relaxation response through sensory awareness. Spend a few minutes every day focusing on a specific sense, for example close your eyes and concentrate on the sounds around you.

The power of lifestyle

Healthy behaviours, including exercise, good nutrition, safe exposure to sunlight and adequate sleep, are associated with improved mental well-being [34]. They are described in detail throughout this book.

Action: Prioritise healthy behaviours, for example try and achieve 8 h sleep every night.

Conclusion

Happiness is a personal emotion. But despite its subjective nature, it can still be measured in a meaningful and reliable way. This is important, as it allows research into factors such as its cause and effect. Happiness is not just about feeling good, but it also encompasses feeling fulfilled.

Being happy has widespread effects and significant health and social implications: Happy individuals may have an increased life expectancy, a reduced risk of cardiovascular disease, exhibit more productivity and creativity at work and have stronger social connections.

Happiness is not predetermined or a fixed emotional state. There are several steps which can be taken to improve an individual's happiness, including spending time with people that are cared about, focusing on good things, being mindful and showing kindness and compassion.

If you are happy, you are more likely to make others happy. Caring about the happiness of others moves us in the direction of a happier society.

> *The purpose of our lives should be to create as much happiness in the world as possible. Richard Layard.*

Summary

- Happiness is a subjective emotion.
- Most definitions of happiness include elements such as pleasure and purpose.
- Happiness is mediated through dopamine, serotonin and endorphins.
- Happiness has significant health and social implications: Happy individuals may have an increased life expectancy, a reduced risk of cardiovascular disease, exhibit more productive behaviors at work and have stronger social connections.
- Strategies to improve happiness include nurturing relationships, helping others, having a purpose and working towards a goal and practising gratitude and mindfulness.

References

[1] Lyubomirsky S, et al. The benefits of frequent positive affect: does happiness lead to success? Psychol Bull November 2005;131(6):803—55.
[2] The World database of happiness. https://worlddatabaseofhappiness.eur.nl/hap_quer/hqi_fp.htm.
[3] Lyubomirsky S. The how of happiness — a practical guide to getting the life you want. USA: Penguin Press; 2007.
[4] Kováč L. The biology of happiness: chasing pleasure and human destiny. EMBO Rep 2012;13(4):297—302.
[5] Ryan RM, et al. On happiness and human potentials: a review of research on hedonic and eudaimonic well-being. Annu Rev Psychol 2001;52:141—66.
[6] Hills P, Argyle M. The Oxford Happiness Questionnaire: a compact scale for the measurement of psychological well-being. Pers Indiv Differ 2002;33:1073—82.
[7] Lyubomirsky S, Lepper H. A measure of subjective happiness: preliminary reliability and construct validation. Soc Indicat Res 1999;46:137—55.
[8] Diener E, Emmons RA, Larsen RJ, Griffin S. The satisfaction with life scale. J Pers Assess 1985;49:71—5.
[9] Watson D, Clark LA, Tellegen A. Development and validation of brief measures of positive and negative affect: the PANAS scales. J PersSocPsychol 1988;54(6):1063—70.
[10] World happiness report 2019. Available at: https://worldhappiness.report/ed/2019/.
[11] Gallup world poll cantril ladder. Available at: https://worldhappiness.report/faq/.
[12] Lawrence EM, Rogers RG, Wadsworth T. Happiness and longevity in the United States. Soc Sci Med 2015;145:115—9.
[13] Steptoe A, et al. Positive affect and health-related neuroendocrine, cardiovascular, and inflammatory processes. Proc Natl Acad Sci Unit States Am 2005;102(18):6508—12.
[14] Boehm JK, Kubzansky LD. The heart's content: the association between positive psychological well-being and cardiovascular health. Psychol Bull 2012;138(4):655—91.

[15] Kim E, et al. Prospective study of the association between dispositional optimism and incident heart failure. Circ Heart Fail 2014;7:394—400.

[16] Claypool K. Organizational success: how the presence of happiness in the workplace affects employee engagement that leads to organizational success. Pepperdine University, ProQuest Dissertations Publishing; 2017.

[17] Oswald AJ, et al. Happiness and productivity. J Labor Econ 2015;33(4):789—822.

[18] Azizi M, et al. The effect of individual factors, socioeconomic and social participation on individual happiness: a cross-sectional study. J Clin Diagn Res 2017;11(6):VC01—4.

[19] Guven C. Are happier people better citizens? Kyklos. Int Rev Soc Sci 2011;64(2):178—92.

[20] AAFP/CompHealth physician happiness survey. 2019. https://comphealth.com/resources/physician-happiness-survey/.

[21] UK doctors' professional satisfaction survey. 2018. https://www.medscape.com/slideshow/uk-doctors-satisfaction-survey-6009772.

[22] UK doctors' burnout & lifestyle survey. 2018. https://www.medscape.com/slideshow/uk-burnout-report-6011058.

[23] Gerada C. Doctors and suicide. Br J Gen Pract 2018;68(669):168—9.

[24] West M, Cola D. Caring for doctors, caring for patients. GMC November 2019. https://www.gmc-uk.org/-/media/documents/caring-for-doctors-caring-for-patients_pdf-80706341.pdf.

[25] Diener E, Seligman ME. Very happy people. Psychol Sci 2002;13:81—4.

[26] Post SG. Altruism, happiness, and health: it's good to be good. Int J Behav Med 2005;12(2):66—77.

[27] Fowler JH, Christakis NA. Co-operative behaviour cascades in human social networks. Proc Natl Acad Sci Unit States Am 2010;107(12):5334—8.

[28] Davidson RJ, et al. Alterations in brain and immune function produced by mindfulness meditation. Psychosom Med 2003;65(4):564—70.

[29] Csikszentmihalyi M. If we are so rich, why aren't we happy? Am Psychol 1999;54(10):821—7.

[30] O'Connor J, Graham CL. More optimistic lives: historic optimism and life expectancy in the United State. J Econ Behav Organ 2019;168(C):374—92. Elsevier.

[31] Lai JCL, et al. Optimism, positive affectivity and salivary cortisol. Br J Health Psychol 2005;10(4):467—84.

[32] Medina J. Brain rules for ageing well: 10 principles for staying vital, happy, and sharp. Pear Press USA; 2017.

[33] Nawijn J, et al. Vacationers happier, but most not happier after a holiday. Appl Res Quality Life 2010;(5): 35—47.

[34] Shin J, Kim JK. How a good sleep predicts life satisfaction: the role of zero-sum beliefs about happiness. Front Psychol 2018;(9):1589.

The role a mentor can play in improving well-being

12

Nita Maha

General Practitioner, Primary Care, Bristol, United Kingdom

Introduction

A mentor can be defined as 'an experienced and trusted advisor' [1], but there are many roles that a mentor can play. One such role is in supporting individuals to achieve their goals, including those in the workplace, in family life and in health. This chapter will look at what mentoring is and how it can be beneficial to both the mentor and the mentee.

Mentorship models

The concept of 'mentorship' dates back to 800 BCE. Homer's classic poem 'The Odyssey' describes a time when the King of Ithaca, Odysseus, was preparing to leave for Troy. He wanted to be sure there was someone who could take care of his son in his absence; someone who would act as a teacher, advisor and friend. The guardian's name was Mentor [2].

Since that time, there have been thousands of articles published about mentorship, which have looked at the important relationship between mentor and mentee and the beneficial effects that such a relationship can bring. Different people regard mentorship in different ways: to some, it means having a positive role model, and to others, it is someone who can help to build supportive networks in business. The mentor—mentee relationship can last for years, or just for a day. Every relationship is different.

Anderson and Shannon [3] define mentoring as *'a nurturing process in which a more skilled or more experienced person, serving as a role model, teaches, sponsors, encourages, counsels, and befriends a less skilled or less experienced person for the purpose of promoting the latter's professional and personal development'*. This 'protégé model' sees the mentee, or protégé, as having a role model, while the mentor nurtures and cares for the mentee.

Some organisations use a 'reverse mentoring' model. With the digital revolution in the 2000s, many businesses paired up younger individuals with more senior colleagues. The younger generation was seen to be more experienced with the latest technology, which was useful for the older generation [4].

A Prescription for Healthy Living. https://doi.org/10.1016/B978-0-12-821573-9.00012-6

Benefits of the mentoring relationship

There are several advantages for those who have a mentor. In the workplace, individuals who engage in a mentoring relationship are reported to have increased awareness of their profession, enhanced assimilation into an organisation, increased job satisfaction and increased likelihood of success [5].

Outside of the workplace, mentoring can also be important. Parent mentors can assist new parents, particularly in vulnerable families. A study by Ayton and Joss [6] evaluated the 'Creating Opportunities and Casting Hope' (COACH) programme, a family mentoring programme for vulnerable parents. The aim of the programme is to break the cycle of generational poverty. It involves mentors providing guidance on parenting strategies, financial management and domestic skills. Mentors are individuals who are chosen by COACH staff who are recognised to have good interpersonal skills, life experience and wisdom. The mentors undergo 14 h of training delivered by COACH staff and are given a manual which includes information on how to communicate effectively, how to assist parents in setting life goals, risk management and how to empower parents to influence their children by implementing behaviour management strategies. Parents engage with their mentor over a minimum of 12 months. The programme was evaluated using a mixture of semistructured interviews with parents, surveys with mentors and reviews of client case reports. It was reported that parents participating in the programme experienced an overall improvement in housing and employment situations. One of the participants stated that:

> ...with the COACH program there is no end date. To be able to talk to somebody about your problems where you wouldn't talk to friends or family ... they are there to be able to help you, listen to you, give you advice when it's needed, um I don't know where I would be actually. I really think the COACH program saved me.

Mentoring in healthcare

Mentoring can have a vital role in the well-being of medical professionals. A randomised controlled trial conducted in Perth, Western Australia [7], looked at the value of peer mentoring for the well-being of junior doctors. In 2015, first-year interns were randomly assigned to either receive a mentor or not. The mentors were resident doctors. 53 interns applied to participate in the programme, out of a total of 79. 26 were mentor–mentee matched, and 27 were in the control group. Semistructured interviews were conducted using validated tools, and qualitative information gained. The interviews explored the degree of communication between members of mentoring pairs, as well as the overall value of the programme for its effects on emotional burnout, organisational engagement, job satisfaction, sense of support and the psychological well-being of the mentee. Interviews and focus groups were conducted at the end of the academic year. Participants with mentors reported high satisfaction with the programme and a positive impact on stress levels, morale, sense of support, well-being and job satisfaction compared with controls.

Mentoring in medicine can have an important role in the career development of physicians. In a review by Mckenna and Straus [8], it was reported that 85% of radiology programme directors strongly agreed that residents should have mentors due to the positive impact this can have. A mentoring relationship was reported to be associated with faculty advancement, research productivity and overall well-being and job satisfaction. However, only 58.7% believed that their trainees had established mentor relationships. Lack of access to mentors has been considered a barrier to ensuring adequate mentor relationships.

In the United Kingdom (UK), Dr Alliott [9], an East Anglian general practitioner (GP), had discussed how mentoring schemes can help doctors experiencing high levels of stress. Stress is a common factor in general practice/family medicine, and many practitioners are reluctant to consult their own doctors to discuss their difficulties. Mentoring is a way of reaching GPs in a nonthreatening way to discuss problems.

Sayan and Ori [10] have demonstrated that strong mentorship can improve mentee productivity, clinical skills, medical knowledge and career progression. A study looking at 126 radiation oncology residents found that those residents who were part of a formal mentoring programme had significantly higher rates of job satisfaction compared with those who were not (90% vs 9%, $P < .001$).

Some of the major benefits of having a mentor include the following:

Support: A mentor can support a mentee through challenges and can act as a confidential 'sounding board'. Mentors can offer guidance, encouragement and support, for example if a mentee lacks confidence or is finding it difficult to manage work politics. A mentor can use their own experiences to help a mentee to work through their problems and to improve their overall performance, which may lead to them staying in a job they may have otherwise left.

Networking: A mentor who is experienced in a relevant industry can aid introductions to like-minded individuals. Growing a network can be invaluable, particularly when building a business.

Preventing Burnout: Sometimes, it can be difficult for individuals to motivate themselves, or conversely to recognise when they are close to 'burning out'. The Yerkes-Dodson law [11] refers to an individual's performance under stress. Overall, increasing stress levels result in increasing performance, but there is a point at which too much stress reduces effective performance. As an individual, it can be difficult to recognise where on this curve you lie, but an effective mentor can identify when an individual needs motivating or when they are close to burnout. Mentors can work with a mentee to implement strategies to ensure that performance is maximised without compromising well-being.

Knowledge: A mentor can act as an experienced guide on a particular topic. Many professions use an *'apprentice model'* of education. This originated in the middle ages and was used by craftsmen to teach a trade to juniors. The knowledge and skills of experienced individuals were passed on to those learning. Apprentices often develop a strong bond with their supervisors.

Interest: Mentors are priceless. They are usually not motivated by money, but instead by the satisfaction of helping others. Having someone with a genuine interest in an individual and their development is very important in helping that individual to feel valued and significant. Passing on wisdom and knowledge is beneficial for the mentee, and mentors can feel fulfilled from sharing information. A good mentor will provide constructive comments, not just simple praise. The mentoring relationship also benefits the mentor. Mentors are reported to experience revitalised interest in their work, increased job satisfaction, a sense of pride in what their mentee has accomplished and enhanced quality of life through improved relationships [12].

Characteristics of mentors and mentees

The most important characteristics and attributes of mentors include good listening skills, being open, warm and enthusiastic, providing feedback, being experienced in their field, being nonjudgemental and having counselling skills when needed [13].

It is recognised that mentees have certain obligations and that mentors have expectations of their mentees. It is desirable that mentees are willing to learn and have a positive attitude, a capacity for professional leadership, commitment and initiative and a capacity for joint decision-making [14].

A key factor needed by both parties is time, so that both individuals can commit to the new relationship. It is also very important to ensure the right combinations of people are selected, to ensure they work well together. A poor relationship would not be beneficial to either party. Geographical considerations are also important. A relationship is likely to be more effective if both parties are able to meet regularly. To ensure adequate recruitment of mentors and mentees, any formal schemes need to be an attractive option for both parties. Mentors are often senior members of staff with plenty of diary commitments. They may need to be persuaded that a mentoring scheme is beneficial for their organisation, their department and for themselves before pursuing this.

Becoming a mentor can be achieved informally, and no qualifications are necessarily required. However, many organisations have training programmes to develop appropriate skills. Courses can range from introductory short courses to master's programmes. They can also vary according to delivery approach and style [15].

How to find a mentor

There are various methods for identifying a mentor. It is useful to identify what skills are required in a mentor, as this can aid in the search for the most suitable and appropriate individual. Finding a mentor through a network is often an option, for example through work or an online community. It is useful to send correspondence to all contacts making it known that a mentor is being searched for. Once potential mentors have been found, it is important to establish that they are compatible. Having a list of questions to determine compatibility can be helpful. When establishing the relationship, important areas to explore and discuss are the expected frequency of meetings, the method of communication, the duration of the relationship and how to go about ending the relationship if either party feels this is necessary. Setting ground rules, such as when it is inappropriate to call, can be helpful [16].

Conclusions

The concept of mentorship dates back to as early as 800 BCE. The mentor–mentee relationship is often unique and can last from a few days to a lifetime. Studies have shown that it can be beneficial to both mentor and mentee by increasing job satisfaction and success in an organisation. Finding a mentor can be challenging, but networks are helpful when searching for appropriate individuals. It is of paramount importance to find the right combination of mentor–mentee. To aid success having 'ground rules' can be helpful, for example clarifying frequency of visits and where these would take place. Certain organisations run training programmes for mentors and mentees, which can include guidance regarding how to set up a mentor–mentee relationship.

Summary

- Mentorship is a historic relationship.
- Most mentoring relationships have an element of support and guidance.
- In the workplace, mentorship can increase job satisfaction and success.
- Mentoring can benefit vulnerable families, especially regarding housing and employment.
- In the medical profession, having a mentor may have a positive impact on stress levels, morale, sense of support, well-being and job satisfaction.
- Each mentor—mentee relationship is unique.

References

[1] oed.com.

[2] study.com/academy/lesson/history-of-mentoring.html.

[3] Anderson EM, Shannon AL. Toward a conceptualization of mentoring. In: Kerry T, Mayes AS, editors. Issues in mentoring. Routledge; 1995.

[4] forbes.com/sites/work-in-progress/2011/01/03/reverse-mentoring-what-is-it-and-why-is-it-beneficial.

[5] warwick.ac.uk/study/cll/courses/professionaldevelopment/wmcett/resources/practitionerarea/mentoring/what/purposes/.

[6] D. Ayton and N. Joss, Empowering vulnerable parents through a family mentoring program, Aust J Prim Health 22 (4), 109-112.

[7] Chanchlani S, Chang D, Ong JS. The value of peer mentoring for the psychosocial wellbeing of junior doctors: a randomised controlled study. Med J Aust November 5, 2018;209(9):401—5.

[8] Mckenna A, Straus. Charting a professional course: a review of mentorship in medicine. J Am Coll Radiol 2011:109—12.

[9] bmj.com/content/313/7060/S2-7060.

[10] frontiersin.org/articles/10.3389/fonc.2019.01369/full.

[11] oxfordreference.com/view/10.1093/oi/authority.20110803125332105.

[12] hbr.org/1979/01/much-ado-about-mentors.

[13] Paetow G, Zaver F, Gottlieb M, et al. Online mastermind groups: a non-hierarchical mentorship model for professional development. Cureus July 20, 2018;10(7):e3013. https://doi.org/10.7759/cureus.301.

[14] coachingnetwork.co.uk/become-a-coachmentor.aspx.

[15] qiconcepts.co.uk/wp-content/uploads/2010/07/Qi-Concepts-Guide-on-How-to-Set-up-a-Mentoring-Programme.pdf.

[16] monster.com/career-advice/article/womans-guide-to-finding-a-mentor.

Physical activity and physical health

Fit for life: the health benefits of cardiovascular activity

13

Caroline Deodhar

Senior Resident Medical Officer Obstetrics and Gynaecology, Westmead Hospital, Sydney, NSW, Australia

Introduction

Physical activity (PA) is an important consideration when addressing the risk factors for many chronic diseases [1]. In middle-income and high-income countries, inadequate PA is thought to contribute to up to 30% of the disease burden from ischaemic heart disease and type 2 diabetes mellitus. Most individuals have some awareness that exercise is beneficial to health, but numerous barriers exist to exercise uptake.

This chapter reviews the physiology of exercise, the health benefits of PA, specific disease reduction rates and the barriers to PA. Guidance shall be offered regarding counselling individuals to increase their activity levels, and the role of wearable activity trackers shall be discussed.

The chapter aims to equip healthcare professionals with an overview of the health-related aspects of PA so that clinically relevant information can be utilised in future consultations.

What is exercise?

Exercise is defined as any activity which causes physical exertion, and it is important to recognise that activities such as housework and gardening make a valid contribution to total time spent on physical exertion. The broad types of exercise are cardiovascular exercise, muscular strength and endurance, and flexibility [2].

Exercise is also categorised according to whether it is primarily aerobic or anaerobic in nature, referring to the underlying energy system that is stimulated [2]. Aerobic exercise activates large muscle groups, using slow twitch/type 1 muscle fibres, and can be maintained continuously, with examples including jogging, cycling and dancing [2,3]. Anaerobic exercise activates fast twitch/type 2 muscle fibres and includes high-intensity interval training and power lifting [2,3].

The physiology of exercise

Exercise results in an immediate increase in energy demand. This is met by metabolism of adenosine triphosphate (ATP), the molecule used to drive intracellular energy transfer. One of the phosphate bonds in the molecule is hydrolysed to form adenosine diphosphate (ADP) plus a free phosphate ion,

A Prescription for Healthy Living. https://doi.org/10.1016/B978-0-12-821573-9.00013-8

and this process yields approximately 7 kcal energy per molecule [3]. Immediate hydrolysis of ATP provides energy to fuel around 1 s of activity, so the body must resynthesise ATP rapidly to continue to meet ongoing energy demands [3]. This occurs by one of three pathways which act as a continuum, with all three pathways being stimulated to some degree during PA [4].

The first system is the phosphocreatine system. During the first 10 s of exercise, hydrolysis of phosphocreatine, a compound stored in skeletal muscle, is used to provide the energy to convert ADP back to ATP under anaerobic conditions [3,4]. Once muscular phosphocreatine stores are depleted, further metabolic processes must be utilised to continue to meet energy demands [3,4].

The second and third energy systems operate via glycolysis, the process by which glycogen is converted to pyruvate, which in turn is used to produce more ATP [3]. In the absence of oxygen, pyruvate is converted to lactic acid [3]. This results in a yield of only three molecules of ATP per molecule of glycogen, so it is relatively inefficient, and the build-up of lactic acid that results causes a drop in pH, resulting muscles 'burn' which limits the degree to which this pathway can be utilised [3,4].

After about 2 min of activity, the aerobic system is fully activated, and in the presence of oxygen, pyruvate is processed via the Krebs cycle, with a yield of 39 molecules of ATP per molecule of glycogen [3]. Fatty acids are also used to produce ATP; however, the process by which this occurs, β-oxidation and entry to the Krebs cycle, is too slow for fats to be regarded as a meaningful energy substrate during high-intensity exercise. In endurance events after depletion of glycogen stores, fat metabolism may become more relevant as an energy source [3].

The increased oxygen demands placed on the body during PA require an increase in cardiac output. Both heart rate and stroke volume increase to achieve this [3]. Forcible systolic contraction causes an increase in stroke volume. With repeated activity sessions, this causes hypertrophy of the cardiac muscle, further increasing the stroke volume [3]. Consequently, at rest, the athletic heart is able to beat at a much slower rate. Although PA mandates increased perfusion of skeletal and cardiac muscle, cerebral perfusion remains constant, and a minimum essential blood flow is provided to the rest of the vital organs [1,3].

Repeated PA leads to an improvement in the elasticity of the smooth muscle wall of the blood vessels and also reduces peripheral vascular resistance, thus leading to a decrease in blood pressure at rest [1].

Increased oxygen requirements demand an increase in respiratory rate and tidal volume. Resting oxygen consumption is around 250 mL/min, and resting ventilation is around 5–6 L/min [3]. During exercise, this rises to as much as 5000 mL/min oxygen consumption and ventilation of over 100 L/min. The initial rise in ventilation in response to exercise is rapid but then plateaus out to a much more gradual increase. After cessation of exercise, the respiratory rate may remain elevated for some time during a process termed *postexercise oxygen consumption*. This has received much interest in the fitness community due to the presumed elevated rate of energy expenditure during this phase compared with rest [3].

The mechanisms underpinning the control of ventilation during exercise are not fully understood but are likely to be related to minute fluctuations in Pao_2 and $PaCo_2$ at peripheral chemoreceptors [3]. Other mechanisms might be mediated through changes in core temperature [3].

Muscle fibres in skeletal muscle are classified as type 1 slow twitch or type 2 fast twitch fibres [3]. Slow twitch fibres contract at a slower, less forceful rate and mainly synthesise ATP under aerobic conditions. Fast twitch fibres produce more forceful contractions and utilise ATP under anaerobic conditions [3]. The proportions of these fibres found in individuals vary according to genetics and type of exercise undertaken, although repeated exercise improves skeletal muscle oxygen utilisation [2,3].

The health benefits of exercise

An inadequate level of exercise is a major risk factor for chronic disease and premature mortality. It has been reported that middle-aged women who complete less than 1 hour of PA per week have a doubling of both all-cause mortality and cardiovascular mortality and a 20% increase in cancer related mortality [1]. Similar risks are seen for hypertension, hyperlipidaemia and obesity. To put this into context, these figures are similar to the increase in risk caused by moderate cigarette smoking [1].

Regular exercise is associated with a reduction in all causes of mortality for those with a good baseline level fitness and for those who improve their fitness over time. The effects are seen irrespective of bodyweight and gender [5–7]. Even when an individual falls short of meeting the current exercise recommendations, there is still an overall reduction in mortality, particularly if the individual was sedentary to start with [7]. Therefore, patients should be counselled that even making small changes is much better than doing nothing at all.

Box 13.1 summarises some of the key health benefits of exercise.

As primary prevention, there is a dose–response relationship between PA conducted during leisure time and cardiovascular disease risk reduction, up to a maximal reduction in risk of around 20%–30% [5]. This is consistent across both genders [5]. As secondary prevention following a cardiac event, cardiac rehabilitation has been shown to reduce both premature death and death from cardiac disease in particular, and increasing levels of exercise convey increased benefits [8]. Expending 1600 kcal per week halts the progression of coronary artery disease. Expending 2200 calories per week reverses plaque formation in those with a diagnosis of heart disease [8,9]. Current research is being carried out to investigate the type and combination of exercise that yields the most superior cardiac benefits [10].

PA is effective in the prevention of hypertension and the management of stage I hypertension due to the improvement in smooth muscle tone, changes to vascular wall function and reduction of peripheral vascular resistance [1,11]. It alters the lipid profile, increasing high-density lipoprotein cholesterol and reducing low-density lipoprotein cholesterol [1,12].

Box 13.1 Health benefits of exercise

- Reduction in cardiovascular disease by 20%–30%
- Reduced formation of atherosclerotic plaques
- Reduced risk of hypertension
- May reverse stage I hypertension (systolic blood pressure of 130–139 mmHg or diastolic blood pressure of 80–89 mmHg)
- Improved blood lipid profile
- Reduced risk of breast and colon cancer, and improved prognosis for those diagnosed with these diseases
- Improved symptoms of depression and anxiety
- Reduced stress
- Improved cognition into old age, reduced risk of dementia
- Reduced risk of osteoporosis
- Reduced rate of fragility fractures in those diagnosed with osteoporosis

There is an inverse ratio between aerobic fitness and stroke mortality, with one study showing a 68% reduction in stroke risk for men considered high fitness [11,12]. This is not surprising as exercise reduces all of the known risk factors for stroke.

An increase in weekly activity rate reduces the incidence of type 2 diabetes mellitus by around 6% for every additional 500 kcal expended, and this effect was even more pronounced for those with raised body mass index (BMI) [1,13]. Weight loss in combination with dietary improvements and PA reduced the risk of developing diabetes among high-risk individuals by 40%—60%, and adding in 150 min of aerobic activity a week in combination with a 7% reduction in body weight was more effective than metformin in control of blood glucose in those with an existing diagnosis [1,14,15].

Regular PA is associated with a reduction in the risks of developing breast and colon cancer [1]. A metaanalysis of studies looking at PA and breast cancer found a reduction in breast cancer deaths by 34%, all-cause mortality reduction of 41% and breast cancer reoccurrence by 24% [16]. A study looking at PA in women with colorectal cancer found a reduction in all-cause mortality and cancer-related mortality when women commenced exercise after cancer diagnosis [17].

Furthermore, the calories expended during exercise have a beneficial effect on weight control. Increased abdominal circumference is associated with adverse health outcomes, whereas a greater fat-free mass reduces all-cause mortality [1]. Regular PA has also been shown to reduce chronic inflammation, which is thought to be one of the major processes behind many chronic diseases and causes of early mortality [1].

PA has a beneficial effect on mental health, having been shown to improve mood and promote a more positive outlook, social interaction, and perceived better physical health [18]. It also reduces the symptoms of anxiety and stress [19]. PA may have beneficial effects in alleviating some of the physical symptoms which can occur as side effects of the medications used in the treatment of mental illness.

Remaining physically active reduces the likelihood of the adverse effects of ageing. Around 30% of European and American women develop osteoporosis [20]. Regular weight-bearing exercise is protective against the development of osteoporosis, slowing the loss of bone mass that occurs with ageing, and thus reducing the risk of osteoporotic fracture [1]. Of women diagnosed with osteoporosis, 40% will go on to have a fragility-related fracture [20]. A fractured neck of femur carries a mortality rate of 33% in the year postfracture, so reduction of fracture risk factors is an important public health consideration [20]. Regular exercise has been found to reduce vertebral fracture rate by 44% and total fracture rate by 51% [20]. Among elderly women with a diagnosis of osteoporosis, exercise commencement has been shown to improve bone mineral density [20,21]. It is important to note that the type of exercise performed is vital as aerobic exercise does not confer the same benefits as resistance work, which ideally needs to be performed at a load of around 70%—55% of the maximal muscle strength [20]. Exercise also reduces likelihood of falls due to the promotion of muscle strength, power, balance and coordination [20]. Exercise reduces the risk of cognitive decline associated with ageing and has been shown to reduce the risk of developing dementia by 28% and Alzheimer's disease specifically by 45% [21]. Research has shown that in Alzheimer's disease regular exercise improves mood, cognitive function, executive function and mobility [21,22].

Physical activity guidelines

Government recommendations in both the United Kingdom (UK) and the United States (US) are that adults should spend 150 min every week doing moderate-intensity aerobic activity, or 75 min doing vigorous activity, or a mixture of both [1,10]. Moderate-intensity exercise is defined as exercise which makes the participant feel warmer with faster breathing and increased heart rate, such as walking and vacuuming [23]. Vigorous activity is defined as making the participant have a very fast heart and respiratory rate and should leave them unable to hold a conversation, such as running or cycling [23]. The benefits of resistance training are becoming increasingly recognised, and it is recommended in both the United Kingdom and the United States that adults incorporate resistance work on at least 2 days per week [10,23].

The British Heart Foundation reports that around 39% of UK adults are failing to meet government recommendations for PA. This equals about 20 million people. The statistics are worse for women than men — around 11.8 million women across the United Kingdom are not active enough compared with around 8.3 million men. Overall, women are 36% more likely to be classified as physically inactive [24]. In the United States, around 50% of adults meet the guidelines for aerobic exercise, but only 20% meet the guidelines for resistance exercise, and even fewer meet both [10,24].

In Australia, numbers of adults who meet the PA recommendations are similarly poor. At age 35 years, only 44% of women meet the exercise guidelines, but by age 55 years, only 37.5% of women are sufficiently active [25].

Barriers to exercise

Given the low uptake rates of exercise across the world, it is clear that current strategies to promote PA are failing. In aiming to better counsel patients, it is worth considering the barriers to becoming more active that individuals may face. These are listed in Box 13.2.

One study which asked individuals about their exercise habits found a difference in barriers to exercise according to socioeconomic status. Wealthier individuals were more likely to complain of lack of time or motivation, whereas lower earning individuals cited reasons such as cost, lack of transport and illness or disability [26]. Middle-aged Australian women cited family reasons, caring responsibilities, lack of time and fatigue [25].

Culture, defined as a set of beliefs, behaviours and values common to a group of people and passed between generations, may also have a bearing on activity participation [27]. A review of cultural barriers to exercise found that language barriers, gender norms, perspectives on heath, lack of role models, lack of culturally appropriate facilities and cultural perspectives on body type and appearance

---

Box 13.2 Commonly cited barriers to exercise

- Time constraints
- Economic constraints
- Caring responsibilities
- Disability
- Lack of encouragement or guidance from doctor
- Stigma
- Lack of social support
- Fear of not fitting in
- Fear of embarrassment
- Body image goals do not tie in with aesthetic promoted by activity
- Lack of culturally sensitive facilities
- Not viewing exercise as integral to health

all influenced PA participation [27]. Some of these factors were explored in more depth in a study that interviewed African American women about their thoughts on yoga. Study participants described feeling ostracised that they were the wrong body type/aesthetic and were put off by the belief that women typically attending yoga classes were very fashion conscious [28]. The study also found that African American women were more likely to respond to suggestions of weight control rather than weight loss [28]. This raises important considerations around tailoring advice and language used according to the individuals, their values and their own determinants of health [28].

Individuals suffering from chronic health conditions experience their own barriers to exercise. Patients with chronic kidney disease found that lack of care provider guidance or counselling was as an issue and reported feeling frustrated by this [27]. Those with epilepsy reported fear of injury, fear of provoking seizures, stigma and concerns over lack of social support [18]. Lack of discussion of exercise by their neurologist or epilepsy specialist nurse was also cited as reasons for not exercising [18]. Individuals poststroke cited fear of stigma, disability and lack of confidence as barriers [29]. Individuals with mental health problems reported poor physical health, fatigue, embarrassment, disorganisation, medication side effects, lack of self-confidence, outwardly visible symptoms of mental illness and lack of money as reasons not to exercise [19]. Individuals with anxiety disorders may be less likely to participate in exercise due to anxiety over joining classes, the social nature of classes, concerns over being able to 'escape' the class and discomfort during exercise due to the physical effects of activity mirroring the symptoms of anxiety [19]. The majority of physicians will be involved in some form of chronic disease management, and it is worth considering that this is a conversation worth having in the context of an ongoing relationship and chronic disease course modification [27].

It is clear that many perceived barriers to exercise exist and that these vary according to gender, socioeconomic group, culture and physical and mental health. Despite this, some individuals will still choose to exercise, and others will not. What differentiates these two groups? One study of middle-aged Australian women found that the impact of perceived barriers could be predicted by just a few factors: Women with poorer physical or mental health, poorer self-motivation and low perceived benefits of exercise were much less likely to engage in activity compared with others who did not have these characteristics [25].

Safety when commencing exercise

For those unaccustomed to exercise, the main risks are muscle soreness, musculoskeletal injury and discontinuation of the exercise regime [7]. The likelihood of injury increases in line with the vigorousness and competitiveness of the activity [7]. Therefore, individuals should be advised to start gradually and slowly build their routine in accordance with increasing fitness.

Sudden vigorous exercise in sedentary individuals can on occasion trigger sudden cardiac death, and this normally occurs on a background of coronary heart disease, other vascular risk factors and when under extreme stress [7]. Hence, the absolute contraindications to exercise are recent myocardial infarction, complete heart block, blood pressure of over 180/110 mmHg and unstable angina [30]. Screening tools such as the Physical Activity Readiness Questionnaire are useful for individuals contemplating a new exercise regime to complete to determine whether they need further medical evaluation [31].

Asymptomatic patients with a blood pressure below 180/100 mmHG intending to participate in mild-to-moderate exercise usually do not require medical evaluation [31]. Of those with hypertension and comorbidities such as ischaemic heart disease or diabetes, consideration should be given to taking antihypertensives before beginning an exercise regime [31].

Wearable activity trackers

Activity trackers have become increasingly popular in recent years. These devices attach to the individual, usually at the wrist, and are able to track step count, exercise sessions, calories burned and sleep patterns [32]. They have the potential to modify behaviour by encouraging the wearer to set goals, reinforcing the desire for change and holding the individual accountable [32]. Various studies have shown that participants using these devices increase their step count and increase the total number of minutes spent in moderate to vigorous activity per week [32]. Studies looking at weight loss showed that over 6 months participants using these devices lost significantly more weight than the control group [32]. There appears to be little disadvantage to using these devices, and for technologically minded individuals, they may further increase motivation for change.

Counselling patients

Whenever possible, PA should be discussed with patients during a consultation. The National Institute for Clinical Excellence (NICE) recommends that activity levels are objectively assessed by use of a tool such as the General Practice Physical Activity Questionnaire [23].

Patients should be made aware that exercise recommendations can be reached in many different ways and that it is vital they find something enjoyable within the spectrum of relevant activities, as pleasure is generally associated with longer-term adherence [7].

Individuals may be more likely to adhere to moderate-intensity exercise over vigorous exercise, particularly if they are generally sedentary at baseline. Indeed, participating in an exercise that pushes the individual beyond their ventilatory threshold is associated with a higher likelihood of discontinuation. Therefore, when counselling patients, it is advisable to ensure they start off gradually and slowly increase duration and intensity of activities as fitness increases [7].

While the type of exercise chosen has little bearing on whether an individual continues with the exercise in question, research has shown that when the exercise session is led by an experienced instructor, adherence is improved, which helps to explain the popularity of the group personal training format gyms [33]. Home-based programs are also associated with better adherence than self-directed activities [7].

Counselling patients on PA using established behavioural intervention techniques can improve exercise uptake [7]. Goal setting, problem-solving/troubleshooting and relapse support techniques, alongside social support, all have been shown to help initiation and adherence to an exercise regime [7,33]. For those with chronic disease, motivation to exercise is much improved if it is considered a part of not letting the illness take over their life [18]. Specific adaptations related to the illness should also be discussed, for example patients with epilepsy often found they were able to reduce the chance of exercise provoking a seizure if they dialled back the intensity of exercise or monitored their heart rate, particularly in warmer weather if that was a trigger for them [18]. On a wider scale, promotion of

---

Box 13.3 Supporting patients to become more active
- Objectively assess activity levels when appropriate within the consultation.
- Provide a brief summary as to the health benefits of exercise.
- Explain current activity-level recommendations and discuss how this target could be reached.
- Tailor advice to the individual's social, cultural, economic and health-related factors.
- Use a preactivity screening tool for cardiac disease.
- Counsel the individual to start gradually and then increase, and that some activity is better than none.
- Wearable activity trackers convey distinct advantages to those who are technologically minded.

PA in schools, provision of community exercise spaces and prompts at the point of decision-making, for example signs suggesting one take the stairs rather than the lift, have also been shown to improve exercise uptake and increase aerobic fitness as a result [34].

When considering specific health parameters, some improve with activity regardless of body weight, whereas others, such as systolic blood pressure, are more heavily affected by degree of obesity [6]. Therefore, when counselling patients, it is useful to caution that the adverse health effects of obesity cannot be completed negated by exercise and that it is imperative to consider weight control in addition to increasing activity levels as part of optimising health.

Box 13.3 summarises the key steps that healthcare professionals can take in supporting their patients to increase their activity levels.

Conclusion

Commencement of PA in the sedentary individual or increasing activity rates in the lightly active individual is one of the most beneficial interventions for the promotion of good health, reduction of premature mortality and reduction of chronic disease burden. Regular PA benefits most organ systems and improves the outcome of many disease processes. Barriers to exercise vary according to age, race, culture, gender and disability and may require gentle empathetic discussion and exploration to inspire change. Individuals should be counselled to start gradually, choosing enjoyable activities, and should consider exercising in a class or group when possible.

Summary
- Adults should participate in 150 min per week total of moderate-intensity exercise.
- An inadequate level of exercise is a major risk factor for chronic disease and premature mortality.
- Meeting exercise guidelines results in a reduction in cardiovascular disease by 20%−30%.
- Exercise is associated with significantly improved mental health, reduction in stress and improved cognition into old age.
- 39% of UK adults are failing to meet government recommendations for PA.
- Barriers to exercise include lack of social support, time constraints, work and family commitments, fear of not fitting in and lack of culturally appropriate facilities.
- When counselling patients on activity uptake, tailor advice to the individual's social, cultural, economic and health-related factors.
- Wearable activity trackers are a useful tool for those who are technologically minded.
- Encouraging PA uptake is a highly beneficial intervention for the promotion of good health, reduction of premature mortality and reduction of chronic disease burden.

References

[1] Warburton DE, et al. Health benefits of physical activity: the evidence. CMAJ 2006;174:801−9.

[2] Patel H, et al. Aerobic vs anaerobic exercise training effects on the cardiovascular system. World J Cardiol 2017;9:134−8.

[3] Burton DA, et al. The physiological effects of exercise. Cont Educ Anaesth Crit Care Pain 2004;4:185−8.

[4] Colberg S. Exercise energy systems: a primer. Accessed at: http://www.diabetesincontrol.com/exercise-energy-systems-a-primer/. (Accessed 29 November 2019).

[5] Li J, et al. Physical activity and risk of cardiovascular disease—a meta-analysis of prospective cohort studies. Int J Environ Res Public Health 2012;9:391−407.

[6] Jakicic JM. The effect of physical activity on body weight. Obesity 2009;17:S3−8.

[7] Garber CE, Blissmer B, Deschenes MR, et al. American College of Sports Medicine position stand. Quantity and quality of exercise for developing and maintaining cardiorespiratory, musculoskeletal, and neuromotor fitness in apparently healthy adults: guidance for prescribing exercise. Med Sci Sports Exerc 2011;43: 1334−59.

[8] Hambrecht R, Niebauer J, Marburger C, et al. Various intensities of leisure time physical activity in patients with coronary artery disease: effects on cardiorespiratory fitness and progression of coronary atherosclerotic lesions. J Am Coll Cardiol 1993;22:468−77.

[9] Franklin BA, Swain DP, Shephard RJ. New insights in the prescription of exercise for coronary patients. J Cardiovasc Nurs 2003;18:116−23.

[10] Brellenthin AG, et al. Comparison of the cardiovascular benefits of resistance, aerobic, and combined exercise (CardioRACE): rationale, design, and methods. Am Heart J 2019;217:101−11.

[11] Prior PL. Exercise for stroke prevention. Stroke Vasc Neurol 2018;3.

[12] Lee CD, Blair SN. Cardiorespiratory fitness and stroke mortality in men. Med Sci Sports Exerc 2002;34: 592−5.

[13] Gregg EW, Gerzoff RB, Caspersen CJ, et al. Relationship of walking to mortality among US adults with diabetes. Arch Intern Med 2003;163:1440−7.

[14] Williamson DF, Vinicor F, Bowman BA. Primary prevention of type 2 diabetes mel−litus by lifestyle intervention: implications for health policy. Ann Intern Med 2004;140:951−7.

[15] Knowler WC, Barrett-Connor E, Fowler SE, et al. Reduction in the incidence of type 2 diabetes with lifestyle intervention or metformin. N Engl J Med 2002;346:393−403.

[16] Ibrahim EM, et al. Physical activity and survival after breast cancer diagnosis: meta-analysis of published studies. Med Oncol 2011;28. 753−265.

[17] Meyerhardt JJA, et al. Physical activity and survival after colorectal cancer diagnosis. J Clin Oncol 2006;24: 3527−34.

[18] Collard SC, et al. How do you exercise with epilepsy? Insights into the barriers and adaptations to successfully exercise with epilepsy. Epilepsy Behav 2017;70:66−71.

[19] Mason JE, et al. Exercise anxiety: a qualitative analysis of the barriers, facilitators and psychological processes underlying exercise participation for people with anxiety related disorders. Ment Health Phys Act 2019;16:128−39.

[20] Daly RM, et al. Exercise for the prevention of osteoporosis in post menopausal women: an evidence based guide to the optimal prescription. Braz J Physiother 2019;23:170−80.

[21] Guitar NA. The effects of physical exercise on executive function in community-dwelling older adults living with Alzheimer's type dementia: a systematic review. Ageing Res Rev 2018;47:159−67.

[22] Van der Wardt V, et al. Adherance support strategies for exercise interventions in people with mild cognitive impairment and dementia: a systematic review. Prev Med Rep 2017;7:38−45.

[23] National Institute of Clinical Excellence. Physical activity: brief advice for adults in primary care. 2013. Accessed at: https://www.nice.org.uk/guidance/ph44/chapter/1-Recommendations. [Accessed 25 November 2019].

[24] Carlson SA, Fulton JE, Schoenborn CA, et al. Trend and prevalence estimates based on the 2008 physical activity guidelines for Americans. Am J Prev Med 2010;39:305−13.

[25] McGuire A, et al. Factors predicting barriers to exercise in midlife Australian women. Maturitas 2016;87: 61−6.

[26] Chinn DJ, et al. Barriers to physical activity and socioeconomic position: implications for health promotion. J Epidemiol Community Health 1999;53:191−2.

[27] Kendrick J, et al. Exercise in individuals with CKD: a focus group study exploring patient attitudes, motivations and barriers to exercise. Kidney Med 2019;1:131−8.

[28] Tenfelde SM, et al. "Maybe black girls do yoga": a focus group study with predominantly low-income African-American women. Complement Ther Med 2018;40:230−5.

[29] Young RE, et al. Experiences of venue based exercise interventions for people with stroke in the UK: a systematic review and thematic synthesis of qualitative research. Physiotherapy 2019. https://doi.org/ 10.1016/j.physio.2019.06.001. Available online 14 June 2019.

[30] Cardinal BJ, Esters J, Cardinal MK. Evaluation of the revised physical activity readiness questionnaire in older adults. Med Sci Sports Exerc 1996;28(4):468−72.

[31] Ghadieh AS. Evidence for exercise training in the management of hypertension in adults. Can Fam Physician 2005;61:233−9.

[32] Shin G, et al. Wearable activity trackers, accuracy, adoption, acceptance and health impact: a systematic literature review. J Biomed Inf 2019;93:103153.

[33] Seguin RA, Economos CD, Palombo R, Hyatt R, Kuder J, Nelson ME. Strength training and older women: a cross-sectional study examining factors related to exercise adherence. J Aging Phys Act 2010;18:201−18.

[34] Kahn EB, Ramsey LT, Brownson RC, et al. The effectiveness of interventions to increase physical activity. A systematic review. Am J Prev Med 2002;22:73−107.

Sedentary behaviour and adverse health outcomes

Emma Short

Department of Cellular Pathology, Division of Cancer and Genetics, Cardiff University,
University Hospital of Wales, Cardiff, United Kingdom

Introduction

The health benefits of being physically active are long established and well defined, as described throughout this book. Regular exercise reduces the overall risk of death [1] and the risks of cardio-vascular disease [2], type 2 diabetes [2], high blood pressure [3], certain cancers [1], falls [2], dementia [2], back pain [2] and osteoporosis [1]. Exercise has a positive impact on mood and reduces symptoms of anxiety and depression.

But, as well as being physically active, it is also important not to spend too much time sitting. Evidence is accumulating that sedentary behaviour is an independent factor that is associated with adverse health outcomes [4]. It is possible that an individual may fulfil their exercise requirements, but they can also be sedentary if the rest of their time is spent sitting.

Physical inactivity and sedentary behaviour can therefore be viewed as separate entities, each with their own health consequences [4].

What is sedentary behaviour?

The Sedentary Behaviour Research Network defines sedentary behaviour as '*any waking behaviour characterised by an energy expenditure of less than, or equal to, 1.5 metabolic equivalents while in a sitting or reclining posture*'. One metabolic equivalent is the energy expended while sitting at rest, or the standard of 3.5 mL of oxygen per kilogram of body weight per minute [4].

In other words, sedentary behaviour is sitting or lying while not doing much else. It includes watching television, working at the computer, driving, listening to music or reading—activities to which many individuals devote a significant amount of time.

Sedentary behaviour patterns and socioeconomic factors

Human beings are designed to move [5]. The human body is built to be active, and to engage in manual labour, this was essential for survival as a species [5]. But data show that sedentary behaviour has increased since the 1960s [4]. In the United Kingdom (UK), sedentary time increased from 30 h per week in 1960 to 42 h per week in 2005 [4].

A Prescription for Healthy Living. https://doi.org/10.1016/B978-0-12-821573-9.00014-X

A large study in the United States (US) looking at nearly 52,000 individuals found that the estimated prevalence of watching television or videos for at least 2 h a day was 62% for children, 59% for adolescents, 65% for adults aged 20—64 years and 84% for adults aged 65 years and above [6]. The study reported that from 2001 to 2016, the estimated prevalence of sitting watching television or videos remained high and stable, the estimated prevalence of computer use during leisure time increased in all age groups and the estimated total sitting time increased among adolescents and adults.

It is currently estimated that adults spend an average of between 9 and 10 h every day sitting [7]. One of the key reasons for this is that there has been a change in the patterns of employment—there has been a move from a construction, farming and manufacturing workforce to one which is increasingly office based [5]. There has also been a rise in screen time. Between 1989 and 2009, the number of households with a computer and Internet access increased from 15% to 69% [5]. More than 4 in 10 adults aged between 40 and 60 years in the United Kingdom do not even achieve 10 min of continuous brisk walking over the course of a month [8].

There are a variety of socioeconomic and environmental factors which are associated with an individual's sedentary behavioral patterns. Most studies have reported that sedentary behaviour is associated with unhealthy eating habits, including high calorie snacking, smoking and low levels of physical activity [9]. The relationship with alcohol consumption is controversial [9]. There is a strong correlation between increased body mass index (BMI) and higher levels of sedentary behaviours [9], and there is a positive correlation between sedentary behaviour and increasing age, so that as an individual gets older, sedentary behaviour generally increases [9].

The impact that education, employment and socioeconomic status (SES) have on sedentary behaviour patterns is interesting and is dependent upon how sedentary behaviour is measured, for example total sitting time, occupational sitting time or specifically screen time. Education level and television and screen entertainment time (TVSE) are inversely correlated, whereas education level and total sedentary time, both self-reported and objectively measured, are positively correlated, which is likely due to increased occupational sitting time [9]. Overall, employment is inversely related to TVSE [9]. The type of employment impacts behaviour patterns, with manual workers showing more sedentary behaviour outside of work and office-based employees being less sedentary during leisure time [9]. Generally, work days correspond to more sedentary time, and full-time workers are more sedentary than part-time workers [9]. Household income, a reflection of an individual's SES, is positively associated with occupational sitting time and total sitting time but negatively associated with TVSE [9]. Many adverse health behaviours are associated with a lower SES; therefore, this is one of the exceptions.

Individuals living in close proximity to green areas are less likely to exhibit sedentary behaviours, whereas those inhabiting regions with air or noise pollution, or living in areas with adverse weather conditions, are more sedentary [9].

Health risks of sedentary behaviour

Sitting still for prolonged periods of time has a negative impact on health and well-being.

The health risks of sedentary behaviour have been apparent for over 50 years. It was in 1953 that the famous London Transport Workers Study reported that bus drivers had a much higher incidence of coronary heart disease compared with bus conductors, with an annual incidence of 2.7/1000 versus 1.9/1000, respectively. Furthermore, it was observed that if the conductors did develop heart disease, it was of later onset than the drivers and was less likely to be fatal [10].

Increased sedentary time is associated with a poorer health-related quality of life and increased depressive symptoms [4]. Depression, tension and anxiety are all positively related to total screen time, as are perceived stress levels and perceived tiredness [9]. But it is not clear whether this is a cause or an effect—do people sit because they feel low or are they low in mood because they spend too long sitting still? The same question arises when considering cognitive impairment. Patients with probable mild cognitive impairment exhibit lower physical activity levels and higher sedentary behaviour than patients without such impairment [11]. It is not evident, however, whether the cognitive impairment is related to a lifetime of inactivity, or whether there is a reciprocal association, so that diminished executive function impacts the individual's functionality, resulting in impaired decision-making about engaging in physical activities [11].

There is a substantial body of data which confirms that sedentary behaviour increases the risk of developing metabolic syndrome, type 2 diabetes, cardiovascular disease and the overall risk of death [4,12]. Significant hazard ratio (HR) associations have been found between prolonged sedentary behaviour and **all-cause mortality** (HR: 1.220, 95% confidence interval [CI] 1.090−1.410), **cardiovascular disease mortality** (HR 1.150, CI 1.107−1.195), **cardiovascular disease incidence** (HR 1.143, CI 1.002−1.729), **cancer mortality** (HR 1.130, CI 1.053−1.213), **cancer incidence** (HR 1.130, CI 1.053−1.213) and **type 2 diabetes** (HR 1.910, CI 1.642−2.222) [13]. Specific cancers associated with sedentary behaviour include lung cancer, uterine cancer and colon cancer [14]. Adverse health outcomes are generally more pronounced at lower levels of physical activity than at higher levels [13].

For individuals who already have a diagnosis of type 2 diabetes, total sedentary time and the number of breaks in sedentary time are associated with various metabolic health parameters [15]. A study of 66 individuals found that, regardless of cardiovascular fitness, total sedentary time is positively associated with HbA1c measurements ($P = .044$). However, once moderate-to-vigorous physical activity levels are adjusted for, total sedentary time is only related to fasting blood glucose levels ($P = .037$). However, breaks in sedentary time have a favorable association with metabolic parameters such as HOMA-IR ($P = .047$), which gives an indication as to the degree of insulin resistance, and fasting glucose ($P = .046$), even after adjustments have been made for physical activity levels. Once cardiovascular fitness has been taken into consideration, associations remain between breaks in sedentary time and with HOMA-IR ($P = .036$), the Matsuda index ($P = .036$), which measures insulin sensitivity, and fasting glucose ($P = .038$). The study concluded that sedentary time and patterns are relevant for glycemic control in patients with type 2 diabetes. While high levels of physical activity and cardiovascular fitness can counteract most of the adverse metabolic outcomes associated with total sedentary time, they do not negate the positive effects of breaks in sedentary time [15]. In other words, physical activity and cardiovascular fitness may not offset the adverse health effects of *prolonged* sedentary periods [16].

In adults aged 60 years and older, it has been found that self-reported sitting time is associated with biomarkers of frailty, namely leukocyte count and C-reactive protein (CRP) concentration, with critical sedentary times being >257 min per day for males and >330 min per day for females [17]. In a study of nearly 2000 adults aged 40−75 years (n = 1932), increasing sedentary times were associated with a shorter distance covered in a 6-min walk test and lower relative elbow extension strength, indicating poorer physical performance [18]. Furthermore, increasing breaks in sedentary time were associated with a faster chair rise performance, whereas longer durations of sedentary periods correlated with a slower chair rise performance and lower strength of knee extension [18].

Both the total amount of time being sedentary and prolonged uninterrupted sedentary bouts are associated with an increased risk of death [7]. Individuals who sit for more than 13 h a day have a 200% greater risk of death than those who sit for less than approximately 11 h a day [19]. Those who sit for stretches of less than 30 min have a 55% lower risk of death than those who sit for more than 30 min [19].

Data from over 1 million individuals have shown that to eliminate the increased mortality risks associated with high total sitting time, individuals would need to engage in 60–75 min of moderate-intensity physical activity *every day* [20]. For individuals who watch television for five or more hours daily, even this level of activity does not completely eliminate the increased risk [20].

The mechanisms that mediate the increased risk of adverse health outcomes are not fully understood and need further investigation but are likely due to the effects on metabolic parameters such as insulin sensitivity, increased body fat, altered production of sex hormones and systemic inflammation [4,14].

Guidelines regarding sedentary behaviour

At the present time, there is insufficient evidence on which to base specific public health recommendations regarding appropriate limits on sedentary time [4]. The general advice is currently to 'sit less, move more'. The Australian government advises people to *minimise the amount of time spent in prolonged sitting and to break up long periods of sitting as often as possible* [21], whereas the UK government recommends that *all adults should minimise the amount of time spent being sedentary (sitting) for extended periods* [22].

Taking this on board, there are some easy ways to encourage your patients to be less sedentary:

How to be less sedentary: patient recommendations

- You can suggest to patients that they set an alarm on their phone or watch to go off at regular intervals, ideally every 30 min but, if this is not possible, every hour. Each time the alarm goes off, they must make themselves move—for example march on the spot, walk up and down a flight of stairs or perform an exercise such as squats or lunges. The longer this is done for the better, but it is likely than anything is better than nothing.
- If individuals have a desk-based job, ask them to consider whether it might be possible to swap the desk to a standing one rather than a standard sitting one. Some businesses are starting to hold standing meetings. Other ideas for things that can be done at work are to move bins away from desks, so people have to get up to throw their litter away and for individuals to choose workstations furthest from the kettle, so they have to walk a greater distance to make a cup of coffee.
- If people are required to sit for long periods of time, there are ways to stay active even though they are seated. One option could be to get an underdesk bike—they can be bought cheaply, and they keep people moving while they would otherwise be sedentary.
- When individuals go to the toilet at work, they should be encouraged to make a conscious effort to go to a bathroom which is further away than the most convenient one.

- When people are watching the television, they should try and minimise their sitting or lying periods. They could use this time to do a chore such as ironing, so they are standing or practise an activity like yoga. Each time there is an advert break, they can walk up and down the stairs or get up and make themselves a drink. If they want to change the channel, they should get up to do it rather than using the remote control.
- Limits should be set on screen time. This way sedentary periods will be minimised, and people could also find that they have got more time to do other activities such as exercise, which will be further beneficial for their health.
- When talking on a mobile phone, individuals should walk and talk rather than sit still.

Conclusions

Sedentary behaviour is an independent risk factor associated with many adverse health outcomes. Levels of sedentary behaviour throughout the developed world are dangerously high and are increasing with time. This is largely explained by a shift in patterns of employment and by increasing screen time. It is currently estimated that adults spend an average of between 9 and 10 h every day sitting and that more than 4 in 10 adults do not even achieve 10 min of continuous brisk walking over the course of a month.

It is important that healthcare professionals are aware of the health risks associated with sedentary behavior and of its significance as a risk factor for chronic disease. There are many simple interventions which can be carried out to reduce sedentary periods, and these should be discussed with patients at all available opportunities.

Summary

- Sedentary behaviour describes behaviours characterised by an energy expenditure of less than, or equal to, 1.5 metabolic equivalents.
- Sedentary behaviour refers to sitting or lying and includes activities such as watching the television, working at a computer or driving.
- Sedentary behaviour increases the risk of
  - all-cause mortality
  - cardiovascular disease
  - certain cancers
  - type 2 diabetes
- Sedentary behaviour is also associated with depression, tension, anxiety, perceived stress levels and perceived tiredness.
- Various socioeconomic and environmental factors are associated with an individual's sedentary behavioral patterns, including their eating habits, smoking status, levels of physical activity, body weight and proximity to green areas.
- Both the total amount of time being sedentary and prolonged uninterrupted sedentary bouts are associated with an increased risk of death.
- To eliminate the increased mortality risks associated with high total sitting time, individuals would need to engage in 60–75 min of moderate-intensity physical activity every day.
- It is vital that healthcare providers encourage their patients to make a conscious effort to incorporate physical activities into their daily routine and try to move as often as possible.

References

[1] Warburton DE, et al. Health benefits of physical activity: the evidence. CMAJ 2006;174(6):801—9.

[2] https://assets.publishing.service.gov.uk/government/uploads/system/uploads/attachment_data/file/541233/Physical_activity_infographic.PDF.

[3] Ghadieh AS. Evidence for exercise training in the management of hypertension in adults. Can Fam Physician 2015;61(3):233—9.

[4] Young DR, et al. Sedentary behaviour and cardiovascular morbidity and mortality. Circulation 2016;134: e262—79.

[5] Owen N, et al. Sedentary behaviour: emerging evidence for a new health risk Mayo. Clin Proc 2010;85(12): 1138—41.

[6] Yang L, et al. Trends in sedentary behaviour among the US population, 2001—2016. JAMA 2019;321(16): 1587—97.

[7] Diaz KM, et al. Patterns of sedentary behaviour and mortality in US middle aged and older adults. Ann Intern Med 2017;167:465—75.

[8] Iacobucci G. Sedentary lifestyle is putting middle ages health at risk, PHE warns. BMJ 2017;358:j3995.

[9] O'Donoghue G, et al. A systematic review of correlates of sedentary behaviour in adults aged 18—65 Years: a socio-ecological approach. BMC Publ Health 2016;16:163. https://doi.org/10.1186/s12889-016-2841-3.

[10] Morris JN, Heady JA, Raffle PAB, et al. Coronary heart disease and physical activity of work. Lancet 1953; 265(6795):1053—7.

[11] Falck RS, et al. Cross-sectional relationship pf physical activity and sedentary behaviour with cognitive function in older adults with probable mild cognitive impairment. Phys Ther 2017;97(10):975—84.

[12] Wilmot EG, et al. Sedentary time in adults and the association with diabetes, cardiovascular disease and death: systematic review and meta-analysis. Diabetologia 2012;55:2895—905.

[13] Biswas A, et al. Sedentary time and its associations with risk for disease incidence, mortality, and hospitalization in adults: a systematic review and meta-analysis. Ann Intern Med 2015;162(2):123—32.

[14] Heath R. Sitting Ducks — sedentary behaviour and its health risks: part one of a two part series BSJM blog series. 2015. https://blogs.bmj.com/bjsm/2015/01/21/sitting-ducks-sedentary-behaviour-and-its-health-risks-part-one-of-a-two-part-series/.

[15] Sardinha LB, et al. Sedentary patterns, physical activity and cardiorespiratory fitness in association to glycaemic control in type 2 diabetes patients. Front Physiol 2017. https://doi.org/10.3389/fphys.2017.00262.

[16] Bailey DP. Editorial: sedentary behavior in human health and disease. Front Physiol 2017. https://doi.org/10.3389/fphys.2017.00901.

[17] Virtuosa Junior JS, et al. Time spent sitting is associated with changes in biomarkers of frailty in hospitalized older adults: a cross sectional study. Front Physiol 2017. https://doi.org/10.3389/fphys.2017.00505.

[18] Van der Velde JHPM, et al. Sedentary behaviour is only marginally associated with physical function in adults aged 40—75 years — the Maastricht study. Front Physiol 2017. https://doi.org/10.3389/fphys.2017.00242.

[19] https://edition.cnn.com/2017/09/11/health/sitting-increases-risk-of-death-study/index.html.

[20] Ekelund U, et al. Does physical activity attenuate, or even eliminate, the detrimental association of sitting time with mortality? A harmonised meta-analysis of data from more than 1 million men and women. Lancer 2016;388:1302—10.

[21] www.health.gov.au.

[22] https://assets.publishing.service.gov.uk/government/uploads/system/uploads/attachment_data/file/213740/dh_128145.pdf.

The gut microbiome

15

Venita Patel

Community Paediatrician, Guy's & St Thomas NHS Trust & Registered Nutritional Therapist, London, United Kingdom

What is the gut microbiota and microbiome?

The gut microbiota is a dense and diverse community of around 100 trillion microorganisms, which includes thousands of different species of bacteria and other microbes such as fungi, viruses and protozoa [1]. The colon houses the most concentrated populations of these organisms [1].

There are over 35,000 species of bacteria in a healthy bowel. It is understood that there is an association between having a greater number and diversity of bacteria in the gut and improved overall health and resistance to illness.

The term gut 'microbiome' actually refers to the combined genetic material or genomes of all of the microbes present through the gastrointestinal tract, which is now regarded by some as an organ in its own right [1]. The gut microbiome contains over 3 million genes, compared to the human genome which has only 23,000 genes. These genes produce numerous metabolites responsible for vital functions in systemic health [1].

In healthy human adult guts, around 90% of the microbes are in the Bacteroidetes and Firmicutes phyla, with the rest in various other phyla.

Each individual has a unique profile of gut microbiota, which differs from person to person, and is thought to be fairly stable. Its composition is influenced by factors which begin before birth, and which are encountered throughout life, in particular environmental factors [2].

There were initially thought to be three main groups of bacterial profiles or 'enterotypes' with predominant bacterial species of the following:

1. Bacteroides *or*
2. Prevotella *or*
3. Ruminococcus

However, studies have shown that enterotypes are more complex than this and are subject to variability [3].

Functions of the gut microbiota relating to human health

The gut microbiota performs numerous essential roles in the functioning of the body, including digestion, immune defence and homeostasis [4]. For example, 70% of the immune system resides in the gut, with microbes being important in host defence and immune modulation [1]. The gut microbiota is also linked to the control of appetite, body weight and composition, cholesterol and lipid metabolism, vitamin and nutrient absorption and production, hormonal modulation, brain health, mood, neurodevelopment and behaviour [4].

Gut microbes produce an array of metabolites and proteins which have numerous effects on the host. Many are derived from dietary sources, underlining the importance of human nutrition [5]. Gut bacteria break down and ferment dietary fibre and intestinal mucus, to produce short-chain fatty acids (SCFA), namely butyrate, propionate and acetate. *Butyrate* acts as a primary fuel for intestinal epithelial cells and is important in cellular proliferation, differentiation, the maintenance of the integrity of the barrier function of the gut epithelium and gut motility. It also helps maintain a level of oxygen which deters microbial imbalance or 'dysbiosis'. *Propionate* is involved in glucose metabolism and the feeling of satiety, and *acetate* supports the growth of other bacteria, cholesterol metabolism and appetite regulation. Furthermore, butyrate and propionate appear to influence the activity of the gut hormones [5] and SCFAs also stimulate the production of the protective mucus layer.

The gut—brain link

There are several ways in which the gut communicates with the brain, and such communication occurs in a bidirectional manner. There are physical connections through the autonomic nervous system, and blood-borne connections via hormones and other chemicals produced by gut bacteria [6]. This relationship has been termed the 'microbiome—gut—brain axis' [7]. Through these connections, the gut microbiome can modulate neurogenerative activity, such as blood—brain-barrier formation, neurogenesis, myelination and maturation of the microglia. The diversity of the microbiome is known to be involved in mood, mental health disease and behaviour [6,7].

Microbial diversity and dysbiosis

Microbial diversity describes the number of different species of microbes present and their distribution. A reduction in diversity is called a microbial imbalance or dysbiosis [8]. Reduced microbial diversity has been observed in various chronic health conditions, including inflammatory bowel disease and Crohn's disease, type 1 and 2 diabetes, psoriatic arthritis, cardiovascular disease and obesity [9]. However, the relationship between cause and effect is not always apparent. Dysbiosis has also been reported in elderly populations, smokers and following antibiotic treatment [1].

There is a hypothesis that dysbiosis can lead to disruption of the gut lining, causing enhanced intestinal permeability or 'leaky gut'. The tight junctions between epithelial cells usually prevent the movement of bacteria and molecules across the gut wall. It has been proposed that loss of this integrity may allow protein to enter the bloodstream and interact with the immune system, potentially leading to adverse health consequences such as allergies and autoimmunity [10]. The clinical significance of this currently remains under investigation.

Factors influencing the gut microbiome

A multitude of factors are reported to influence the gut microbiome, with one population study describing 126 exogenous factors, 31 intrinsic factors, 12 health conditions, 60 dietary factors, 19 pharmaceutical drug groups and 4 smoking categories [11].

Early-life influences

The population of human gut bacteria begins to be shaped in early life, starting in utero, largely under the influence of genetics and maternal health during pregnancy. Later contributing factors include the gestational age at delivery, the mode of delivery, method of infant feeding, the external environment and the weaning process (summarised in Table 15.1). There are two major transitions in early life, one at birth and early feeding, and the next at weaning with the introduction of solid foods [12].

The assumption that the uterus is a sterile environment is now being questioned, with evidence suggesting that there are microbial DNA and RNA in the amniotic fluid and in the preterm gut. Maternal-to-foetal transmission of microbial genetic material, and subsequent seeding of the gut in the prenatal period, may impact foetal and infant physiology and immune priming. The foetus swallows amniotic fluid from the second trimester onwards, and it is reported that there are shared microbiota between the amniotic fluid and meconium [13].

Babies born by vaginal delivery will acquire similar microbiota to those forming the maternal vaginal flora, whereas those born by caesarean section will have flora similar to those on the maternal skin [14]. Vaginal delivery and breastfeeding are both shown to be associated with a healthy cohort of microbes. Birth by caesarean section appears to reduce the diversity of microbe species in the neonatal period, but this pattern may be improved with breastfeeding, even for a short period [12].

Oligosaccharides found in breast milk are largely undigested until they reach the colon, at which point bacteria such as *Bifidobacterium* ferment them to produce SCFAs [13,15]. Antibiotic exposure during the neonatal period and early cessation of breastfeeding are both linked to significant disruption to the establishment of healthy populations of microbial species [15].

The foods given to children from weaning onwards can cause a shift in the composition of the bacteria present in the gut. For example, a reduction in milk intake is associated with a reduction in lactase-producing bacteria. The nature of the weaning foods given tends to follow cultural norms and is

Table 15.1 Early life factors which can influence the microbiome.

Foetus	Newborn	Infant feeding method	Weaning
Genetics	Age at delivery	Breast milk	Composition of diet
Maternal health	Mode of delivery	Formula milk	
Uterine environment	Vaginal microbiome		
Amniotic fluid	Maternal skin microbiome		
Predominant bacterial species ->	*Enterobacteriaceae*	*Bifidobacterium*	*Firmicutes*

affected by socioeconomic factors. Diets higher in fibre and carbohydrate-rich meals favour particular bacterial diversity. There is a difference in gut microbiota between developed and developing countries, and also between urban and rural areas [14].

The composition of the gut microbiome is thus dynamic in the infancy period. It becomes more static and stable from around the age of 3 years, and then remains relatively constant throughout life.

Later life influences

From the age of 3 years onwards, multiple factors are influencing the number and diversity of gut microbes. These include nutrition and diet, exposure to antibiotics, lifestyle and cultural factors, exercise, body weight, stress, sleep, gastrointestinal infections and other illnesses [16]. They are highlighted in Table 15.2. Several of these factors have a bidirectional relationship with gut microbiome.

Diet and nutrition

Dietary factors, including the ratios of macronutrients and micronutrients, along with eating patterns, appear to be the most influential factors in shaping and modulating microbial populations. The balance of beneficial and harmful bacteria can respond rapidly to modifications of diet, with significant changes possible in five to 6 days. However, this effect is transient, and it takes longer-term dietary manipulation to establish stable changes to the microbiome enterotype [18]. One of the most important dietary factors appears to be the variety of plant-based foods consumed, with studies showing that individuals who eat 30 or more different plants in a week have the most diverse populations of gut bacteria [5].

Plant carbohydrates which are not digested or broken down by the time they reach the large bowel are classed as dietary fibre and are fermented by the gut microbes. They are collectively known as prebiotic compounds. Diets rich in fibre content and diversity favour the growth of beneficial bacterial species, such as Bifidobacteria and Lactobacilli. A high fibre intake will also reduce the ratio of

---

Table 15.2 Factors that influence the composition of the gut microbiome [1,2,16,17].

1. Race, ethnicity, genetics
2. Socio-economic status, geographic location and local environment
3. Age and sex
4. Overall diet, diversity of foods
5. Home environmental exposure, including housing conditions, family members and animals
6. Gut infections and gastrointestinal disease
7. Antibiotics and other medications
8. Smoking and alcohol consumption
9. Physical and psychological stresses
10. Body mass index and obesity
11. Injury, surgery or illness
12. Exercise
13. Sleep quality and quantity

Firmicutes: Bacteroidetes species (F:B ratio), which is associated with a lower risk of obesity. Inulin, a plant polysaccharide, and other resistant starches consistently promote the proliferation of beneficial microbial populations [19].

Protein in the diet also has an impact on the composition of microbiota, in particular its type, quality and quantity. For example, whey protein and mung bean protein seem to be beneficial, whereas high concentrations of casein and soy can have a detrimental effect. Different cooking methods of meat can affect bacterial profiles, particularly if food-borne pathogens are ingested, with enterotoxins being harmful to commensals [19].

It is generally reported that diets high in saturated fat and in total fat content have negative effects on the gut microbiome. A high-fat diet appears to increase the F:B ratio and can reduce diversity and numbers of healthy microbial species. In contrast, the ingestion of unsaturated fats and poly-unsaturated fats (PUFAs) appears to have a positive effect on commensal bacteria, along with a reduction in detrimental bacteria. For example, olive oil is said to reduce the F:B ratio [19].

Polyphenols are compounds such as flavonoids, phenolic acids, stilbenes and lignans. They are derived from a variety of fruits, vegetables and cereals, as well as tea, coffee, and wine. They have been suggested to have antioxidant, anti-inflammatory and anticarcinogenic effects, and there is also evidence that they promote beneficial gut microbes, reduce detrimental microbes and reduce the F:B ratio, appearing to have 'prebiotic-like' actions. Specific compounds and foods reported to mediate positive effects are berry anthocyanins, pomegranate, green tea and cocoa polyphenols and grape and red wine polyphenols [19].

Vitamins have a close relationship with gut microbiota. Some vitamins are produced by gut microbes, whilst some vitamins modulate their activity and composition. For instance, vitamins A, D and E impact the F:B ratio. The B complex vitamins are a collection of eight water-soluble compounds. Commensal gut microbes produce some B vitamins, such as B12 and B6, which promote bacterial colonisation, and are involved in host—pathogen interactions [19].

Table 15.3 summarises the dietary components which impact the gut biome.

Dietary eating patterns and fasting

There has been significant interest in the health effects of short-term fasting, time-restricted feeding (TRF) and calorie restriction. Studies have shown that TRF has beneficial effects on metabolism, intestinal inflammation and even central nervous system disorders, which are partly mediated through changes in gut microbiome composition [20].

Short fasting periods, for example overnight, appear to allow replenishment of bacterial populations. However, prolonged fasting or unhealthy dietary restriction, such as that seen in eating disorders, may lead to disruption of the integrity of the protective mucus layer in the gut and thus have negative effects [20].

Body weight and obesity

Obesity is a chronic disorder with a multifactorial aetiology. In recent years, the gut microbiome has been studied as a regulator of weight gain and obesity. It is proposed that the biome impacts host

Table 15.3 Factors influencing the balance of beneficial and detrimental gut bacteria.
Dietary factors promoting beneficial bacteria
Fruit/Wine polyphenols
Tea polyphenols
Flavonoids
Phenolic acid
Vitamin A
Fibre/unrefined carbohydrates
PUFAs
Lower fat diets
Whey protein
Dietary factors promoting detrimental bacteria
Low mineral diets
High fat diets
Saturated fats
Fried meats
Casein protein

metabolism through modulation of energy balance, low-grade inflammation and gut wall integrity. In obesity, overall microbial diversity is low and the F:B ratio is increased, which has been suggested to lead to increased energy extraction from food and storage in adipose tissue [21]. Furthermore, the regulation of appetite, energy intake and satiety is also under the influence of the gut microbiota via SCFAs [5]. In obese individuals, the appetite drive is generally high, with a preference for higher energy foods. Consumption of these foods has adverse effects on the composition of the gut microbiota, and a reinforcing feedback cycle is perpetuated.

Exercise and the gut

Exercise and the gut microbiome have a bidirectional relationship. Aerobic exercise has been shown to influence both the composition of the microbiome and its functions [22].

The key phyla which increase in diversity and abundance in response to exercise are *Firmicutes* and *Actinobacteria*, along with some SCFA-producing species [7]. In turn, the biome impacts the body's adaptation to the physiological stress of training. Therefore, it may play a role in athletic performance and the variation in sporting achievements following training. In the future, targeting the gut microbiome could be an important factor in personalised sports nutrition [22].

Sleep and circadian rhythm

Several studies have demonstrated that sleep deprivation and sleep fragmentation have a negative impact on microbiota populations, which is associated with an adverse modulation of metabolic parameters. This includes reduced insulin sensitivity and impaired glucose tolerance. These abnormalities may be normalised following a return to normal sleeping patterns [23,24].

The gut microbiome has its circadian rhythm, with different populations active at different times of the day and night. Disruption of the host circadian rhythm causes shifts in microbiota activity patterns, particularly if the timing of feeding changes [23].

Interestingly, the gut microbiome has also been implicated in the regulation of sleep, with some evidence showing that depletion of microbiota is associated with reduced sleep quality [24].

Fig. 15.1 illustrates the bidirectional relationship that the gut microbiome has with various lifestyle factors and Table 15.4 summarises practical tips to optimise the microbiome.

Prebiotics, probiotics and supplementation

Prebiotics and probiotics are dietary-based compounds which can modulate the composition of the colonic microbiota. A prebiotic is defined as 'a substrate that is selectively utilized by host micro-organisms conferring health benefit' [25].

These are the compounds, particularly fermentable dietary fibre and nondigestible carbohydrates, which 'feed' beneficial bacteria and help to increase their numbers. They are mostly vegetable foods containing soluble and insoluble fibre, including onion, leek, garlic, Jerusalem artichokes, chicory root, apples, bananas, oats, bran, and other wholegrains and legumes [16,18].

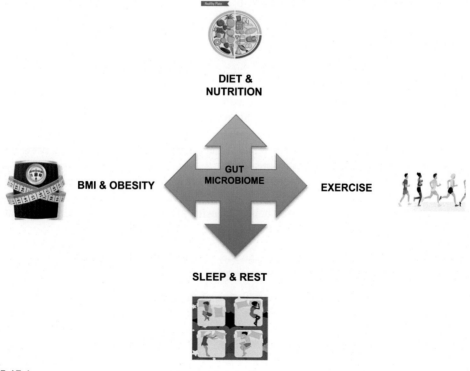

FIGURE 15.1

Factors with a bidirectional relationship with the gut microbiome.

Table 15.4 How to optimize the gut microbiota.

1. Increase the number and variety of vegetables, wholegrains, legumes and fibre-rich foods consumed
2. Consider including prebiotic-rich foods and probiotic fermented foods/drinks in the diet
3. Limit intake of high sugar and refined foods
4. Allow a window of time between dinner and breakfast of at least 10–12 h and limit snacks between meals
5. Drink adequate fluids, especially water
6. Spend time in natural environments as often as possible
7. Regular aerobic exercise, outside if possible
8. Stroke pets and interact with animals
9. Reduce stress levels
10. Optimise sleep

Probiotics are defined as 'live microorganisms which, when administered in adequate amounts, confer a health benefit on the host'. Probiotics may be consumed as foods or as supplements. However, it is debatable how many live bacteria reach the large bowel. The most effective dietary sources include fermented foods and drinks, such as yoghurt, kefir, miso, natto, tempeh, sauerkraut, kimchi, beet kvass, apple cider vinegar, and kombucha [25]. There are many commercially available probiotic supplements, but their efficacy is unclear, and they are not recommended in seriously unwell or immune-compromised individuals [25].

The gut microbiome and disease

Early life factors play a major role in influencing the composition of the gut microbiome, which in turn has an important impact on the long-term health of an individual. Variations in microbiome composition observed in early life have been associated with specific diseases and conditions. For example, there is a close link between the nature of the microbiome in infancy and early childhood and the development of a healthy immune system. Bacterial colonisation of the gut is required for normal immune functioning and healthy development and metabolism, which is mediated via host–bacteria interactions, and bacteria–bacteria interactions [26].

The lack of a healthy bacterial microbiome leads to an increased risk of developing atopy and allergies, gastrointestinal diseases such as inflammatory bowel disease, irritable bowel syndrome and colorectal cancer, and metabolic disease such as obesity [17,26]. It has also been implicated in neurodevelopmental disorders, mental health disease and mood [6].

Clinical interventions and future research

Several strategies are being explored for manipulation of the gut microbiome in the management of certain diseases and conditions [20]. For example, faecal microbiota transfer is the transfer of healthy faecal bacteria from one person to another. It has been investigated as a therapeutic intervention in *Clostridium difficile*-related disease, ulcerative colitis and irritable bowel syndrome, with mixed results [20].

Prebiotics and probiotics may have beneficial effects in conditions such as childhood diarrhoea, antibiotic-associated diarrhoea, necrotising enterocolitis and inflammatory bowel disease. However, studies addressing the clinical utility of these are limited due to wide interindividual variability in gut microbiome composition [25].

Pharmabiotics is the term used to describe any products obtained from the microbiota that confer a potential benefit to health [20]. Currently, research regarding the role of pharmabiotics in the management of human health and disease is ongoing.

Conclusion

The gut microbiome describes the vast collection of bacteria and microbes, primarily in the large bowel, which has numerous physiological and metabolic functions in the human body. The composition of the microbiome is influenced by a wide range of factors, which begin in utero and continue throughout life, including diet and nutrition, exercise and sleep patterns. The lack of a healthy biome is associated with diseases such as atopy, inflammatory bowel disease, colorectal cancer and neuropsychiatric disorders. The manipulation of the gut biome to optimise health and treat disease is an exciting field of research currently underway.

Summary

- The human body is composed of over 37 trillion cells, but this figure is dwarfed when the microbiome is considered — the 100 trillion microorganisms which live on and inside the human body, including bacteria, fungi and parasites. A significant proportion of the microbiome lives in the gut, predominantly in the colon.
- There are over 35,000 species of bacteria in a healthy bowel. It is beneficial to have a diversity of bacterial species present.
- Microbiome diversity is partly determined by heritable factors, but environmental factors are even more important. Environmental factors include the composition of the diet; early life experiences such as type of birth and mode of infant feeding; interactions with other people and animals; gut infections; antibiotics and other medications; physical and psychological stresses; injury, surgery and illness; exercise; sleep.
- One of the major roles of the gut microbiome is in fermenting nondigestible substrates. For example, gut bacteria break down and ferment dietary fibre and intestinal mucus to produce chemicals called short-chain fatty acids. One of these, butyrate, is the main energy source for gut mucosal cells, and it is important in maintaining the integrity of the gut lining. Others are involved in appetite regulation.
- The gut flora is also involved in immune regulation, metabolism, vitamin production, hormone production and mucus formation.
- The gut is connected with the nervous system and brain. As well as direct connections through nerves, the gut can communicate with the central nervous system through chemicals produced by the gut bacteria.
- Dysbiosis refers to an imbalance of the microorganisms present in the biome. It has been implicated in diseases as diverse as rheumatoid arthritis, psoriatic arthritis, coeliac disease, obesity, diabetes, cardiovascular disease, colorectal cancer, inflammatory bowel disease and neuropsychiatric disorders such as depression.
- Diet is one of the key determinants of bacterial diversity in the gut. It is important to eat a varied diet, full of fibre, as well as proteins, fats and carbohydrates.
- *Probiotics* are actual live bacteria. Many drinks and yoghurts are marketed as containing probiotics, but they vary in the quantity of bacteria present and their quality. It is debatable how many of these bacteria actually reach the gut alive, but certain whole fermented foods do seem to be effective

Summary—cont'd

- *Prebiotics* are the compounds which feed the beneficial bacteria and help to increase their numbers. These are mostly, but not exclusively, vegetable products which contain soluble and insoluble fibre, including onion, leek, garlic, Jerusalem artichokes, chicory root, apples, bananas, oats, bran, other wholegrains and legumes.
- General advice to improve the diversity of the gut flora includes the following:
  1. Increase the variety of vegetables, wholegrains, legumes and fibre-rich foods consumed
  2. Consider eating prebiotic-rich/probiotic-rich foods
  3. Limit high sugar and processed foods
  4. Try to avoid eating snacks in between meals, and eat over a defined time period during the day, for example between 8 a.m. and 6 p.m.
  5. Drink enough fluids, especially water
  6. Spend time in nature
  7. Exercise regularly, outdoors if possible
  8. Stroke pets
  9. Reduce stress levels
  10. Optimise sleep

References

[1] Valdes AM, Walter J, Segal E, Spector TD. Role of the gut microbiota in nutrition and health. BMJ June 13, 2018;361:k2179.

[2] Rinninella E, Raoul P, Cintoni M, Franceschi F, Miggiano GAD, Gasbarrini A, et al. What is the healthy gut microbiota composition? A changing ecosystem across age, environment, diet, and diseases. Microorganisms January 10, 2019;7(1):E14.

[3] Cheng M, Ning K. Stereotypes about enterotype: the old and new ideas. Dev Reprod Biol 2019 02;17(1): 4—12.

[4] Jandhyala S. Role of the normal gut microbiota. World J Gastroenterol 2015;21(29):8787.

[5] Martin-Gallausiaux C, Marinelli L, Blottière HM, Larraufie P, Lapaque N. SCFA: mechanisms and functional importance in the gut. Proc Nutr Soc 2020. https://doi.org/10.1017/S0029665120006916.

[6] Sharon G, Sampson TR, Geschwind DH, Mazmanian SK. The central nervous system and the gut microbiome. Cell November 3, 2016;167(4):915—32.

[7] Dalton A, Mermier C, Zuhl M. Exercise influence on the microbiome-gut-brain axis. Gut Microb 2019; 10(5):555—68.

[8] Carding S, Verbeke K, Vipond D, Corfe B, Owen L. Dysbiosis of the gut microbiota in disease. Microb Ecol Health Dis 2015;26(0).

[9] Ganal-Vonarburg SC, Duerr CU. The interaction of intestinal microbiota and innate lymphoid cells in health and disease throughout life. Immunology January 2020;159(1):39—51.

[10] Mu Q, Kirby J, Reilly C, Luo X. Leaky gut as a danger signal for autoimmune diseases. Front Immunol 2017;8.

[11] Zhernakova A, Kurilshikov A, Bonder MJ, Tigchelaar EF, Schirmer M, Vatanen T, et al. Population-based metagenomics analysis reveals markers for gut microbiome composition and diversity. Science April 29, 2016;352(6285):565—9.

[12] Kashtanova DA, Popenko AS, Tkacheva ON, Tyakht AB, Alexeev DG, Boytsov SA. Association between the gut microbiota and diet: fetal life, early childhood, and further life. Nutrition June 2016;32(6):620—7.

[13] Stinson LF, Boyce MC, Payne MS, Keelan JA. The not-so-sterile womb: evidence that the human fetus is exposed to bacteria prior to birth. Front Microbiol 2019;10:1124.

[14] Tanaka M, Nakayama J. Development of the gut microbiota in infancy and its impact on health in later life. Allergol Int October 2017;66(4):515−22.

[15] Yasmin F, Tun HM, Konya TB, Guttman DS, Chari RS, Field CJ, et al. Cesarean section, formula feeding, and infant antibiotic exposure: separate and combined impacts on gut microbial changes in later infancy. Front Pediatr 2017;5:200.

[16] Conlon MA, Bird AR. The impact of diet and lifestyle on gut microbiota and human health. Nutrients December 24, 2014;7(1):17−44.

[17] Royston KJ, Adedokun B, Olopade OI. Race, the microbiome and colorectal cancer. World J Gastrointest Oncol October 15, 2019;11(10):773−87.

[18] Yang Q, Liang Q, Balakrishnan B, Belobrajdic DP, Feng QJ, Zhang W. Role of dietary nutrients in the modulation of gut microbiota: a narrative review. Nutrients January 31, 2020;12(2):E381.

[19] Zhang N, Ju Z, Zuo T. Time for food: the impact of diet on gut microbiota and human health. Nutrition July−August 2018;51−52:80−5.

[20] Quigley EMM, Gajula P. Recent advances in modulating the microbiome. F1000Res 2020;9. F1000 Faculty Rev-46.

[21] Kim B, Choi HN, Yim JE. Effect of diet on the gut microbiota associated with obesity. J Obes Metab Syndr December 2019;28(4):216−24.

[22] Hughes RL. A review of the role of the gut microbiome in personalized sports nutrition. Front Nutr 2019;6: 191.

[23] Kaczmarek JL, Thompson SV, Holscher HD. Complex interactions of circadian rhythms, eating behaviors, and the gastrointestinal microbiota and their potential impact on health. Nutr Rev September 1, 2017;75(9): 673−82.

[24] Parkar SG, Kalsbeek A, Cheeseman JF. Potential role for the gut microbiota in modulating host circadian rhythms and metabolic health. Microorganisms January 31, 2019;7(2):E41.

[25] O'Connell TM. The application of metabolomics to probiotic and prebiotic interventions in human clinical studies. Metabolites March 24, 2020;10(3):E120.

[26] Stiemsma LT, Michels KB. The role of the microbiome in the developmental origins of health and disease. Pediatrics April 2018;141(4). e20172437.

Cigarettes: the facts, strategies for smoking cessation, e-cigarettes and vaping

16

Adam Douglas[1] and Arfa Ahmed[2]

[1]*Department of Cellular and Anatomical Pathology, Derriford Hospital, Plymouth, United Kingdom;*
[2]*General Practitioner, National Health Service, Manchester, United Kingdom*

Introduction

Smoking kills. Clichéd, yes, but true nonetheless. Tobacco smoking is responsible for more than 7 million deaths worldwide per year, and this is predicted to rise to over 8 million by the year 2030 [1,2]. Smoking is a major modifiable risk factor for numerous diseases including cancer and cardiovascular disease and is an ongoing public health concern.

Tobacco is produced from the leaves of plants of the genus Nicotiana, which are cured after harvest to make the smoke easily inhalable [3]. The leaves are primarily packaged as cigarettes or burned in pipes (Figs. 16.1 and 16.2). Smokeless tobacco consumption methods are less common, but include *snuff*, which describes powdered tobacco inhaled into the nasal cavity, *snus* and *dipping tobacco*, which refers to powdered tobacco placed in between the gum and lip, and *chewing tobacco*.

Tobacco is currently used mainly as a recreational drug, but previously tobacco enjoyed popularity as a medicinal product for centuries and had a reputation as a panacea of sorts. Notable uses included tobacco salves for wound healing, and tobacco smoke enemas for drowning victims. Unsurprisingly these interventions were ineffective and potentially dangerous [4].

Nicotine is the constituent responsible for tobacco addiction and dependence although its potential role as a carcinogen is beginning to be understood [5]. The only genuine medical indication for nicotine is in the form of nicotine replacement therapy (NRT), assisting individuals to quit smoking.

Tobacco leaves release more than 5,000 chemicals when they are burned, with 98 considered hazardous to human health. The chemicals include at least 70 carcinogens [6,7]. Tobacco smoke consists of nicotine, hydrogen cyanide, lead, arsenic, ammonia, carbon monoxide and many other chemicals.

This toxic combination makes the cigarette the single most deadly invention in the history of mankind, but what is a life worth to the manufacturers of such products? If it is assumed that the cigarette-producing industry generates a penny of profit per cigarette and that the death rate is one death per million cigarettes smoked, Robert Proctor of Stanford University estimates that manufacturers generate just US $10,000 profit for every death caused [8]. Given everything we know, who is still smoking?

A Prescription for Healthy Living. https://doi.org/10.1016/B978-0-12-821573-9.00016-3

FIGURE 16.1

Cigarette poster from 1923.

Ludwig Hohlwein. Public Domain.

FIGURE 16.2

Pipe alongside tobacco.

Adam Douglas.

Epidemiology in the United States

Tobacco use is responsible for nearly 20% of deaths in the United States (US) [9]. In 2017, 14% of all adults in the US were cigarette smokers, equating to 34.3 million individuals, with more men smoking than women (15.8% vs 12.2%) [10]. Comprehensive control programs at both state and federal level have been successful in decreasing the rates of smoking, from 20.9% of adults in 2005 to 14% in 2017, in turn leading to declining mortality rates for diseases caused by smoking, including lung cancer and cardiovascular disease [7]. These programs utilise multiple elements including the creation of

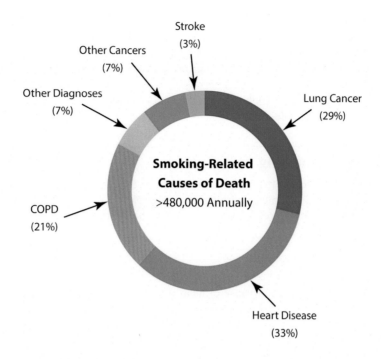

FIGURE 16.3

Average annual numbers of smoking-related deaths for adults aged 35 or older in United States, 2005–09.

2014 Surgeon General's Report. Table 11.4. page 660.

smoke-free environments, public education through media and schools, preventing youth access to tobacco products and increasing the availability of smoking cessation services [11]. Despite this, smoking remains a major public health concern, with almost 90% of lung cancers attributable to smoking (Fig. 16.3).

There are significant racial and socioeconomic determinants of tobacco use. Statistics from 2017 showed current cigarette smoking highest amongst non-Hispanic American Indians/Alaska natives (24%), and lowest in non-Hispanic Asians (7.1%). Tobacco smoking broadly mirrors the level of educational attainment, with 36.8% of current cigarette smokers holding a general educational development certificate only (high school academic standard), compared with 4.1% of adults with a graduate degree. In tandem with education, cigarette smoking was more prevalent amongst persons with a low annual household income (21.4% with an income less than $35,000) compared to higher annual incomes (7.6% with an income greater than $100,000).

Adults with a disability were more likely to be current smokers compared to those without (20.7% vs 13.3%), and adults with serious psychological distress as measured by the Kessler scale were more than twice as likely to smoke as those unaffected (35.2% vs 13.2%).

Smoking is often initiated in adolescence, and reducing tobacco use amongst young people remains an important public health goal. Each day around 2,000 children under the age of 18 smoke their first cigarette, and over 300 become daily smokers [12].

Cigarette smokers are not the only group to suffer the effects of tobacco smoke. Second-hand smoke exposure is responsible for more than 41,000 deaths per annum in the US, and accounts for approximately 2.5 million deaths since 1964 [7,13]. Second-hand smoke exposure is also higher amongst people with low incomes [14], and in some occupation groups, including blue collar, service, and construction workers [15].

It is therefore observed that the smoking-related health crisis will disproportionately affect lower socioeconomic groups and the vulnerable.

Pathophysiology of smoking

Tobacco smoking has widespread effects on the human body. Smoking increases the risk of coronary heart disease and stroke by 2—4 times, and lung cancer by 25 times [7,13]. The effects of cigarette smoking on the cardiovascular and respiratory system shall be reviewed, along with its carcinogenic effects and other health impacts.

Cardiovascular disease

Exposure to tobacco smoke is a major cause of heart disease, peripheral arterial disease, stroke and aortic aneurysm. Even low levels of cigarette smoke exposure, including passive smoking, increase the risk of developing cardiovascular disease. The effects of smoking are partially mediated through damaging blood vessels, therefore playing a role in the pathogenesis of atherosclerosis. Furthermore, nicotine stimulates catecholamine release, and other products of tobacco combustion cause endothelial injury and dysfunction.

Smoking also leads to increased oxidation of low-density lipoproteins, which in turn stimulate arterial intimal cells, again leading to the development of atherosclerosis. Plasma fibrinogen concentrations are increased in smokers, which alters platelet activity and potentiates thrombosis [16]. Atherosclerosis in turn can lead to myocardial ischaemia, myocardial infarction and stroke.

Smoking increases the risk of developing type 2 diabetes by 30%—40% compared to non-smokers [7], and smoking is independently associated with higher haemoglobin A1c (HbA1c) concentration [17]. The association is not fully understood, but data suggest smoking and nicotine can affect insulin sensitivity, body composition, and pancreatic β-cell function.

Carbon monoxide in tobacco smoke combines with haemoglobin to form carboxyhaemoglobin, which reduces the oxygen-carrying capacity of blood and impairs the delivery of oxygen to the tissues. Other deleterious cardiovascular effects of smoking include coronary artery spasm, increased blood clotting, and susceptibility to arrhythmias [18].

Non-malignant respiratory disease

Smoking causes 79% of chronic obstructive pulmonary disease (COPD) cases in the US [7]. COPD, encompassing emphysema, that is permanent dilation of the alveoli and wall destruction without fibrosis, and chronic bronchitis, namely a chronic productive cough, is the primary non-malignant respiratory condition attributed to cigarette smoking. COPD is a rising cause of morbidity and mortality [19]. Smoking has also been linked to asthma and idiopathic pulmonary fibrosis although the evidence is not sufficiently strong to imply causation.

Tobacco smoke acts on the respiratory system in multiple ways. The entire respiratory tract, from mouth and upper airways down to the terminal bronchioles and alveoli, are affected by the carcinogens and toxins in smoke. The components of cigarette smoke cause irritation to the bronchioles and alveoli which can result in chronic inflammation and scarring through lung remodelling [19]. Smoking can also damage elastin, an essential protein for lung elasticity and ventilatory function, impairing gas exchange.

Smoke exposure can cause genetic mutations potentiating the effect of endogenous proteases, resulting in tissue damage and, if unchecked, emphysema. Cigarette smoke also contains a significant number of oxidants which can cause oxidative injury throughout the lung. Toxic components of cigarette smoke can paralyse and destroy cilia on the surface of epithelial cells lining the respiratory tract, damaging an important part of the innate immune response and increasing susceptibility to infection.

Cancer

Cigarette smoking is the biggest preventable cause of cancer, accounting for one in three cancer deaths in the US [7]. Smoking has been causally linked to many cancers, including cancer of the lung, oropharynx, larynx, oesophagus, bladder, and cervix. Tobacco smoke and smokeless tobacco products contain several carcinogens, including tobacco-specific N-nitrosamines and polycyclic hydrocarbons [5].

Carcinogens in cigarette smoke bind to DNA, forming DNA adducts. There are mechanisms within cells to remove DNA adducts and repair DNA, but if the repair enzymes are not functioning efficiently or are overwhelmed by DNA damage, the DNA adducts remain and increase the chance of developing somatic mutations. Such mutations may lead to unchecked cell growth and proliferation, hallmarks of carcinogenesis [19].

The link between lung cancer and tobacco smoking was first investigated in the 1930s [8], but it did not gain widespread acceptance in the medical community until the 1960s. It was even longer before the message filtered through to the general public, having contended with denial and propaganda spread by the tobacco industry.

Smoking is strongly associated with small cell lung cancer, squamous cell carcinoma (SqCC), and adenocarcinoma [7]. As cigarette manufacturers moved to producing low-tar, filtered cigarettes, the predominant histological subtype of lung cancer has shifted from SqCC to adenocarcinoma (Fig. 16.4). Filtered cigarettes allow deeper inhalation of smoke than unfiltered varieties, allowing tobacco-derived carcinogens to travel more distally to the bronchoalveolar junction [3]. Damage to more distal parts of the lung is more likely to result in adenocarcinoma than SqCC.

Other health risks

The effects of tobacco smoking on the cardiovascular and respiratory systems are well known, but research has found causal relationships between smoking and rheumatoid arthritis, immunodeficiency, ectopic pregnancy, cleft palate in babies of women who smoke during pregnancy, age-related macular degeneration, and erectile dysfunction [7].

Smoking can affect sperm, reducing fertility and increasing the risk of miscarriage and birth defects [20].

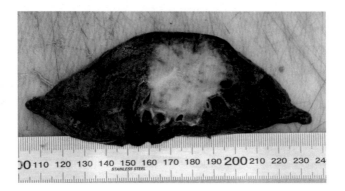

FIGURE 16.4

Cross-section of lung with a large tumour. Histology confirmed the diagnosis of adenocarcinoma.

Adam Douglas.

Passive smoking

Second-hand smoke has been responsible for the death of 2.5 million adult American non-smokers since the 1964 Surgeon General's Report, which stated the link between smoking and disease [7]. Adult non-smokers exposed to second-hand smoke have a higher risk of developing stroke, coronary heart disease, and lung cancer.

Adults are not the only group affected by passive smoking. Mothers who smoke during pregnancy increase the risk of sudden infant death syndrome (SIDS) in their child, and infants who are exposed to second-hand smoke are at higher risk of suffering SIDS [7,14]. Higher concentrations of nicotine and cotinine, the main metabolite of nicotine and a biomarker for exposure, have been found in the lungs of infants who die from SIDS. There are a number of postulated mechanisms for the increase in risk, including changes in brain chemistry affecting regulation of breathing, but the mechanism remains poorly understood.

Second-hand smoke exposure can also increase the risk of bronchitis and pneumonia in children and result in more frequent ear infections compared to unexposed children. Children with asthma exposed to tobacco smoke experience more frequent and severe asthma attacks [7].

Addiction

Addiction is defined by the American Society of Addiction Medicine as a 'treatable, chronic medical disease involving complex interactions among brain circuits, genetics, the environment, and an individual's life experiences' [21].

The complexity of the definition hints at the multifactorial nature of addiction. People engage in behaviours or use substances in a repeated and compulsive way despite harmful and potentially dangerous consequences [22].

The addictive component of tobacco is nicotine, and cigarettes are a highly efficient nicotine delivery system. Regular cigarette smoking leads to addiction in many users, who can become reliant on smoking for mood regulation, relief of withdrawal symptoms, and positive reinforcement in certain situations.

Nicotine is easily absorbed through the lungs into the bloodstream. Once absorbed, nicotine acts on nicotinic cholinergic receptors, triggering the release of dopamine in the nucleus accumbens, and other neurotransmitters. These neurotransmitters reduce stress and anxiety and provide a feeling of pleasure. Dopamine is important in the reward-related neuronal system, reinforcing further tobacco use. As the body adapts to nicotine, individuals tend to increase the amount of tobacco use. Dopamine, glutamate, and γ-aminobutyric acid play important roles in the development of nicotine dependence [22].

Tobacco use typically begins in childhood or adolescence, and the risk of nicotine dependence is higher when people start smoking at a young age. Smoking is also more prevalent in individuals with psychiatric or substance abuse disorders than the general population, suggesting an increased susceptibility to nicotine addiction.

Smoking cessation often leads to nicotine withdrawal, which describes physical and psychological symptoms including low mood, anxiety, anger, tiredness, irritability, change in behaviour, mood swings, nervousness, headaches and sleeping problems [23]. The symptoms are usually most marked in the first week and will gradually taper off over several weeks.

Nicotine is metabolised to cotinine in the liver by the CYP2A6 enzyme, and the rate of nicotine metabolism influences the severity of withdrawal symptoms. Individuals with a slow nicotine metabolism smoke fewer cigarettes than those with a faster metabolism and may be more likely to successfully quit smoking [22].

Benefits of stopping smoking

It is never too late to stop smoking. There are immediate and long-term health benefits for smokers who quit. Within 20 minutes, blood pressure drops to the level it was before the last cigarette was smoked. Eight hours after stopping, carbon monoxide levels in the blood return to normal. Lungs regain normal ciliary function anywhere from 1 to 9 months, reducing the risk of infection [24].

Long-term health benefits are also seen in the cardiovascular and respiratory systems. After a year of smoking cessation, the risk of coronary heart disease is halved compared to a smoker, and at 5 years of abstinence the risk of stroke is that of a non-smoker. Respiratory symptoms including cough, wheeze and shortness of breath tend to improve, and the rate of decline in lung function, and hence the risk of developing COPD decreases after stopping smoking as measured by a slowing in the decline of forced expiratory volume (FEV1) [7,18,25]. In the Lung Health Study, participants who quit smoking experienced a decline in FEV1 of 34 mL/year, which was almost half of those who continued to smoke who had a decline of 63 mL/year.

A decade after cessation, the risk of lung cancer is about half of a smoker. The risk of developing head and neck cancer falls by 40% after 10 years of smoking cessation, and the risk of pancreatic cancer by around 30% [26].

Benefits from smoking cessation are felt at any age, but the most marked effects are seen in younger individuals. People who stop smoking at age 30 gain around 10 years in life expectancy compared to lifelong smokers. Individuals who stop smoking at age 40, 50 or 60 add approximately 9, 6 or 3 years, respectively, to life expectancy compared to those who continue to smoke [27].

Male sexual function has been shown to improve following smoking cessation, with the enhancement of physiological and self-reported indices. This aspect could provide a novel motivation for smoking abstinence in men [28].

Effectiveness of different smoking cessation strategies

Almost 70% of cigarette smokers in the US wanted to stop smoking in 2015, and more than half had attempted to quit in the last year. Only a third of cigarette smokers who attempted to quit used evidence-based smoking cessation treatments [29]. Even PEZ Candy, a popular sweet in the US, was originally invented as a smoking cessation aid [30].

A variety of psychosocial and pharmacological interventions have been developed for smoking cessation which have been shown to increase the rate of long-term smoking abstinence. Despite this, the majority of smokers fail to quit. Motivation is a major determinant in quitting smoking, regardless of the aid used.

Target audience

In terms of public health strategies, identifying cigarette smokers is the first step. Clinic-based protocols to encourage assessment and documentation of tobacco use, for example the use of a pop-up or prompt on a computer system in primary care, increase the rates of intervention by clinicians, though it does not increase the rates of smoking cessation by itself [23].

Once identified, assessing an individual's willingness to make a quit attempt is the next step, and it influences the intervention chosen.

Brief interventions and counselling

Never underestimate the power of time. Behavioural support helps people to quit smoking and can work well in conjunction with other therapies. Even minimal counselling, involving sessions lasting less than 3 minutes, increases smoking abstinence rates, and there appears to be a dose-dependent relationship between session length and success. Longer sessions provide better outcomes for quitting [23,31]. A meta-analysis of session contact time showed a rise in abstinence rates from 14% for 1−3 minutes of contact time to 27% for 31−90 minutes [23].

Both individual and group counselling are more effective than self-help and other less intensive smoking cessation support such as physician or nurse advice and health education [31,32]. Group therapy educates individuals about techniques for smoking cessation whilst offering mutual support from other members. Intensive individual counselling and group counselling appear to have a similar efficacy [32,33].

Telephone counselling can provide information and support to smokers trying to quit and takes the form of smokers calling 'quitlines' or opting to receive calls from counsellors. Proactive telephone counselling has been shown to increase quit rates in smokers although evidence was inconclusive on reactive telephone counselling [34].

Nicotine replacement therapy

Nicotine replacement therapy (NRT) comes in the form of gum, nasal sprays, transdermal patches, and sublingual tablets. Providing an alternative source of nicotine helps to ease the transition from smoking by reducing cravings and withdrawal symptoms. All forms of NRT can increase the rate of quitting by 50%−70%. This effect is robust and fairly independent of the level of additional support given to the individual [35]. A recent meta-analysis of the available evidence showed that combination NRT, the use of a fast-acting form in combination with a patch, also increased the chances of smokers quitting successfully compared with single NRT therapy [36].

Other pharmacotherapy

Several non-NRT pharmacotherapies have been developed to improve quit rates amongst people who smoke. Varenicline is a nicotine partial agonist which acts on the nicotinic acetylcholine receptors to stimulate dopamine release in the nucleus accumbens, mimicking the effects of nicotine and helping to reduce cravings and withdrawal. Varenicline is effective for smoking cessation, increasing the rates of smoking abstinence at 6 months by 2–3 times compared to placebo, and it is more effective than NRT single therapy and as effective as NRT combination therapy [37]. Nausea is a common side effect, and although there were post-marketing reports of suicidal thoughts, subsequent systematic reviews have found no evidence for an increase in suicidal thoughts or behaviour with varenicline use.

Bupropion is an antidepressant medication that is also marketed for smoking cessation. Bupropion acts as a norepinephrine–dopamine reuptake inhibitor and partial nicotinic receptor antagonist [38]. Bupropion is as effective as NRT single therapy, with smokers almost twice as likely to quit long term when compared with placebo, but it is less effective than varenicline [37].

The combination of counselling and pharmacotherapy is more effective than using either alone and should be considered as first line in smokers who want to quit [23].

E-cigarettes and vaping

Electronic cigarettes (e-cigarettes) are handheld, battery-powered vaporisers, designed to simulate smoking without exposing the user to the harms of tobacco. The use of e-cigarettes is known colloquially as 'vaping'. There were 6.9 million current e-cigarette users in the US in 2017 [10]. They have become virtually ubiquitous, and vaping has become a multibillion-dollar industry, but what is really known about the safety of e-cigarettes?

E-cigarettes utilise a heating element to heat and vaporise liquid mixtures of propylene glycol or glycerine, nicotine and flavourings, allowing the user to inhale the resultant vapour. Liquids for use with e-cigarettes are available in a variety of flavours and concentrations of nicotine, including none. E-cigarettes can resemble ordinary cigarettes, pipes, cigars, or even USB sticks or pens (Fig. 16.5).

E-cigarettes and smoking cessation

Available evidence on the efficacy of e-cigarette usage for smoking cessation is limited, and of variable quality. Results from a recent large meta-analysis including 1,007 randomised control trial (RCT) participants and 13,115 cohort participants were mixed [39]. E-cigarettes with and without nicotine were the interventions, and the comparator groups were: no smoking cessation aid, alternative smoking cessation aid, including NRT, behavioural, and pharmacological aids, and alternative e-cigarette products. The outcomes examined were tobacco smoking cessation measured at 6 months (preferentially biochemically validated, for example blood carbon dioxide level), reduction of cigarette use of at least 50%, and serious (for example pneumonia and myocardial infarction) and nonserious (for example nausea and vomiting) adverse events measured at 1 week or longer follow-up. Results from two of the included RCTs suggested an increase in smoking cessation with nicotine-containing e-cigarettes compared to non-nicotine-containing e-cigarettes (RR 2.03, 95% CI 0.94–4.38; $P = .07$, I2 = 0%, risk difference (RD) 64/1,000 over 6–12 months), but this evidence was considered low quality due to imprecision and risk of bias. Very low-certainty evidence provided by analysis of eight

FIGURE 16.5

Variety of e-cigarette devices.

'Teen vaping is now getting popular due to discreet vaping devices', by Sarah Johnson, licensed under CC BY 2.0.

cohort studies suggested a reduction in smoking cessation rates with those using nicotine-containing e-cigarettes compared to not using e-cigarettes (OR 0.74, 95% CI 0.55–1.00; $P = .051$; I2 = 56%).

Another systematic review showed an improvement in smoking cessation rates with nicotine-containing e-cigarette usage compared to placebo e-cigarettes, but again this was a low confidence result given the paucity of trial data [40]. Only two completed RCTs contributed follow-up data on smoking cessation at 6 months or longer, including 662 participants. Smokers who used nicotine-containing e-cigarettes had higher quit rates than those in the placebo group (RR 2.29, 95% CI 1.05–4.96), but the quality of evidence is considered low overall given the small number of trials.

High-quality evidence demonstrating e-cigarette efficacy in smoking cessation is needed before they can be recommended over evidence-based interventions that are known to be effective, such as NRT.

Safety and regulation

E-cigarettes are a rapidly changing and evolving product, and with a wide variety of technologies and composition of e-cigarette liquids, making research into the harms and potential benefits, and regulation, a challenge. To date, no long-term vaping safety or toxicological studies have been conducted in humans, and without these, it cannot be definitively said that e-cigarettes are safer than conventional tobacco smoking.

In 2016, the US Food and Drug Administration (FDA) decided to regulate e-cigarettes as tobacco-containing products, as the nicotine is derived from tobacco. Purchasers of e-cigarettes have to be at least 18 years old in the US and United Kingdom (UK), in an attempt to curb the increasing usage among the youth [41,42]. Extension of legislation allows the regulation of the manufacture, import, packaging, labelling, advertising, promotion, sale, and distribution of e-cigarettes by the FDA.

Governmental perspectives on e-cigarette products are divided internationally. The US FDA recognise that e-cigarettes may not be safe and will not endorse the substitution of e-cigarettes for traditional tobacco products [43]. Public Health England (PHE) in the UK has recognised e-cigarette usage as safer than tobacco smoking and endorses them for harm reduction and smoking cessation [44]. The UK-centric view focuses on harm reduction to the smoker, whereas the US view prioritises minimising involuntary exposure of bystanders (children and non-smokers) to nicotine and other chemicals. The conflicting approaches reflect the novelty of e-cigarettes and the overall lack of evidence around long-term health impacts.

E-cigarette or vaping product use associated lung injury

Although initial research findings have suggested that e-cigarettes are less harmful than tobacco smoking, cases are emerging which suggest that e-cigarette usage can be dangerous. By the end of 2019, 2290 cases of e-cigarette, or vaping, product use associated lung injury (EVALI) were reported in the US, with 47 deaths. The patients' ages ranged from 17 to 75 years [45].

Tetrahydracannabinol (THC), the main psychoactive component of cannabis, was identified in most of the samples tested by FDA, and most of the patients reported a history of THC product use. The Centre for Disease Control (CDC) undertook testing of bronchoalveolar lavage samples from 29 of the patients with EVALI, all of which contained vitamin E acetate. Vitamin E acetate is used as a thickening agent in THC-containing e-cigarette products and is now listed as a chemical of concern in EVALI. Ordinarily vitamin E is found in many foods and is available as a dietary supplement and an additive in cosmetic products, but when inhaled it may interfere with lung functioning. There is not yet sufficient evidence to rule out other chemicals as contributory to EVALI and investigation is ongoing.

CDC recommendations at the time of writing are to avoid the use of e-cigarette/vaping products containing THC, and if possible, as the causative agent of EVALI is not yet known, refrain from the use of all e-cigarette or vaping products.

Conclusion

Tobacco smoking is a dangerous pursuit. Despite the overall decline in smoking rates, cigarette smoking remains a major cause of morbidity and mortality across the globe. It is a major modifiable risk factor for cancer and puts a significant burden on healthcare systems worldwide, being responsible for seven million deaths every year. Smoking remains a public health emergency with significant racial and socioeconomic determinants, disproportionately affecting certain ethnic groups, and individuals with lower educational attainment and lower household incomes.

Smoking cessation has wide-ranging benefits including a reduced risk of developing many cancers, improved respiratory function, improved sexual function, and an increased life expectancy. People who quit smoking live longer, healthier lives.

Most cigarette smokers want to stop. Despite this, the majority do not successfully quit smoking. Motivation is a major factor in quit attempts and is essential for sustained abstinence, but effective evidence-based therapies are available to assist individuals who smoke. Simple measures including brief interventions and counselling improve quit rates, and a number of pharmacological options have been developed. A combination of counselling and nicotine replacement therapy should be considered first line and has proven efficacy over either therapy alone.

Electronic cigarettes are showing some promise for smoking cessation, providing an alternative nicotine delivery method emulating cigarette smoking but with fewer harmful chemicals, but they remain a controversial subject. Given recent public health scares with EVALI, it is clear that more studies exploring the long-term safety and efficacy as a smoking cessation aid are required before e-cigarettes can be endorsed as a safe and effective adjunct to stop smoking.

Never underestimate the power of time. Do not be afraid to discuss smoking with patients. A short conversation highlighting the benefits of quitting and the assistance available might be enough to make someone reconsider a lifelong habit and make a change for the better.

Summary

- Smoking is a significant cause of morbidity and mortality worldwide, responsible for 7 million deaths annually.
- There are significant racial and socioeconomic determinants of tobacco use.
- The major effects of smoking are to increase the risk of cardiovascular disease, respiratory disease and malignancy.
- Additional adverse health effects include increasing the risk of rheumatoid arthritis, immunodeficiency, ectopic pregnancy, cleft palate in babies of women who smoke during pregnancy, age-related macular degeneration and erectile dysfunction.
- The benefits of smoking cessation can be felt at any age, including a significant reduction in cancer risk.
- Effective, evidence-based pharmacological and non-pharmacological smoking cessation therapies are available including nicotine replacement therapy and counselling.
- Even minimal counselling is proven to increase smoking cessation rates.
- Combination therapy with counselling and nicotine replacement is more effective than monotherapy and should be offered first line.
- E-cigarettes may be useful as a smoking cessation aid, but little is known regarding their long-term safety and efficacy, and more research is needed.

References

[1] WHO report on the global tobacco epidemic. Geneva: World Health Organisation; 2017.

[2] WHO report on the global tobacco epidemic. 2014.

[3] Furrukh M. Tobacco smoking and lung cancer: perception-changing facts. Sultan Qaboos Univ Med J 2013; 13(3):345–58.

[4] Charlton A. Medicinal uses of tobacco in history. J R Soc Med 2004;97(6):292–6.

[5] Sanner T, Grimsrud TK. Nicotine: carcinogenicity and effects on response to cancer treatment - a review. Front Oncol 2015;5:196.

[6] Talhout R, Schulz T, Florek E, van Benthem J, Wester P, Opperhuizen A. Hazardous compounds in tobacco smoke. Int J Environ Res Publ Health 2011;8(2):613–28.

[7] The health consequences of smoking - 50 years of progress: a report of the surgeon general. Atlanta: U.S. Department of Health and Human Services, Centres for Disease Control and Prevention, National Center for Chronic Disease Prevention and Health Promotion, Office on Smoking and Health; 2014.

[8] Proctor RN. The history of the discovery of the cigarette-lung cancer link: evidentiary traditions, corporate denial, global toll. Tobac Contr 2012;21(2):87–91.

[9] Centers for Disease C, Prevention. Smoking-attributable mortality, years of potential life lost, and productivity losses–United States, 2000–2004. MMWR Morb Mortal Wkly Rep 2008;57(45):1226–8.

[10] Wang TW, Asman K, Gentzke AS, Cullen KA, Holder-Hayes E, Reyes-Guzman C, et al. Tobacco product use among adults - United States, 2017. MMWR Morb Mortal Wkly Rep 2018;67(44):1225–32.

[11] Wakefield M, Chaloupka F. Effectiveness of comprehensive tobacco control programmes in reducing teenage smoking in the USA. Tobac Contr 2000;9(2):177−86.

[12] 2017 National Survey on Drug Use and Health: Detailed tables. Substance Abuse and Mental Health Services Administration; 2017.

[13] Reducing the health consequences of smoking: 25 years of progress: a report of the surgeon general. U.S. Department of Health and Human Services, Public Health Service, Centers for Disease Control, National Center for Chronic Disease Prevention and Health Promotion, Office on Smoking and Health; 1989.

[14] Homa DM, Neff LJ, King BA, Caraballo RS, Bunnell RE, Babb SD, et al. Vital signs: disparities in non-smokers' exposure to secondhand smoke–United States, 1999−2012. MMWR Morb Mortal Wkly Rep 2015; 64(4):103−8.

[15] Arheart KL, Lee DJ, Dietz NA, Wilkinson JD, Clark 3rd JD, LeBlanc WG, et al. Declining trends in serum cotinine levels in US worker groups: the power of policy. J Occup Environ Med 2008;50(1):57−63.

[16] Powell JT. Vascular damage from smoking: disease mechanisms at the arterial wall. Vasc Med 1998;3(1): 21−8.

[17] Maddatu J, Anderson-Baucum E, Evans-Molina C. Smoking and the risk of type 2 diabetes. Transl Res 2017; 184:101−7.

[18] Novello AC. Surgeon General's report on the health benefits of smoking cessation. Publ Health Rep 1990; 105(6):545−8.

[19] How tobacco smoke causes disease: the biology and behavioral basis for smoking-attributable disease: a report of the surgeon general. Atlanta (GA): Publications and Reports of the Surgeon General; 2010.

[20] Dai JB, Wang ZX, Qiao ZD. The hazardous effects of tobacco smoking on male fertility. Asian J Androl 2015;17(6):954−60.

[21] Definition of Addiction [cited 2019 Nov 20th]. Available from: https://www.asam.org/resources/definition-of-addiction.

[22] Benowitz NL. Nicotine addiction. N Engl J Med 2010;362(24):2295−303.

[23] Phs Guideline Update Panel L, Staff. Treating tobacco use and dependence: 2008 update U.S. Public Health Service Clinical Practice Guideline executive summary. Respir Care 2008;53(9):1217−22.

[24] Shah RS, Cole JW. Smoking and stroke: the more you smoke the more you stroke. Expert Rev Cardiovasc Ther 2010;8(7):917−32.

[25] Anthonisen NR, Connett JE, Kiley JP, Altose MD, Bailey WC, Buist AS, et al. Effects of smoking intervention and the use of an inhaled anticholinergic bronchodilator on the rate of decline of FEV1. The Lung Health Study. J Am Med Assoc 1994;272(19):1497−505.

[26] Ordonez-Mena JM, Schottker B, Mons U, Jenab M, Freisling H, Bueno-de-Mesquita B, et al. Quantification of the smoking-associated cancer risk with rate advancement periods: meta-analysis of individual participant data from cohorts of the CHANCES consortium. BMC Med 2016;14:62.

[27] Doll R, Peto R, Boreham J, Sutherland I. Mortality from cancer in relation to smoking: 50 years observations on British doctors. Br J Canc 2005;92(3):426−9.

[28] Harte CB, Meston CM. Association between smoking cessation and sexual health in men. BJU Int 2012; 109(6):888−96.

[29] Babb S, Malarcher A, Schauer G, Asman K, Jamal A. Quitting smoking among adults - United States, 2000-2015. MMWR Morb Mortal Wkly Rep 2017;65(52):1457−64.

[30] About Us - PEZ 2019 [cited 2020 Jan 9th]. Available from: https://us.pez.com/pages/history.

[31] Stead LF, Buitrago D, Preciado N, Sanchez G, Hartmann-Boyce J, Lancaster T. Physician advice for smoking cessation. Cochrane Database Syst Rev 2013;(5):CD000165.

[32] Stead LF, Carroll AJ, Lancaster T. Group behaviour therapy programmes for smoking cessation. Cochrane Database Syst Rev 2017;3:CD001007.

[33] Lancaster T, Stead LF. Individual behavioural counselling for smoking cessation. Cochrane Database Syst Rev 2017;3:CD001292.

[34] Matkin W, Ordonez-Mena JM, Hartmann-Boyce J. Telephone counselling for smoking cessation. Cochrane Database Syst Rev 2019;5:CD002850.

[35] Hartmann-Boyce J, Chepkin SC, Ye W, Bullen C, Lancaster T. Nicotine replacement therapy versus control for smoking cessation. Cochrane Database Syst Rev 2018;5:CD000146.

[36] Lindson N, Chepkin SC, Ye W, Fanshawe TR, Bullen C, Hartmann-Boyce J. Different doses, durations and modes of delivery of nicotine replacement therapy for smoking cessation. Cochrane Database Syst Rev 2019; 4:CD013308.

[37] Cahill K, Lindson-Hawley N, Thomas KH, Fanshawe TR, Lancaster T. Nicotine receptor partial agonists for smoking cessation. Cochrane Database Syst Rev 2016;(5):CD006103.

[38] Khan SR, Berendt RT, Ellison CD, Ciavarella AB, Asafu-Adjaye E, Khan MA, et al. Bupropion hydrochloride. Profiles Drug Subst Excipients Relat Methodol 2016;41:1—30.

[39] El Dib R, Suzumura EA, Akl EA, Gomaa H, Agarwal A, Chang Y, et al. Electronic nicotine delivery systems and/or electronic non-nicotine delivery systems for tobacco smoking cessation or reduction: a systematic review and meta-analysis. BMJ Open 2017;7(2):e012680.

[40] Hartmann-Boyce J, McRobbie H, Bullen C, Begh R, Stead LF, Hajek P. Electronic cigarettes for smoking cessation. Cochrane Database Syst Rev 2016;9:CD010216.

[41] FDA. FDA's deeming regulations for E-cigarettes, cigars, and all other tobacco products. 2019 [cited 2019 Nov 29th]. Available from: https://www.fda.gov/tobacco-products/rules-regulations-and-guidance/fdas-deeming-regulations-e-cigarettes-cigars-and-all-other-tobacco-products.

[42] GOV.UK. Rules about tobacco, e-cigarettes and smoking: 1 October 2015. 2015 [cited 2019 Nov 29th]. Available from: https://www.gov.uk/government/publications/new-rules-about-tobacco-e-cigarettes-and-smoking-1-october-2015/new-rules-about-tobacco-e-cigarettes-and-smoking-1-october-2015.

[43] Washington (DC). In: Eaton DL, Kwan LY, Stratton K, editors. Public health consequences of e-cigarettes; 2018.

[44] PHE. PHE publishes independent expert e-cigarettes evidence review. 2018 [cited 2019 Nov 29th]. Available from: https://www.gov.uk/government/news/phe-publishes-independent-expert-e-cigarettes-evidence-review.

[45] CDC. Outbreak of lung injury associated with the use of e-cigarette, or vaping, products: the centers for disease control and prevention. 2019 [cited 2019 Nov 28th]. Available from: https://www.cdc.gov/tobacco/basic_information/e-cigarettes/severe-lung-disease.html#latest-outbreak-information.

Alcohol: its impact on wellbeing, morbidity and mortality

17

Hayley S. McKenzie

Medical Oncology, University Hospital Southampton NHS Foundation Trust, Southampton, United Kingdom

Introduction

> *Modern life, too, is often a mechanical oppression and liquor is the only mechanical relief.*
>
> — **Hemingway**

Alcohol is a drug and is one of the most widely used drugs around the world. The effects of alcohol have long been appreciated — research shows that communities were making a type of rice wine in China around 9000 years ago [1]. In more recent times, for many alcohol consumers, the quantity ingested means that the negative effects outweigh the positive.

Absorption and metabolism

Alcoholic drinks contain ethanol, C_2H_5OH, which is produced by the fermentation of a sugar source such as grains or fruit. Alcohol is largely absorbed in the small intestine, with a small amount absorbed through the oral and gastric mucosa. Once consumed, blood alcohol levels peak after 15—90 min. Absorption is slowed down by the presence of food in the gut [2]. The elimination of ethanol occurs in the liver, with a minor amount being excreted through breath and sweat. The liver breaks down ethanol at a standard rate that cannot be altered [2]. Blood alcohol content (BAC) therefore increases when consumption exceeds the rate of excretion.

Ethanol is metabolised in a number of steps by several enzymes. Alcohol dehydrogenase is the hepatic enzyme that oxidises ethanol into acetaldehyde. This substrate is then oxidised to acetate by aldehyde dehydrogenase. Acetate is activated in peripheral tissues to Acetyl CoA, which can be converted into fatty acids, ketone bodies, cholesterol and carbon dioxide [3]. The rate at which this occurs in an individual is dependent on their gender, age, ethnicity, medications and other environmental or metabolic factors.

There is no way to enhance the elimination of alcohol once it has been consumed.

Different physiological and physical effects are experienced as BAC increases. The initial effects of alcohol include increased talkativeness, relaxation and confidence. If the BAC level continues to rise, inhibitions are lowered, judgment is impaired and speech becomes slurred. There are negative effects on balance and coordination, and reflexes become slower. Nausea and vomiting can follow, and individuals may reach the stage when they are unable to walk, have a reduced level of consciousness, reduced respiratory rate and incontinence. Ultimately there is a risk of death.

Units are used to explain how much alcohol is in a drink. One unit is 10 mL or 8 g of pure alcohol. This is equal to the amount an adult of average weight can eliminate from their system per hour. The number of units contained in a drink depends on the size and strength of the drink. For instance, a pint of beer can be two or three units depending on its strength. The number of units can be calculated by multiplying the percentage strength by the volume in millilitres, divided by 1000. For example, a 25 mL single shot of spirits is one unit, a 330 mL bottle of 5% beer or cider is 1.7 units and a 175 mL glass of 12% wine is 2.1 units [4].

Alcohol consumption and socioeconomic factors

Overall, the economic wealth of countries is correlated with higher alcohol consumption [5]. Within countries, socioeconomic status is positively related to alcohol use although this is moderated by other variables including ethnicity and gender [6].

Culture and environment also have an impact, for instance African Americans and Latinos consume less alcohol than Whites, in line with a more conservative attitude towards drinking. There are likely disparities within communities too; for example, there is a higher rate of alcohol dependence amongst Puerto Rican men compared to Cuban Americans. Cultural norms are further affected by environment and context including peer group, gender and neighbourhood [7]. In places where drinking alcohol is seen as an act of socialisation, different ethnic groups may modify their alcohol intake as they integrate into their new culture [8].

Despite individuals with a lower socioeconomic status overall drinking less alcohol, they are disproportionately more likely to suffer negative alcohol-related health effects [6]. This is termed the 'alcohol harm paradox'. The causes of this paradox are unclear but may include comorbidities in addition to nutritional factors, smoking and obesity, health access issues and/or more extreme drinking within a lower number of overall drinkers in the lower socioeconomic status groups [9].

Drinking patterns around the world

Globally, alcohol use is increasing. The per adult capita consumption rose from 5.5 L of pure alcohol in 2005 to 6.4 L in 2016. However, the percentage of current drinkers has decreased, indicating that those who are drinking have increased levels of consumption. Although the region with the highest alcohol intake is Europe, consumption there is decreasing over time.

Two countries that partially account for the worldwide increase are China and India. In India, consumption increased from 2.4 L in 2005 to more than double that at 5.7 L in 2016 [5].

Worldwide, heavy episodic drinking, defined as the consumption of ≥60 g pure alcohol on at least one occasion at least once a month, decreased from 22.6% of adults in 2000 to 18.2% in 2016. However, it remains high in some areas, particularly in Eastern Europe and some sub-Saharan African countries [5].

In the United Kingdom (UK), 28.7% of men and 25.6% of women admit to binge drinking, which is defined as men drinking eight units a day and women drinking six [10]. Binge drinking is highest amongst 16—24-year-olds although this group tends to drink a smaller amount of alcohol overall than other age groups.

The group most at risk from drinking too much is the middle-aged, as they are more likely to drink every day and consume more alcohol overall.

In England, 31% of men and 16% of women drink more than 14 units in an average week. The highest level of drinking is amongst 55—64-year-olds [11].

Short- and long-term effects of alcohol

Alcohol is classified as a depressant, a drug that lowers levels of neurotransmitters in the central nervous system. Alcohol has a complex pharmacology and interacts with a number of neuromodulator systems in the brain's stress and reward circuits, which is partly mediated through its interaction with $GABA_A$ and NMDA receptors. Alcohol appears to affect almost all neurotransmission, resulting in the signs and symptoms of intoxication [12].

Alcohol also increases the levels of the reward hormone, dopamine, in the brain. This explains why, after a few drinks, individuals can feel happy and energised. However, continued and excessive alcohol consumption, over a prolonged period of time, results in the dopamine effects being attenuated. Individuals may, therefore, drink increasing volumes to try and achieve the 'feel good' effect.

Chronic alcohol misuse impacts neuronal functioning, and this can lead to the development of an addiction, which can encompass tolerance and withdrawal symptoms [13].

Excess alcohol consumption is associated with an increased risk of physical injury. A person is up to five times more likely to injure themselves if they drink around five to seven units over a 3—6-h period than if they have not drunk [14]. Using Emergency Department data from 18 countries, it was determined that 16.4% of all injuries were estimated to be attributable to alcohol, with males sustaining twice as many injuries as females [15]. Injuries can include fractures and head injuries.

Fractures are four times more common in those who drink alcohol chronically, compared to age-matched controls. The causes for this include lower bone mass density, more falls and poor nutrition [16]. In one study, 54% of men with alcohol misuse had at least one fracture on surveillance [17]. Fractures of the hip, vertebrae and ribs are particularly common [16].

There is a dose—response relationship between alcohol and injury. Injury risk is 2.3—2.7 times more likely after one drink and 4.6 times more likely after three drinks, increasing to 12.8 times more likely for men and 28.6 times more likely for women after 20 drinks [18].

The outcome can be very serious for those who decide to drive after drinking. Even low BAC levels impair driving ability. Globally, 370,000 deaths occurred secondary to road injuries attributable to alcohol in 2016 [5]. In most countries, there is also a risk of conviction, resulting in imprisonment, a large fine or a driving ban.

There is a strong relationship between alcohol consumption and violence, including domestic abuse and assault. Moreover, 65% of the victims of stranger violence and 39% of victims of domestic violence believed the perpetrator was under the influence of alcohol [19]. The rate of partner violence in alcoholic men in the year before starting treatment for their substance misuse is 50%—60%

compared to 12% for non-alcoholics [20]. It appears that alcohol lowers the threshold for committing violence in individuals already predisposed to being aggressive [21]. Binge drinking is more likely to be associated with partner abuse than drinking alcohol per se [20].

Intoxication can also affect sexual health. When drunk, a person is more likely to engage in unprotected sex. One review concluded that for each 0.1 mg/mL increase in BAC, the risk of engaging in unprotected sex increased by 5% [22]. This clearly exposes an individual to the risks of unplanned pregnancy and sexually transmitted infections.

The negative physiological and psychological effects of alcohol can last for more than 24 h, if a large enough amount is consumed. Typical symptoms of a 'hangover' can include a dry mouth, dizziness, headache, fatigue, nausea, vomiting and anxiety. There is no evidence that any medication is effective at preventing or treating a hangover. The number of days lost because of alcohol-related sickness is estimated to cost the UK economy £1.9 billion per year [23]. Worldwide, in 2016, alcohol caused 132.6 million disability-adjusted life years (DALYs), which accounted for 5.1% of all DALYs for that year [5]. Concentration and focus are likely to be diminished the day following alcohol consumption due to a reduction in memory and reaction speed [24]. Exercise performance is also affected. Alcohol impairs protein synthesis following exercise, leading to slower recovery and less muscle growth [25].

In the long term, moderate or heavy alcohol consumption is associated with an increased risk of developing a range of illnesses including heart disease, stroke, liver disease, cancer, brain and nerve damage [14].

Most organs can be damaged by alcohol, but usually symptoms do not develop until it is too late for the effects to be reversed. Cardiomyopathy and pancreatitis are common consequences. As the liver is primarily responsible for the metabolism of alcohol, it is one of the most common organs affected by repeated exposure to this drug. Liver cirrhosis is irreversible and once it occurs, complications can include gastrointestinal haemorrhage secondary to oesophageal varices, hepatic encephalopathy and an increased risk of developing hepatocellular carcinoma. Once cirrhosis develops, 5 year survival is only 50% [26].

Alcohol increases the risk of cancer, including oral cancer, oesophageal cancer, bowel cancer, breast cancer and liver cancer [27]. The risk of malignancy is dose dependent. Worldwide, there were 337,400 deaths from alcohol-attributable cancers, with liver cancer being the most common.

In 2010, 4.2% of all cancer deaths were attributable to alcohol. Alcohol is more strongly associated with some cancers than others, for example it is estimated to cause 9.0% of breast cancer cases but 29.3% of nasopharyngeal cancers [28] (Fig. 17.1). The mechanisms for the carcinogenic effects of alcohol include the generation of reactive oxygen species, toxic metabolites, carcinogenic contaminants, inhibition of nutrient absorption and increased oestrogen levels [29]. Alcohol and tobacco have a synergistic effect, particularly in the development of cancers of the head and neck and oesophagus [29].

Overall, worldwide in 2016, around 3 million deaths, which equates to 5.3%, occurred as a result of the harmful use of alcohol [5] (Fig. 17.2).

Alcohol and stigma

There is often stigma surrounding individuals who are labelled as alcoholics. Alcoholism is less likely to be judged as a mental illness and more likely to be seen as a voluntary condition compared to

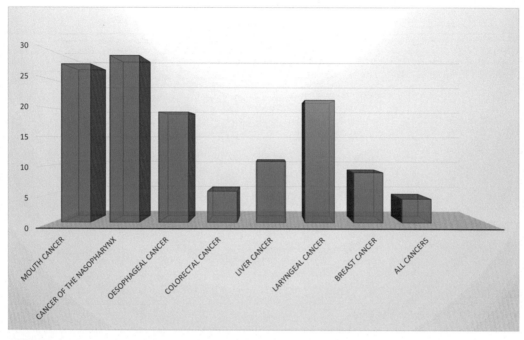

FIGURE 17.1

Percentage of cancer deaths attributable to alcohol.

Data from IACR World Cancer Report 2014 International Agency for Research on Cancer, World Health Organization, B.W. Stewart
& C. Wild. World cancer report 2014.

depression or schizophrenia. An individual is more likely to be blamed for their alcoholism compared to those with other mental health disorders. Generally, alcoholics are deemed to be dangerous and unpredictable. In many studies, it has been shown that people desire more social distance from alcoholics than those with other mental health problems including schizophrenia and panic disorder [30]. Unfortunately, this stigma is likely to isolate those suffering from alcohol dependence and worsen the condition, as they are less likely to seek help. This is compounded by health professionals generally having a negative attitude towards patients with substance use disorders [31]. In the United States of America (USA), only 5% of patients with alcohol dependence seek treatment in the first year of their condition [32].

How much alcohol is 'safe' to drink?

The benefits of alcohol are frequently overstated, and it is common to hear people say, 'But a glass of red wine a day is good for your heart!' A recent study revealed that people who consumed an average of 0.5 drinks per day, or up to five drinks per week, had a lower combined risk of cancer or death than those who never drank [33]. Therefore, a small amount of alcohol appeared to be protective in some

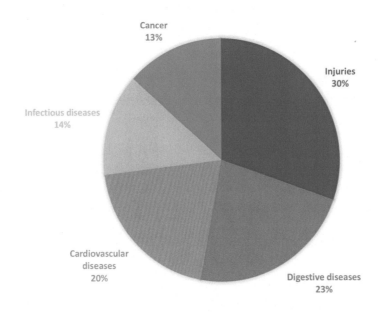

FIGURE 17.2

Causes of alcohol-related death, 2016.

Data from WHO Global status report on alcohol and health 2018 WHO. reportGlobal status report on alcohol and health 2018. (2018).

way, possibly related to a reduction in heart disease. This requires further research. However, those who had two or more drinks per day had a significantly increased risk of early death. When the study evaluated the risk of cancer alone, this seemed to increase with each drink per day, so any amount of alcohol was detrimental. The link between light drinking and an increased risk of cancer of the oral cavity, pharynx, oesophagus and breast has been shown in a large meta-analysis [34].

If individuals are concerned about developing cancer, they should be advised to consider that no alcohol is the best policy. Otherwise, up to five drinks a week appears to be safe.

Current guidelines

Guidelines regarding alcohol consumption vary in different countries. In the USA, the Centre for Disease Control's Dietary Guidelines, 2015–20, recommend as follows [35]:

- Up to one drink per day/14 g pure alcohol for women and up to two per day for men;
- It is advised that people who do not currently drink do not start to drink;
- Those under the age of 21 should not drink.

The National Health Service in the UK advises [4]:

- Not to drink more than 14 units a week on a regular basis;

- Spread drinking over three or more days for individuals who regularly drink as much as 14 units a week;
- Try to have several drink-free days each week.

Other regions have different guidelines. For example, in Spain, men are advised to drink a maximum of 40 g per day or 17 drinks per week and women 25 g per day or 28 drinks per week [36]. This contrasts with the gender-neutral policy in the UK.

Working with patients with alcohol problems

If there are concerns a patient may be drinking too much, there are various screening tools that can be used to gain a greater understanding of their situation. There are different tools in different countries, but they include the AUDIT-C score in the UK and the Alcohol Use Disorders Identification Test in the USA.

It is important to be empathetic and nonjudgmental, and consideration may need to be given to harm reduction strategies, counselling or detoxification and abstinence programmes.

For individuals who do not necessarily have an alcohol problem but who want to reduce their intake, several strategies can be explored:

- It is beneficial to explain the benefits of reducing alcohol intake — these include more energy, improved skin and weight loss, better mood and sleep, improved immunity and improved health.
- Prompt patients to consider what barriers are preventing them from changing their behaviour, and encourage them to address these.
- It is worth advising the patient to inform friends and family of their plan to cut down on their alcohol intake. This means the patient is more likely to have social support during what might be a challenging time.
- You might consider suggesting the individual plans social activities that do not revolve around alcohol — for instance going on a hike, instead of meeting in a pub.
- If the patient is trying to reduce their intake rather than stop altogether, they could be advised to try low or no alcohol alternatives, or a smaller serving size.

All patients should be advised to have two to three completely alcohol-free days per week. A month off may be even more beneficial if alcohol consumption has been excessive. In one small study, one month of abstinence resulted in improvements in parameters such as blood pressure, bodyweight and insulin resistance in moderate-to-heavy drinkers [37]. The longevity of this effect is however unknown. Furthermore, it could be helpful to refer a patient to local support groups.

Conclusions

Despite being one of the most widely used and harmful drugs worldwide, the negative effects of alcohol remain unknown to many members of the public. A small amount of alcohol consumption, three to four drinks per week, is likely to be safe. However, increased levels of intake are associated with a multitude of adverse health consequences, including cancer, cardiovascular disease, physical injury and gastrointestinal disease. Healthcare providers have a responsibility to inform their patients of these risks and to support patients in reducing their alcohol intake.

Summary

- Alcohol is one of the most widely used drugs worldwide.
- Globally, alcohol use is increasing although decreasing in Europe.
- The middle-aged are those most likely to be consuming an excessive quantity of alcohol.
- Alcohol increases the risk of road traffic accidents, injury, violence, heart disease, stroke, liver and pancreatic disease and multiple types of cancer.
- Small amounts of alcohol may be associated with a reduced risk of heart disease.
- Overall, alcohol increases the risk of death from multiple causes.
- Lower socioeconomic status groups drink less but have more adverse outcomes associated with alcohol.
- Healthcare providers should always consider whether their patients could have a problem with alcohol misuse, and should support them in obtaining the appropriate help.

References

[1] McGovern PE, et al. Fermented beverages of pre- and proto-historic China. Proc Natl Acad Sci U S A 2004; 101:17593−8. https://doi.org/10.1073/pnas.0407921102.

[2] Dasgupta A, Ebrary - York University. 1 online resource. Lanham: Rowman & Littlefield Publishers; 2011.

[3] Cederbaum AI. Alcohol metabolism. Clin Liver Dis 2012;16:667−85. https://doi.org/10.1016/j.cld.2012.08.002.

[4] NHS. Alcohol Units. 2018. https://www.nhs.uk/live-well/alcohol-support/calculating-alcohol-units/.

[5] WHO. Global status report on alcohol and health 2018. 2018.

[6] Collins SE. Associations between socioeconomic factors and alcohol outcomes. Alcohol Res 2016;38: 83−94.

[7] Sudhinaraset M, Wigglesworth C, Takeuchi DT. Social and cultural contexts of alcohol use: influences in a social-ecological framework. Alcohol Res 2016;38:35−45.

[8] Castaldelli-Maia JM, Bhugra D. Investigating the interlinkages of alcohol use and misuse, spirituality and culture - insights from a systematic review. Int Rev Psychiatr 2014;26:352−67. https://doi.org/10.3109/09540261.2014.899999.

[9] Lewer D, Meier P, Beard E, Boniface S, Kaner E. Unravelling the alcohol harm paradox: a population-based study of social gradients across very heavy drinking thresholds. BMC Publ Health 2016;16:599. https://doi.org/10.1186/s12889-016-3265-9.

[10] Office for National Statistics. Adult drinking habits in great Britain. 2017. https://www.ons.gov.uk/peoplepopulationandcommunity/healthandsocialcare/drugusealcoholandsmoking/bulletins/opinionsandlifestylesurveyadultdrinkinghabitsingreatbritain/2017.

[11] NHS. Health survey for England. 2016. https://digital.nhs.uk/data-and-information/publications/statistical/health-survey-for-england/health-survey-for-england-2016.

[12] Vengeliene V, Bilbao A, Molander A, Spanagel R. Neuropharmacology of alcohol addiction. Br J Pharmacol 2008;154:299−315. https://doi.org/10.1038/bjp.2008.30.

[13] Gilpin NW, Koob GF. Neurobiology of alcohol dependence: focus on motivational mechanisms. Alcohol Res Health 2008;31:185−95.

[14] Department of Health. UK chief medical officers' alcohol Guidelines review. 2015. https://assets.publishing.service.gov.uk/government/uploads/system/uploads/attachment_data/file/489795/summary.pdf.

[15] Cherpitel CJ, et al. Alcohol attributable fraction for injury morbidity from the dose-response relationship of acute alcohol consumption: emergency department data from 18 countries. Addiction 2015;110:1724−32. https://doi.org/10.1111/add.13031.

[16] Kelly KN, Kelly C. Pattern and cause of fractures in patients who abuse alcohol: what should we do about it? Postgrad Med 2013;89:578−83. https://doi.org/10.1136/postgradmedj-2013-131990.

[17] Gonzalez-Reimers E, et al. Vitamin D and nutritional status are related to bone fractures in alcoholics. Alcohol Alcohol 2011;46:148−55. https://doi.org/10.1093/alcalc/agq098.

[18] Cherpitel CJ, Ye Y, Bond J, Borges G, Monteiro M. Relative risk of injury from acute alcohol consumption: modeling the dose-response relationship in emergency department data from 18 countries. Addiction 2015; 110:279−88. https://doi.org/10.1111/add.12755.

[19] Institute of Alcohol Studies. Alcohol, domestic abuse and sexual assault. 2014. http://www.ias.org.uk/uploads/IAS report Alcohol domestic abuse and sexual assault.pdf.

[20] Foran HM, O'Leary KD. Alcohol and intimate partner violence: a meta-analytic review. Clin Psychol Rev 2008;28:1222−34. https://doi.org/10.1016/j.cpr.2008.05.001.

[21] Abbey A. Alcohol's role in sexual violence perpetration: theoretical explanations, existing evidence and future directions. Drug Alcohol Rev 2011;30:481−9. https://doi.org/10.1111/j.1465-3362.2011.00296.x.

[22] Rehm J, Shield KD, Joharchi N, Shuper PA. Alcohol consumption and the intention to engage in unprotected sex: systematic review and meta-analysis of experimental studies. Addiction 2012;107:51−9. https://doi.org/10.1111/j.1360-0443.2011.03621.x.

[23] Independent. Hangovers have 'serious consequences' for driving and working, study finds. 2018.

[24] Gunn C, Mackus M, Griffin C, Munafo MR, Adams S. A systematic review of the next-day effects of heavy alcohol consumption on cognitive performance. Addiction 2018. https://doi.org/10.1111/add.14404.

[25] Parr EB, et al. Alcohol ingestion impairs maximal post-exercise rates of myofibrillar protein synthesis following a single bout of concurrent training. PloS One 2014;9. https://doi.org/10.1371/journal.pone.0088384. e88384.

[26] NHS. Alcohol-related liver disease. 2018. https://www.nhs.uk/conditions/alcohol-related-liver-disease-arld/.

[27] Cancer Research UK. Does alcohol cause cancer?. 2018. https://www.cancerresearchuk.org/about-cancer/causes-of-cancer/alcohol-and-cancer/does-alcohol-cause-cancer.

[28] International Agency for Research on Cancer, World Health Organization, B.W. Stewart & C. Wild. reportWorld cancer report 2014.

[29] National Cancer Institute. Alcohol and cancer risk. 2018. https://www.cancer.gov/about-cancer/causes-prevention/risk/alcohol/alcohol-fact-sheet.

[30] Schomerus G, et al. The stigma of alcohol dependence compared with other mental disorders: a review of population studies. Alcohol Alcohol 2011;46:105−12. https://doi.org/10.1093/alcalc/agq089.

[31] van Boekel LC, Brouwers EP, van Weeghel J, Garretsen HF. Stigma among health professionals towards patients with substance use disorders and its consequences for healthcare delivery: systematic review. Drug Alcohol Depend 2013;131:23−35. https://doi.org/10.1016/j.drugalcdep.2013.02.018.

[32] Hammarlund R, Crapanzano KA, Luce L, Mulligan L, Ward KM. Review of the effects of self-stigma and perceived social stigma on the treatment-seeking decisions of individuals with drug- and alcohol-use disorders. Subst Abuse Rehabil 2018;9:115−36. https://doi.org/10.2147/SAR.S183256.

[33] Kunzmann AT, Coleman HG, Huang WY, Berndt SI. The association of lifetime alcohol use with mortality and cancer risk in older adults: a cohort study. PLoS Med 2018;15. https://doi.org/10.1371/journal.pmed.1002585. e1002585.

[34] Bagnardi V, et al. Light alcohol drinking and cancer: a meta-analysis. Ann Oncol 2013;24:301−8. https://doi.org/10.1093/annonc/mds337.

[35] CDC. Alcohol and public health: moderate drinking. 2016. https://www.cdc.gov/alcohol/fact-sheets/moderate-drinking.htm.

[36] Alcohol in moderation. Sensible drinking Guidelines. 2018. http://www.drinkingandyou.com/site/pdf/Sensibledrinking.pdf.

[37] Mehta G, et al. Short-term abstinence from alcohol and changes in cardiovascular risk factors, liver function tests and cancer-related growth factors: a prospective observational study. BMJ Open 2018;8. https://doi.org/10.1136/bmjopen-2017-020673. e020673.

Cancer: how to help your patients to reduce their cancer risk

18

Ailsa Sita-Lumsden

Medical Oncology, Guy's and St Thomas' NHS Foundation Trust, London, United Kingdom

The major lifestyle recommendations to reduce the risk of developing cancer are smoking cessation, maintaining a healthy weight, optimising nutrition, minimising alcohol, engaging in physical activity, participating in screening activities, treating or vaccinating against specific infections and avoiding unnecessary radiation.

Introduction

There have been remarkable improvements in cancer diagnosis and treatment over the last few decades. Cancer incidence has decreased amongst men although it has remained stable for women. Importantly, in the United States (US) cancer mortality has declined overall in recent decades [1]. However, cancer has overtaken cardiovascular disease as the leading cause of death in many parts of the world for example Canada [2]. The highest rates of cancer are in high-income countries. This raises the question as to whether some of the factors that contribute to the development of cancer can be linked to Western lifestyles. Cancer can affect anyone, from infants to octogenarians, but certain individuals are at a higher risk than others. Some of the factors that can increase the risk of cancer developing are fixed, such as inherited genetic mutations, but others can be influenced by lifestyle and the environment.

The implementation of evidence-based interventions to reduce cancer risk factors and increase cancer-screening uptake results in decreased cancer incidence, morbidity and mortality [3]. An estimated 42% of cancer cases and 45% of cancer deaths in the US are attributed to potentially modifiable risk factors [4]. This means that almost half of all new cancers diagnosed are preventable or could be detectable at an earlier stage leading to improved survival [5]. Given these statistics, healthcare professionals must educate patients and the public regarding cancer risk reduction and cancer screening.

Assessing a patient's risk allows the healthcare professional to provide accurate counselling on cancer risk reduction strategies, appropriate screening recommendations and potentially prophylactic or treatment options. It starts with a detailed history, including past medical history, social history and documentation of recent age-appropriate screening tests. Family history is critical in establishing the risk of hereditary cancer syndromes, for example Lynch syndrome. Medication history, diet, levels of physical activity, environmental exposures and smoking and alcohol use are important factors when determining cancer risk.

A Prescription for Healthy Living. https://doi.org/10.1016/B978-0-12-821573-9.00018-7

> **Box 18.1 Tumourigenesis in lay terms**
> **What is cancer?**
> Cancer is a disease of cells. Adults have approximately 100 trillion cells in their bodies. There are over 200 different types of cell, which vary in their characteristics and functions, and the majority have the potential to form a cancerous tumour.
>
> Most cells in the body can divide to produce more copies of themselves. One cell grows and then divides into 2, 2 into 4, 4 into 8 and so on. This cell division is crucial for growth and healing, so cell division is a normal and essential process to keep us healthy. New cells must be only made when they are needed. In the development of cancer, cells divide uncontrollably — usually forming a mass of abnormal cells called a tumour.
>
> Cell division goes out of control due to misinformation in the cell's genetic code — its DNA. DNA acts as a recipe, telling the cells what proteins to make. These proteins then play an essential role in determining how the cell behaves. When the DNA becomes 'faulty' and acquires defects called mutations, cancer can result. Just one mutation in one gene is not enough to cause cancer. Several mutations in critical genes need to accumulate before a healthy cell can develop into a cancer cell. This is one of the reasons why cancer is more common in older people — DNA damage gradually accumulates throughout a person's lifetime, so the longer we live, the more likely we are to develop cancer.
>
> DNA mutations can occur during natural cell replication and can result from agents in our environment called 'carcinogens', such as chemicals in tobacco and ultraviolet light from the sun. It is estimated that only 1 in 10 of all cancers is linked to inheritance of particular 'high-risk' genes. This means that a faulty gene has been passed down to an individual from one or both parents. This will increase the risk of cancer, as individuals are essentially a step closer to accumulating all the defects that are needed for the disease to develop. So, inheriting a faulty gene *does not* mean that a person will definitely get cancer. Instead, it increases their risk of the disease.
>
> A lot of the DNA mutations that can cause cancer are due to factors we have been exposed to throughout our life. These include viruses, diet, activity levels, environmental carcinogens or certain medical conditions.

To make the necessary lifestyle adjustments to lower cancer risk, it can help if the patient understands what cancer is. This pathogenesis of tumourigenesis is described in lay terms in Box 18.1. It can be useful to share this in appropriate consultations.

Risk reduction advice

There are many lifestyle measures that individuals can take to reduce their risk of developing malignancy:

1. Avoid tobacco in any form.

It has been well communicated to the general public that cigarette smoking causes lung cancer, following the first US Surgeon General's Report on Smoking and Health in 1964. This message does seem to be having an impact. In 1965, 42% of adults in the US were cigarette smokers, decreasing to 14% in 2017 [5]. However, the burden of smoking-related cancers remains high [6]. Smoking prevalence is higher in some parts of the population than others, for example those with low socioeconomic status or with a mental illness [7]. It is estimated that around 30% of all cancer deaths in the US are still caused by smoking [5].

It is important to encourage smokers to stop. Smoking cessation reduces the risk of developing cancer and improves the outcomes for patient's who already have a cancer diagnosis.

E-cigarettes are battery-powered devices that allow users to inhale an aerosol typically containing nicotine, as well as other ingredients such as flavouring. The use of e-cigarettes or vapes, as they are also known, has rapidly grown. Amongst high school students in the US, e-cigarette usage increased from around 2% in 2011 to 21% in 2018 [5]. E-cigarettes have been praised as a safer alternative to smoking and evidence thus far suggests that e-cigarettes are less harmful than conventional cigarettes. However, the risks associated with long-term use are not clear [8]. Certainly, metals and other hazardous chemicals can seep into the inhaled aerosol through contact with heating coils or wicks, and some commonly used flavouring components are dangerous to the lungs. There are reports of deaths secondary to the use of e-cigarettes in the US.

2. Maintain a healthy weight.

Obesity is associated with an increased risk of developing several types of cancer: endometrial, oesophageal, liver, stomach, kidney, multiple myeloma, pancreas, colorectal, gallbladder, ovary, postmenopausal breast and thyroid [4].

An estimated 18% of cancer cases are attributable to the combined effects of excess body weight, alcohol consumption, physical inactivity and an unhealthy diet [4]. There are several proposed mechanisms by which obesity may be associated with malignancy, including an increase in circulating inflammatory signals such as insulin and insulin-like growth-factor-1. This may promote the development of certain tumours by inhibiting programmed cell death. Excess amounts of oestrogen can increase the risk of female malignancies such as breast and endometrial cancers. Adipose cells have direct and indirect effects on other growth regulators, including mammalian target of rapamycin and adenosine monophosphate-activated protein kinase. These metabolic and hormonal abnormalities lead to a general upregulation of cell growth, a decrease in cell death and tumour-promoting inflammation [9].

Amongst adults, the prevalence of being overweight has remained relatively stable since the early 1960s, but obesity has markedly increased. In 2015−16 in the US, approximately 7 in 10 adults were overweight or obese and around 4 in 10 were obese. From 1971 to 2002, the prevalence of obesity amongst youths aged 2−19 years tripled from 5% to 15%, increasing to 19% in 2015−16 [5]. Maintaining a healthy weight throughout life is one of the most important ways to protect against cancer.

3. Eat an appropriate diet.

A healthy diet includes both avoiding things that cause harm and eating sufficient foods that are known to be beneficial. It has been estimated that around 4% of cancer cases can be attributed to poor diet, predominantly colon cancer [4,10]. It has also been indicated that those individuals with the healthiest diets are between 11% and 24% less likely to die from cancer than those with unhealthy eating habits [11]. This likely to result from a reduced risk of developing a malignancy as well as increased fitness allowing the delivery of treatment for those who do receive a cancer diagnosis.

Minimise processed and red meat consumption

Processed meat has been classified as a carcinogen based on the evidence of association with increased colorectal cancer risk [17]. Although the specific mechanisms are not fully understood, substrates such as nitrates or nitrites used to process meats, and heme iron in red meat, can contribute to the formation of nitrosamines, which are implicated in carcinogenesis [18]. Furthermore, processed meats generally have a high salt content, which may result in damage to the gastric mucosa, leading to inflammation and colonisation by the bacteria *Helicobacter pylori*. *Helicobacter pylori* is implicated in gastric malignancies. Curing, smoking and cooking meat at high temperatures such as grilling can form carcinogenic chemicals, which may also contribute to the increased risk. There is no amount of processed meat that can be confidently assessed as not having an increased colorectal cancer risk [19].

Fruit and vegetables

There is some evidence that increased consumption of non-starchy vegetables and fruits are associated with a lower risk of mouth, laryngeal, oesophageal and stomach cancers. There is also evidence that higher vegetable intake may lower the risk of breast tumours [20]. Fruit and vegetables are a rich source of vitamins and minerals such as selenium, vitamin A, C, D and E that are good antioxidants, so may help in reducing the risk of cancer developing. It may also stem from their replacement of more calorie-dense foods within the whole diet and associated maintenance of optimal weight.

Whole grains

Whole grains, for example, oats, quinoa and wild rice, are an important part of a healthy diet due to their relatively high fibre, vitamin and mineral content when compared to their refined products. Overall, the evidence of association is limited, but studies support that a diet high in whole grains protects against bowel cancer as does dietary fibre, which is found in abundance in plant foods [21].

Supplements are not necessary

High-dose dietary supplements of vitamins and minerals are not recommended for cancer prevention unless there is a confirmed deficiency. The aim is always to meet micronutrient requirements through diet alone. The exception to this in the United Kingdom (UK) is Vitamin D — micronutrients are explored in detail in Chapter 23.

Key dietary advice to offer to patients is listed in Box 18.2.

4. Minimise alcohol intake.

Excess alcohol increases the risk of cancers of the oropharynx, oesophagus, liver and colon. It also increases a woman's risk of breast cancer. Alcohol can be broken down by the bacteria that live in the gut to produce a toxic metabolite called acetaldehyde, which can damage the epithelial cells of the colorectal mucosa to increase the risk of bowel cancer [12,13]. Alcohol is also linked to changes in hormone metabolism and is associated with increased levels of oestradiol, which increases the risk of breast cancer [14,15]. Recent results from the Global Burden of Disease (2016) indicated that the amount of alcohol consumption that minimises health outcomes was zero [16].

5. Stay physically active.

Physical activity has many health benefits, including cancer prevention. Approximately 3% of cancer cases can be attributable to physical inactivity [4]. Physical activity has been associated with a reduced risk of colon cancer, endometrial cancer and postmenopausal breast cancer [22]. Exercise is

---

Box 18.2 Dietary advice for patients to reduce their risk of developing a malignancy

- Consume a healthy diet, with an emphasis on plant sources.
- Choose food and drinks in portions that help achieve and maintain a healthy weight.
- Limit consumption of processed meats and red meats.
- Eat at least 2½ cups/5 portions of vegetables and fruits each day.
- Choose whole grain instead of refined-grain products.

protective in addition to its effects on weight [23]. There are a few purported mechanisms by which this may occur. First, exercise improves the way that the body responds to the hormone insulin and may reduce levels of hormones such as oestrogen [23]. Second, physical activity has been shown to support the immune system, which is essential for immune surveillance to potentially destroy cancer cells [24]. Aerobic exercise can decrease oxidative stress, which can damage DNA, and can also enhance DNA repair mechanisms [25].

It is advised that adults should engage in at least 150 min of moderate-intensity or 75 min of vigorous-intensity physical activity each week. In 2017, it was estimated that only 54% of adults and 26% of high school students met recommended levels of physical activity [5]. Additionally, there is mounting evidence that greater time spent being sedentary increases the risk of colon and endometrial cancers [26].

6. Engage in cancer screening programmes.

The aim of screening for cancer in the general population is to detect cancer when no symptoms are apparent to decrease cancer-related morbidity and mortality. For almost all types of cancer, early detection leads to improved outcomes as treatment is initiated at an earlier stage. For cancer screening to be effective, the test must be able to detect cancer earlier than the stage at which symptoms appear and there must be evidence that earlier treatment improves outcomes.

The government-led screening programmes differ across the world, and multiple organisations have published screening guidelines, such as the National Comprehensive Cancer Network, the American Cancer Society and the National Institute for Clinical Excellence. In general, there is consensus around screening recommendations for breast, cervical, colorectal, lung and prostate cancer across the Western world. However, there are considerable differences in the details regarding the age that screening is initiated and screening intervals.

Cervical cancer

Cervical cancer is the third most common female gynaecological cancer. Over 50 years ago, the Pap smear test was developed, and since then, the incidence and mortality rates of cervical cancer have declined steadily. It is estimated that in England screening has prevented 70% of cervical cancer deaths.

A smear identifies preinvasive atypical epithelial cells, and patients may be offered regular surveillance or an active intervention, depending on the severity of the abnormal findings. When detected early, cervical cancer is one of the most successfully treated cancers. Despite the introduction of the human papillomavirus (HPV) vaccine, cervical screening still has a role. Women should be encouraged to engage in the screening programme.

Colorectal cancer

Colorectal cancer is the third leading cause of cancer death. When colorectal cancer is detected at Stage 1, the 5-year survival rate is 90% [1]. However, less than half of colorectal cancers are diagnosed at this stage. The purpose of screening is to identify precancerous polyps and remove them before they progress to invasive malignancy. Bowel cancer screening differs across different countries but primarily consists of the faecal occult blood test or a faecal immunochemical test and visual screening with sigmoidoscopy or colonoscopy. Across the world, screening programmes start at different ages, use different methods and have different time intervals between tests.

Breast cancer

Breast cancer is the most common female malignancy and the second most common cause of cancer-related death in women. The average lifetime risk for women developing breast cancer is 1 in 8 [1]. Breast cancer screening has been shown to decrease the mortality from breast cancer. For every 1000 women screened, five lives will be saved. The screening test offered across the Western world is the mammogram, but the age range and interval changes depending on the particular country. Magnetic resonance imaging (MRI) may be superior to mammography in high-risk women, but MRI screening has not yet been found to reduce mortality in any group of women [27].

Lung cancer

The most significant cause of cancer-related death is lung cancer, and only 15% are diagnosed at an early stage [5]. In 2011, the National Lung Screening Trial showed a 20% reduction in lung cancer deaths with an annual low dose computed tomography (CT) in smokers with a 30 pack-year history [28]. Since then, some countries, such as the US, have adopted an annual low dose CT into their screening guidelines.

Prostate cancer

Prostate cancer is the most commonly diagnosed cancer in men. The majority of prostate cancers are diagnosed at Stage 1 (93%), and the 5-year survival rate for localised disease approaches 100%. The prostate-specific antigen (PSA) blood test is able to detect prostate cancer at an earlier stage. However, using it as a screening tool is far less straightforward as large studies have demonstrated that more than half of the prostate cancers diagnosed early with PSA screening are low risk and would never have caused clinically significant problems. Additionally, the test has a low specificity and can be raised by other benign conditions of the prostate such as infection or inflammation. Routine screening would lead to overdiagnosis and overtreatment in a large proportion of the screened population [29].

7. Engage in medical interventions if offered

Human papilloma virus vaccine

HPV is a virus that can cause cervical cancer, as well as anal and some mouth and head and neck cancers. More than 200 types of HPV have been identified, 20 of which have been associated with cancer. HPV strains 16 and 18 are most commonly associated with cervical cancer and are targeted by HPV vaccines. The vaccine should be given before a patient is sexually active to be most effective.

Hepatitis B vaccine

Hepatitis B can cause an acute or chronic hepatic infection. Hepatitis B virus is the leading cause of hepatocellular cancer around the world. The hepatitis B vaccine has been available since 1982, it is on the infant immunisation schedule in many countries including the US and the UK. It can also be given to adults in specific high-risk groups, for example healthcare workers or travellers to high-risk regions of the world.

Chemoprevention

The use of Tamoxifen, a selective oestrogen receptor modulator in women at risk of oestrogen receptor-positive breast cancer, is the most common example of chemoprevention. This was approved for preventative use after the results of the Breast Cancer Prevention Trial showed a 49% reduction in invasive breast cancer in a population of 13,000 high-risk women, for example those carrying mutations in the *BRCA1* gene [30]. The decision to start chemoprevention is individualised due to the risks associated with the use and the specific subgroups that benefit and is a decision made by the treating oncologist.

8. Avoid unnecessary radiation

Overexposure to ultraviolet radiation such as UVA and UVB is the most significant risk factor for all types of skin cancer. Some people are at higher risk, for example the immunosuppressed as well as those who burn easily or with red hair. Most types of skin cancers, for example basal cell carcinoma and squamous cell carcinoma, are highly treatable. However, the skin cancer with the highest mortality rate is melanoma, and its incidence is rising in the Western world, where it is the fifth most commonly occurring cancer [5]. There is clear evidence that artificial sources of UV radiation, such as tanning booths, should be avoided. In fact, the risk of melanoma is increased by 75% in those who used tanning beds as young adults [31]. Despite declining use in recent years, 8% of female high school students in 2017 reported the use of indoor tanning in the past year [5]. In addition, it is wise to advise patients to minimise sun exposure with protective behaviours such as suncream use, protective clothing, hats and sunglasses.

Another potential source of radiation to avoid is medical imaging studies. It is essential that only necessary imaging is undertaken with a view given to minimising radiation exposure to the patient.

In terms of electromagnetic radiation from high-voltage power lines or radiofrequency radiation from microwaves and cell phones, there is currently no evidence that they cause cancer.

Summary

Cancer results from multiple mutations in various genes. There are lifestyle modifications that can be made to reduce the risk of cancer developing. In addition to avoiding tobacco use, it is highly beneficial to maintain a healthy weight and to limit alcohol consumption to reduce cancer risk. It is estimated that 18% of cancer cases and 16% of cancer deaths are attributable to the combined effects of excess body weight, alcohol consumption, physical inactivity, and consuming an unhealthy diet [5]. Adults who adopt appropriate healthy lifestyles are 10%−20% less likely to be diagnosed with cancer and 25% less likely to die from cancer compared to those who engage in risky behaviours [4].

Patient education and encouraging early presentation to healthcare professionals for investigation will also confer preventative benefit. Cancer detected earlier is often more amenable to treatment and is linked to improved mortality.

Furthermore, there are lifestyle modifications that can be made for those individuals who have a cancer diagnosis. These are summarised in Box 18.3.

Box 18.3 Health promotion for patients diagnosed with cancer

A cancer diagnosis can be a source of motivation for individuals to make lifestyle changes that will lower the risk of recurrence or lower the side-effect profile of interventions such as surgery or chemotherapy that are offered to treat cancer.

1. Smoking cessation should be encouraged and supported as continuing to smoke decreases the effectiveness of chemotherapy and the risk remains for development of a second primary cancer.
2. It is important to encourage healthy eating: the advice is largely the same as the preventative advice. It is important that patients avoid fad diets frequently advertised as 'cures' for cancer. A balanced and healthy diet is the best way to support the body through a cancer diagnosis.
3. The traditional advice following a cancer diagnosis was rest is best. This is now proven to be wrong. Graded exercise during treatment and recovery is encouraged.
4. It can be difficult for some patients with cancer to maintain their weight due to the catabolic nature of some cancers. Referral to the oncology dietetic team may be appropriate to support the patient's intake. However, it is also not uncommon for patients to put on weight with some cancer diagnoses and during some chemotherapy regimens. The use of steroids as an antiemetic causes excess hunger and often patients reduce their exercise levels resulting in a net weight gain. The hormonal modifications used to treat some cancers like prostate and breast can also encourage weight gain. There is evidence in breast cancer that maintaining a healthy weight throughout treatment confers a lower risk of recurrence. Patients should be supported in weight control.
5. Ensure that alcohol intake is moderated.

Summary

Approximately 42% of cancer cases are preventable.
To reduce the risk of developing a malignancy, basic lifestyle interventions include the following:

1 Avoiding tobacco in all its forms, including exposure to second hand smoke
2 Maintaining a healthy weight
3 Eating a healthy and balanced diet
4 Moderating alcohol intake or avoiding alcohol
5 Exercising regularly
6 Engaging in screening programmes
7 Avoiding or treating infections that contribute to cancer, including hepatitis viruses, HIV, and the human papillomavirus
8 Avoiding unnecessary exposure to radiation
   Such lifestyle interventions may improve the prognosis for those with a cancer diagnosis.

References

[1] Siegel RL, Miller KD, Jemal A. Cancer statistics. CA Cancer J Clin 2019;69(1):7–34. 2019.

[2] Tadayon S, Wickramasinghe K, Townsend N. Examining trends in cardiovascular disease mortality across Europe: how does the introduction of a new European Standard Population affect the description of the relative burden of cardiovascular disease? Popul Health Metrics 2019;17(1):6.

[3] Siegel RL, Jemal A, Wender RC, Gansler T, Ma J, Brawley OW. An assessment of progress in cancer control. CA Cancer J Clin 2018;68(5):329—39.

[4] Islami F, Goding Sauer A, Miller KD, Siegel RL, Fedewa SA, Jacobs EJ, et al. Proportion and number of cancer cases and deaths attributable to potentially modifiable risk factors in the United States. CA Cancer J Clin 2018;68(1):31—54.

[5] Society AC. Cancer facts and figures 2015. 2015.

[6] Lortet-Tieulent J, Goding Sauer A, Siegel RL, Miller KD, Islami F, Fedewa SA, et al. State-level cancer mortality attributable to cigarette smoking in the United States. JAMA Intern Med 2016;176(12):1792—8.

[7] Drope J, Liber AC, Cahn Z, Stoklosa M, Kennedy R, Douglas CE, et al. Who's still smoking? Disparities in adult cigarette smoking prevalence in the United States. CA Cancer J Clin 2018;68(2):106—15.

[8] Dinakar C, O'Connor GT. The health effects of electronic cigarettes. N Engl J Med 2016;375(26):2608—9.

[9] Institute NC. Obesity and cancer risk. 2012.

[10] Grosso G, Bella F, Godos J, Sciacca S, Del Rio D, Ray S, et al. Possible role of diet in cancer: systematic review and multiple meta-analyses of dietary patterns, lifestyle factors, and cancer risk. Nutr Rev 2017; 75(6):405—19.

[11] Liese AD, Krebs-Smith SM, Subar AF, George SM, Harmon BE, Neuhouser ML, et al. The Dietary Patterns Methods Project: synthesis of findings across cohorts and relevance to dietary guidance. J Nutr 2015;145(3): 393—402.

[12] Seitz HK, Becker P. Alcohol metabolism and cancer risk. Alcohol Res Health 2007;30(1):38—41. 4-7.

[13] Albano E. Alcohol, oxidative stress and free radical damage. Proc Nutr Soc 2006;65(3):278—90.

[14] Hankinson SE, Willett WC, Manson JE, Hunter DJ, Colditz GA, Stampfer MJ, et al. Alcohol, height, and adiposity in relation to estrogen and prolactin levels in postmenopausal women. J Natl Cancer Inst 1995; 87(17):1297—302.

[15] Endogenous H, Breast Cancer Collaborative G, Key TJ, Appleby PN, Reeves GK, Roddam AW, et al. Circulating sex hormones and breast cancer risk factors in postmenopausal women: reanalysis of 13 studies. Br J Cancer 2011;105(5):709—22.

[16] Collaborators GBDA. Alcohol use and burden for 195 countries and territories, 1990-2016: a systematic analysis for the Global Burden of Disease Study 2016. Lancet 2018;392(10152):1015—35.

[17] Bouvard V, Loomis D, Guyton KZ, Grosse Y, Ghissassi FE, Benbrahim-Tallaa L, et al. Carcinogenicity of consumption of red and processed meat. Lancet Oncol 2015;16(16):1599—600.

[18] Cross AJ, Pollock JR, Bingham SA. Haem, not protein or inorganic iron, is responsible for endogenous intestinal N-nitrosation arising from red meat. Cancer Res 2003;63(10):2358—60.

[19] Gilsing AM, Fransen F, de Kok TM, Goldbohm AR, Schouten LJ, de Bruine AP, et al. Dietary heme iron and the risk of colorectal cancer with specific mutations in KRAS and APC. Carcinogenesis 2013;34(12): 2757—66.

[20] Farvid MS, Chen WY, Rosner BA, Tamimi RM, Willett WC, Eliassen AH. Fruit and vegetable consumption and breast cancer incidence: repeated measures over 30 years of follow-up. Int J Cancer 2019;144(7): 1496—510.

[21] Song M, Wu K, Meyerhardt JA, Ogino S, Wang M, Fuchs CS, et al. Fiber intake and survival after colorectal cancer diagnosis. JAMA Oncol 2018;4(1):71—9.

[22] Choudhury F, Bernstein L, Hodis HN, Stanczyk FZ, Mack WJ. Physical activity and sex hormone levels in estradiol- and placebo-treated postmenopausal women. Menopause 2011;18(10):1079—86.

[23] Moore SC, Lee IM, Weiderpass E, Campbell PT, Sampson JN, Kitahara CM, et al. Association of leisure-time physical activity with risk of 26 types of cancer in 1.44 million adults. JAMA Intern Med 2016;176(6):816–25.

[24] McTiernan A. Mechanisms linking physical activity with cancer. Nat Rev Cancer 2008;8(3):205–11.

[25] Cormie P, Zopf EM, Zhang X, Schmitz KH. The impact of exercise on cancer mortality, recurrence, and treatment-related adverse effects. Epidemiol Rev 2017;39(1):71–92.

[26] Kerr J, Anderson C, Lippman SM. Physical activity, sedentary behaviour, diet, and cancer: an update and emerging new evidence. Lancet Oncol 2017;18(8):e457–71.

[27] Saslow D, Boetes C, Burke W, Harms S, Leach MO, Lehman CD, et al. American Cancer Society guidelines for breast screening with MRI as an adjunct to mammography. CA Cancer J Clin 2007;57(2):75–89.

[28] Patel AR, Wedzicha JA, Hurst JR. Reduced lung-cancer mortality with CT screening. N Engl J Med 2011;365(21):2035. author reply 7-8.

[29] Schroder FH, Hugosson J, Roobol MJ, Tammela TL, Zappa M, Nelen V, et al. Screening and prostate cancer mortality: results of the European randomised study of screening for prostate cancer (ERSPC) at 13 years of follow-up. Lancet 2014;384(9959):2027–35.

[30] Fisher B, Costantino JP, Wickerham DL, Cecchini RS, Cronin WM, Robidoux A, et al. Tamoxifen for the prevention of breast cancer: current status of the national surgical adjuvant breast and bowel project P-1 study. J Natl Cancer Inst 2005;97(22):1652–62.

[31] Cust AE, Armstrong BK, Goumas C, Jenkins MA, Schmid H, Hopper JL, et al. Sunbed use during adolescence and early adulthood is associated with increased risk of early-onset melanoma. Int J Cancer 2011;128(10):2425–35.

Lifestyle factors and women's health

Alexandra J. Kermack

School of Human Development and Health, Faculty of Medicine, University of Southampton,
Southampton, United Kingdom

Introduction

Diet, exercise and behaviours such as smoking and alcohol consumption all affect women's gynaecological health throughout their life, from menarche to postmenopause. Lifestyle factors also influence conception and pregnancy and have consequences for offspring.

It is essential that women are provided with evidence-based knowledge and advice. This will empower them to make lifestyle changes which could improve their health and wellbeing, manage and improve any adverse symptoms they may experience and give them the optimum chance of conceiving, maintaining healthy a pregnancy and having healthy offspring.

It is important to recognise that many gynaecological conditions should be managed by medical professionals and that although lifestyle factors may modify the course of a disease, this should be regarded as complementary to traditional medical management. Furthermore, lifestyle modifications undertaken while trying to conceive and during pregnancy should be done with the help and advice of midwifery and obstetric teams.

Lifestyle factors and menstruation, dysmenorrhoea and menorrhagia
Menarche

As early as 3 years of age, the diet and exercise levels of females play a role in their future gynaecological, reproductive and general health, by having an effect on the age of menarche [1]. There is a large body of evidence linking early menarche to an increased lifetime risk of breast cancer, possibly due to the prolonged exposure to oestrogens [2]. There is also controversial evidence that early menarche may increase a woman's risk of developing ovarian cancer [3]. However, early menarche has some protective effects on bone mineral density and therefore lowers the risk of osteoporosis [4]. It is reported that girls who have a high fat intake are more likely to have accelerated menarche [5], whereas those with a Mediterranean-style diet [6] and higher levels of physical activity [5] are more likely to have a delay in the onset of their periods. There are concerning statistics regarding poor nutritional intake and declining physical activity levels, particularly in girls, in the United States [7] and in Europe [8,9], both of which may impact the age of menarche. It is reported that there are high levels of snacking and fast food consumption, with less than a quarter of individuals consuming the recommended amounts of fruit and vegetables.

Optimising the health of young and adolescent girls not only provides an opportunity to improve their health but also that of their future offspring. In the United Kingdom (UK), national policymakers and health service providers have been urged to implement policies to work towards this goal [10]. Strategies include school wellness programmes [11], active educational video games [12,13] and reducing sedentary behaviours at home, especially screen-based leisure activities [14]. All of these may have a role to play in reducing rates of childhood obesity and accelerated menarche.

Dysmenorrhoea

Dysmenorrhoea and menorrhagia account for approximately 12% of all gynaecology consultations, and 1 in 20 women have visited their family physician/general practitioner (GP) about these symptoms. A recent systematic review reported that exercise was more effective than simple analgesics in reducing pain intensity in primary dysmenorrhoea, with low intensity exercises such as yoga, pilates and stretching showing the greatest and most consistent benefits. Improvements were also seen with the utilisation of heat and to a lesser extent acupuncture [15].

Endometriosis is a well-recognised cause of dysmenorrhoea and is estimated to affect approximately 10% of women of reproductive age worldwide [16,17]. The aetiology is unknown, and it is difficult to diagnose without direct visualisation of the pelvis at laparoscopy. Research has demonstrated that diet is possibly a modifiable risk factor for the development of endometriosis, with intake of thiamine, folate, vitamin C and vitamin E from food sources inversely related to endometriosis risk [18]. A systematic review also reported a potential protective effect of a diet high in vegetables and omega 3 fatty acids, with low levels of red meat, trans-fats and coffee [19]. However further large population-based studies are required to replicate and evaluate these findings.

Women with endometriosis frequently suffer with severe pelvic pain, dyspareunia, dyschezia and dysuria. These symptoms can have a profound effect on quality of life and well-being. Endometriosis also has a significant adverse impact on the mental health of sufferers, with one study reporting depressive symptoms in 86.5% and anxiety in 87.5% of patients [20]. It has been identified that women are increasingly modifying their lifestyle to try and manage symptoms [21] as they are frustrated with a perceived lack of effectiveness of medical treatments [22]. Changes in diet, including increasing vegetable intake or avoiding gluten, are the most common modifications, and some women opt to reduce levels of work-related stress or increase physical exercise, although evidence for these is limited [22]. In terms of the impact that diet may have, studies are currently under way to evaluate the effectiveness of polyunsaturated omega 3 fatty acids in reducing pain [23].

Menorrhagia

Menorrhagia is a common symptom, with several possible aetiologies, including hormonal factors, such as in polycystic ovarian syndrome (PCOS), mechanical factors such as leiomyomas and endometrial polyps, or related to abnormal clotting. Some of these may be modified by lifestyle changes, for example diets low in carbohydrates have been shown to improve insulin resistance in women with PCOS [24], which may, in turn, lead to weight loss and an improvement in symptoms such as menorrhagia and hirsutism [25]. The evidence for the benefit of exercise in women with PCOS is limited, but it has been shown that aerobic training reduces body mass index (BMI) and may improve reproductive function, ovulation and menorrhagia [26]. A 2014 prospective cohort study examining the medical, surgical and complementary management of leiomyomas demonstrated significant symptom improvement and minimal side effects with diet, exercise, 'herbs' and acupuncture. However, the details of these interventions were not stated, and improvement in pain-related symptoms was qualitative and self-reported [27]. It has been hypothesised that certain vitamins and minerals reduce endometrial prostaglandin production [28] and therefore myometrial contractions [27], which may explain how dietary modifications mediate their beneficial effects.

Lifestyle factors, fertility and early pregnancy
Fertility

There is a growing body of evidence, including large prospective cohort studies, which has demonstrated the impact of female and male preconceptional nutritional status and lifestyle on fertility [29]. Subfertility is defined by the International Committee for Monitoring Assisted Reproductive Technology (ICMART) and the World Health Organization (WHO) as:

> *a disease of the reproductive system defined by the failure to achieve a clinical pregnancy after 12 months or more of regular unprotected intercourse* [30]

Most women who plan their pregnancy report at least one lifestyle change in preparation, the most common being adding nutritional supplements such as folic acid or a multivitamin to their diet [31]. Essential advice for women hoping to conceive is to ensure they are supplementing their diet with folic acid to minimise the risk of neural tube defects (NTDs) [32] and to ensure that those with a raised body mass index (BMI) are taking vitamin D supplements [33].

Obesity is a serious health concern that plays an important role in reproductive disorders. It is associated with anovulation, menstrual disorders such as menorrhagia and infertility. A prolonged time to conceive has been described in women with a BMI greater than 25 [34] and conversely less than 19 [35]. Physical activity is a simple method for improving fertility in women trying to conceive: Higher levels of physical activity are associated with reduced fertility problems. However, this protective effect is only observed in women with a normal BMI [36]. In extreme circumstances, excessive physical activity may have detrimental effects, potentially leading to the 'triad of a female athlete' described by the American College of Sports Medicine: amenorrhoea, osteoporosis and eating disorders [37].

Research has shown that what a woman eats affects the amino acid composition of her endometrial secretions and hence the nutritional environment within the reproductive tract during preimplantation embryo development [38]. A Mediterranean diet, high in vegetable oils, fish, vegetables and legumes and low in carbohydrate-rich snacks, is positively associated with red blood cell folate levels and vitamin B6 in blood and follicular fluid and with a 40% increase in the probability of pregnancy compared with those with a less healthy diet [39]. A 'fertility' diet with high levels of monounsaturated fats compared with trans-fats, vegetable proteins, low glycemic-index carbohydrates, high-fat dairy, multivitamins and iron from plants and supplements is associated with a decreased risk of ovulatory disorder fertility [40]. In addition, a diet supplemented with omega-3 fatty acids has been shown to improve some markers of embryo quality, such as the time the embryo takes to develop from the five-cell to the nine-cell stage, and the synchrony of this growth. There is evidence that this increases the likelihood of blastocyst formation and therefore implantation, but whether this leads to higher live birth rates requires further investigation [41].

Other lifestyle factors that have been investigated to determine their effect on fertility include smoking, alcohol and caffeine intake. Smoking has been shown to alter the morphology of oocytes and to increase the risk of cytoplasmic anomalies, which decrease the chance of pregnancy. In the United Kingdom, women undergoing fertility treatment need to be nonsmokers for the 6 months preceding in vitro *fertilisation* (IVF) and women who are attempting to conceive naturally should be counselled and offered smoking cessation support [42]. Studies examining alcohol intake and fertility have had inconsistent results, with a number of studies demonstrating no correlation [43]. However, a prospective cohort study identified a dose−response inverse relationship between alcohol intake and fecundability and suggested that even moderate alcohol intake had a significant and adverse effect on fertility [44]. Caffeine intake has also been extensively studied. There are numerous reports that caffeine impairs fertility, but a similar number describes no harm [45]. Randomised controlled trials involving alcohol and/or caffeine intake would not be ethical, the evidence from retrospective studies is limited due to recall bias and therefore more prospective cohort trials are required. Interestingly, stress levels may also be involved in fertility levels. Research has demonstrated a relationship between preconception perceived stress levels and a higher risk of anovulation and time to pregnancy [46].

Miscarriage and pregnancy complications

Miscarriage is the most common early pregnancy complication, affecting between 15% and 20% of all pregnancies. Several studies have investigated the relationship between lifestyle factors and miscarriages. Nonmodifiable risk factors include increased maternal age [47] and pregnancy history [48], but increased miscarriage rates are also observed in women with a BMI greater than 25 compared with those with a BMI within the healthy range (20−25) [49]. Remarkably, overweight and obese women with previous pregnancy loss have significantly improved fecundability if they walk for more than 10 min at a time at least once a day [50]. Smoking [51], alcohol and caffeine intake [52] have also been linked to an increased risk of miscarriage, and therefore, preconception counselling should involve discussions around these. Smoking increases miscarriage rates by over double the background rate and exposure to tobacco smoke increases the risk by 1.5 times [52]. Alcohol intake is one of the leading causes of neurodevelopmental deficits in children in the United States (US), and current advice is to consume no alcohol while pregnant. The evidence for caffeine is less clear: caffeine has been found to be significantly associated with fetal loss when more than 300 mg/day is consumed [53].

Recurrent miscarriage, a distressing reproductive outcome, is defined as three or more consecutive miscarriages. It occurs in approximately 1% of the population. Despite thorough investigation, examining for chromosomal, endocrine or uterine abnormalities, half of cases remain unexplained. The study of Lashen et al. showed that obese women, with a BMI>30, had a significantly higher risk of recurrent miscarriage compared with those with a BMI within healthy limits [53]. There has been minimal assessment of the relationship between diet and recurrent miscarriage, although one study hypothesised that a gluten-free diet may improve reproductive outcomes in women with undiagnosed coeliac disease [54]. Studies looking at the effect of tobacco smoking on risk of recurrent miscarriage have been inconsistent, although several studies [55−57] have reported that smoking significantly increased the risk, with smokers having more than a twofold increased risk of pregnancy loss compared with nonsmokers [56]. The effect of alcohol and caffeine is debatable, but as reduction of these may improve outcomes, it may still be advisable that this is suggested to patients.

A further early pregnancy complication is ectopic pregnancy. Lifestyle risk factors which increase the risk of this include smoking [58] and oral contraceptive use prior to the age of 16. The relationship between *Chlamydia trachomatis* infection and the occurrence of ectopic pregnancies is well documented [59] and further promotes the need for good sex education in schools [60] and a national screening programme in people under the age of 25 years [61].

Lifestyle factors, pregnancy and lactation

Prenatal complications often have their origins before conception. In 1983, the American College of Obstetrics and Gynaecology (ACOG) and the American Academy of Paediatrics (AAP) published guidelines for perinatal care, and within these guidelines were recommendations for preconception care to ensure that parents have optimal physical health and are emotionally prepared for parenthood [62]. In the United Kingdom, preconception advice has been specifically encouraged for high-risk groups including those with a BMI of greater than 30, those from low-income families and those with diabetes and other chronic illnesses. Preconception care is considered of particular value in these high-risk groups because waiting until the first antenatal appointment means that major developmental milestones for the fetus have already passed and lifestyle modifications at this point will have less impact on outcomes.

Obesity

Obesity is an increasingly common problem and an important preventable risk factor for adverse pregnancy outcomes. Obesity rates in the United Kingdom doubled between 1989 and 2007 to 15.6% of women of child bearing age [63]. Within Europe, the prevalence of obesity ranges from 7% to 25% of the population [64]. In the Mothers and Babies: Reducing Risk through Audits and Confidential Enquiries across the UK (MBRACE-UK) Confidential Enquiry into Maternal Deaths, 2017, obesity was found to be a risk factor for maternal death, with more than a third of women who died being obese and 19% overweight. Obesity increases the risk of gestational diabetes, hypertensive disorders and thromboembolic complications [65]. In addition to increasing maternal mortality and morbidity, detrimental effects are also observed in the fetus. Women with a BMI greater than 35 have been reported to have a stillbirth rate that is twice that of the general population: 8.6 versus 3.9 per 1000 singleton births respectively, with a relative risk of 16.7 in women with morbid obesity [66]. Furthermore, there is an increased risk of congenital abnormalities and macrosomia and complications during labour, including shoulder dystocia [65]. Some research has suggested that pregnancy is an ideal time to educate and encourage women to participate in healthy lifestyle and weight management strategies [67] as they are in regular contact with healthcare professionals and have the opportunity to make changes that impact on their own and their offspring's long-term health [68]. It has also been demonstrated that if women understand the implications of their periconceptional health and their habits throughout their pregnancy on their offspring, then they are more likely to change their habits [69]. However, some studies have shown that women do not alter their nutritional intake [70] and actually decrease their physical activity during pregnancy [71]. There are a number of barriers to weight loss; the most commonly quoted include a lack of time and/or money [72].

Physical activity

Physical activity has numerous health benefits in pregnant women, including helping them to have greater control over weight gain and reducing the risk of antenatal complications such as gestational diabetes [73], hypertension and related disorders such as preeclampsia [74]. A recent metaanalysis demonstrated an increase in the likelihood of a normal vaginal delivery without obstetric intervention and a reduction in the risk of caesarean section in women who exercised in the second and third trimesters [75]. Despite early concerns that moderate to vigorous exercise might have an adverse impact on offspring [76], recent data have demonstrated no detrimental effects on either the woman or the fetus [77]. In fact, vigorous exercise has been shown to reduce the risk of preterm birth [78] and improve glucose concentrations in women with gestational diabetes [79] and there is no evidence that it reduces birthweight of offspring [77]. However, surveillance of fetal growth is recommended in women who undertake vigorous exercise in pregnancy [80].

The impact of a low body mass index

The risk of being underweight during pregnancy is less clear. Some research has demonstrated that a low BMI is associated with an increased risk of preterm delivery and low birthweight of offspring [81]. However, while other research concurred with the findings that the risk of low birthweight was increased, it also showed that the risk of preeclampsia increased with increasing BMI, that the underweight group was at the lowest risk [82] and that those with a low BMI had a decreased prevalence of gestational diabetes, obstetric intervention and postpartum haemorrhage [83].

Diet and nutrition

As stated previously, the importance of folic acid to minimise NTDs is vital, and the dose should be increased from 400 µg daily to 5 mg in those at increased risk, such as those who have previously had a child with an NTD [84]. There is a change in many of the macro- and micro-nutrient requirements throughout pregnancy. It is important that women are educated about these, as there are many myths about what they should and should not eat. The biggest falsehood is that they should be eating for two: energy requirement actually only increases by 340 and 450 kcal daily in the second and third trimesters, respectively [85].

Many women also take a prenatal multivitamin supplement during pregnancy, and this has been associated with a decreased risk of preterm delivery and very low birth weight infants, along with improvements in maternal folate and ferritin levels [86]. The biggest improvements in health and outcomes were observed in lower-income, urban women, and the UK National Health Service (NHS) now provides multivitamins for these women free of charge as part of its 'healthy start' scheme.

Vitamin D deficiency or insufficiency is common in pregnancy. A Cochrane review concluded that supplementation may reduce the risk of premature delivery, pregnancy-induced hypertension or preeclampsia and reduce the risk of an infant with a low birthweight [87]. Other studies have demonstrated that reduced maternal vitamin D is linked to reduced bone mineral density in the offspring [88]. However, despite the conclusion that the evidence for supplementation of vitamin D during pregnancy remains unclear, and the lack of consensus on the result of high levels, the UK Royal College of Obstetricians and Gynaecologists advises that all women should take 10 µg daily and high-risk women, such as obese or those with reduced exposure to sunlight, should take 2.5 times this amount.

There are also increased iron requirements during pregnancy. Iron is required by the fetus, for the formation of the placenta, for the expansion of maternal red cell mass and to compensate for maternal blood loss during delivery. The recommended iron intake is increased by 9 mg daily during pregnancy, to a total of 27 mg. It may be difficult to obtain this amount from food, and therefore, supplementation may be advised [85]. However, routine use is not endorsed in the United Kingdom because of the potential side effects including abdominal pain and constipation.

Lactation

Lactation requires a significantly higher energy intake than during pregnancy: 500 kcal per day during the first 6 months. However, other than maintaining a healthy diet, knowledge on supplementation during the postpartum period and during lactation is limited. Vitamin D supplementation is recommended for women who avoid milk and other foods fortified with vitamin D, such as cereals, and vitamin B12 is recommended for vegetarians [89].

Lifestyle factors and the menopause

Menopause occurs at a mean age of 51.3 years. There is no evidence that hormone replacement therapy (HRT) is associated with an increased risk of all-cause mortality, including cancer and cardiovascular-related mortality [90]. However, a growing number of women are attempting to manage their menopausal symptoms without its use. This is largely due to previous safety concerns that were been raised regarding increased morbidity resulting from breast cancer [91], ovarian cancer [92] and endometrial cancer, along with cardiovascular disease. The US Preventative Task Force does not recommend the use of HRT for the primary prevention of chronic diseases such as osteoporosis and dementia after the menopause [93] but instead advocates a holistic approach involving a healthy diet and physical activity. 80% of women report being interested in a structured lifestyle program to try to alleviate some or all of their symptoms [94]. The most common and problematic menopausal symptoms are vasomotor — hot flushes and night sweats [95]. Studies have demonstrated that women who consume a vegan diet, or one containing a high intake of vegetables and fruit, report less frequent or severe vasomotor symptoms. It has also been shown that a significant reduction in weight of more than 4.5 kg or 10% of bodyweight has been associated with elimination of symptoms. The trial did not specify whether these women were in the overweight or obese category prior to their weight loss, but it was noted that women with a BMI greater than 40 were more likely to experience symptoms [96]. Other studies have also documented that obese women suffer more with vasomotor and psychological symptoms than women with a normal BMI [97], reinforcing the importance of maintaining a healthy weight during menopause. Another symptom that causes distress in women is the psychological impact of the menopause, which can include symptoms of depression. Interestingly, women who consume a Mediterranean diet may be somewhat protected against this [98,99].

Lifestyle and urogynaecology

Pelvic floor disorders are thought to affect nearly a quarter of all women, with 15% experiencing urinary incontinence, 9% bowel incontinence and 3% pelvic organ prolapse [100]. Women suffering with incontinence and prolapse often feel self-conscious and have concerns about their body image, including feeling less sexually attractive than women who do not have these complaints [101].

The mainstay of treatment for mild urinary stress incontinence and pelvic organ prolapse are pelvic floor exercises. These have been shown to significantly improve quality of life in the majority of patients [102]. Despite 92% of women reporting knowing about these exercises, surprisingly 57% of symptomatic women do not perform them [103]. Evidence has suggested that these exercises are most effective when supervised [104,105] or following proper training [106].

A modifiable risk factor for incontinence [107] and prolapse [108] is obesity. It is thought that obesity mediates harmful effects due to an increase in intraabdominal pressure causing weakening of pelvic floor muscles and fascia [108]. A diet and exercise intervention for urinary incontinence in overweight women demonstrated limited weight loss, with only a mean loss of 3.6 kg. However, there was significant symptomatic improvement [109,110]. In women with prolapse, weight loss is not associated with an anatomical improvement, but research has demonstrated that it improves symptoms and quality of life [108].

A further modifiable risk factor is smoking. Smoking is thought to exacerbate stress incontinence as it causes a chronic cough [111], and laboratory investigations have suggested that vaginal macrophage elastase activation may play a role in the pathogenesis of prolapse in women who smoke [112].

Urgency and urge incontinence are also experienced by many women. The first conservative measure for its management is bladder training and fluid optimisation [113]. There is some suggestion that reducing caffeine intake [114] and reducing the consumption of alcohol and fizzy drinks [115] may improve symptoms; therefore, this should be encouraged as part of a holistic approach to patient care.

Lifestyle factors and gynaecological malignancy

Endometrial cancer is the sixth most common malignancy in women, and there are 320,000 new cases diagnosed worldwide annually [116]. Incidence is higher in the developed world, and this has been attributed to the greater rates of obesity, with 57% of cases in the United States thought to be due to raised BMI [117]. Moreover, as the obesity epidemic continues, it is predicted that rates of endometrial cancer will increase, with one study predicting a 55% rise between 2010 and 2030. This means that the incidence could reach 42.13 cases per 100,000 women per year [118]. The diagnosis and treatment of endometrial cancer has been suggested as a 'teachable moment' to initiate behavioural lifestyle changes, including discussing with patients the importance of a healthy diet and physical activity. A current randomised controlled trial (RCT) is investigating the impact and cost-effectiveness of this [119]. Unfortunately, most women gain weight or maintain the same weight following treatment for endometrial cancer, and most show no changes in their attitude towards weight-related behaviours [120]. Systematic delivery of weight loss interventions should be a priority in all patients as it may reduce morbidity and mortality, both from a risk of cancer recurrence and other diseases associated with obesity.

Another gynaecological cancer in which lifestyle factors play a large role is cervical cancer. It is well established that the most significant risk factor for cervical cancer is infection with high-risk human papillomavirus (HPV) subtypes. HPV is estimated to cause 99.7% of all cases [121]. HPV is transmitted sexually, and therefore, good sex education and increased condom use reduces the risk. Furthermore, the HPV vaccination programme has reduced incidence and mortality by approximately two-thirds [122].

Conclusion

Every interaction with a patient can be regarded as an opportunity to discuss healthy lifestyle choices, including maintaining a healthy weight through diet and exercise, not smoking and limiting alcohol and caffeine intake. All of these factors may influence the reproductive and gynaecological health of females and their offspring, influencing parameters such as the age of menarche, the likelihood of a successful pregnancy, symptoms of the menopause and the risk of certain malignancies.

Summary

- Diet, exercise and behaviours such as smoking and alcohol consumption all impact gynaecological health.
- Diet, exercise and weight affect the age of menarche.
- Exercise is more effective than simple analgesics in reducing pain intensity in primary dysmenorrhoea.
- Diets low in carbohydrates have been shown to improve insulin resistance in women with PCOS, which may lead to weight loss and an improvement in symptoms such as menorrhagia and hirsutism.
- Obesity is a major health concern and plays an important role in reproductive disorders. It is associated with anovulation, menstrual disorders and infertility.
- Physical activity may improve fertility in women with a normal BMI.
- Diet may impact the probability of conceiving, with some evidence that a Mediterranean diet is beneficial.
- Smoking has an adverse impact on fertility. The evidence for alcohol and caffeine consumption is inconsistent.
- A high BMI, smoking, alcohol and caffeine all increase the risk of miscarriage.
- Obesity increases the risk of gestational diabetes, hypertensive disorders and thromboembolic complications during pregnancy, and women with a BMI greater than 35 have an increased risk of stillbirths.
- The most common and problematic menopausal symptoms are hot flushes and night sweats. Symptoms may be improved by maintaining a healthy weight and through ensuring a high intake of fruit and vegetables.
- Obesity is a risk factor for endometrial cancer.

References

[1] Berkey CS, et al. Relation of childhood diet and body size to menarche and adolescent growth in girls. Am J Epidemiol 2000;152(5):446—52.
[2] Coughlin SS. Epidemiology of breast cancer in women. Adv Exp Med Biol 2019;1152:9—29.
[3] La Vecchia C. Ovarian cancer: epidemiology and risk factors. Eur J Cancer Prev 2017;26(1):55—62.
[4] Zhang Q, et al. Age at menarche and osteoporosis: a Mendelian randomization study. Bone 2018;117:91—7.
[5] Merzenich H, Boeing H, Wahrendorf J. Dietary fat and sports activity as determinants for age at menarche. Am J Epidemiol 1993;138(4):217—24.
[6] Szamreta EA, et al. Greater adherence to a Mediterranean-like diet is associated with later breast development and menarche in peripubertal girls. Publ Health Nutr 2019:1—11.
[7] Zapata LB, et al. Dietary and physical activity behaviors of middle school youth: the youth physical activity and nutrition survey. J Sch Health 2008;78(1):9—18. quiz 65—7.
[8] Ruiz JR, et al. Objectively measured physical activity and sedentary time in European adolescents: the HELENA study. Am J Epidemiol 2011;174(2):173—84.
[9] Moreno LA, et al. Trends of dietary habits in adolescents. Crit Rev Food Sci Nutr 2010;50(2):106—12.

[10] Branca F, et al. Nutrition and health in women, children, and adolescent girls. BMJ 2015;351:h4173.

[11] Heo M, et al. Behaviors and knowledge of HealthCorps New York city high school students: nutrition, mental health, and physical activity. J Sch Health 2016;86(2):84—95.

[12] Sun H, Gao Y. Impact of an active educational video game on children's motivation, science knowledge, and physical activity. J Sport Health Sci 2016;5(2):239—45.

[13] Lau PW, Wang JJ, Maddison R. A randomized-controlled trial of school-based active videogame intervention on Chinese children's aerobic fitness, physical activity level, and psychological correlates. Game Health J 2016;5(6):405—12.

[14] Arundell L, et al. Informing behaviour change: what sedentary behaviours do families perform at home and how can they be targeted? Int J Environ Res Publ Health 2019;16(22).

[15] Armour M, et al. The effectiveness of self-care and lifestyle interventions in primary dysmenorrhea: a systematic review and meta-analysis. BMC Compl Altern Med 2019;19(1):22.

[16] Buck Louis GM, et al. Incidence of endometriosis by study population and diagnostic method: the ENDO study. Fertil Steril 2011;96(2):360—5.

[17] Rogers PA, et al. Priorities for endometriosis research: recommendations from an international consensus workshop. Reprod Sci 2009;16(4):335—46.

[18] Darling AM, et al. A prospective cohort study of vitamins B, C, E, and multivitamin intake and endometriosis. J Endometr 2013;5(1):17—26.

[19] Parazzini F, et al. Diet and endometriosis risk: a literature review. Reprod Biomed Online 2013;26(4): 323—36.

[20] Sepulcri RdP, Amaral VFd. Depressive symptoms, anxiety, and quality of life in women with pelvic endometriosis. Eur J Obstet Gynecol Reprod Biol 2009;142(1):53—6.

[21] Vennberg Karlsson J, Patel H, Premberg A. Experiences of health after dietary changes in endometriosis: a qualitative interview study. BMJ Open 2020;10(2). e032321.

[22] Culley L, et al. The social and psychological impact of endometriosis on women's lives: a critical narrative review. Hum Reprod Update 2013;19(6):625—39.

[23] Abokhrais IM, et al. A two-arm parallel double-blind randomised controlled pilot trial of the efficacy of Omega-3 polyunsaturated fatty acids for the treatment of women with endometriosis-associated pain (PurFECT1). PLoS One 2020;15(1). e0227695.

[24] Porchia LM, et al. Diets with lower carbohydrate concentrations improve insulin sensitivity in women with polycystic ovary syndrome: a meta-analysis. Eur J Obstet Gynecol Reprod Biol 2020;248:110—7.

[25] Paoli A, et al. Effects of a ketogenic diet in overweight women with polycystic ovary syndrome. J Transl Med 2020;18(1):104.

[26] Dos Santos IK, et al. The effect of exercise as an intervention for women with polycystic ovary syndrome: a systematic review and meta-analysis. Medicine (Baltimore) 2020;99(16). e19644.

[27] Jacoby VL, et al. Use of medical, surgical and complementary treatments among women with fibroids. Eur J Obstet Gynecol Reprod Biol 2014;182:220—5.

[28] Lloyd KB, Hornsby LB. Complementary and alternative medications for women's health issues. Nutr Clin Pract 2009;24(5):589—608.

[29] Inskip HM, et al. Cohort profile: the southampton women's survey. Int J Epidemiol 2006;35(1):42—8.

[30] Zegers-Hochschild F, et al. The international committee for monitoring assisted reproductive technology (ICMART) and the world health organization (WHO) revised glossary on ART terminology. Hum Reprod 2009;24(11):2683—7. 2009.

[31] Goossens J, et al. Preconception lifestyle changes in women with planned pregnancies. Midwifery 2018;56: 112—20.

[32] Imbard A, Benoist J-F, Blom HJ. Neural tube defects, folic acid and methylation. Int J Environ Res Publ Health 2013;10(9):4352—89.

[33] Wagner CL, et al. Vitamin D and its role during pregnancy in attaining optimal health of mother and fetus. Nutrients 2012;4(3):208−30.

[34] Talmor A, Dunphy B. Female obesity and infertility. Best Pract Res Clin Obstet Gynaecol 2015;29(4): 498−506.

[35] Boutari C, et al. The effect of underweight on female and male reproduction. Metabolism 2020;107: 154229.

[36] Mena GP, Mielke GI, Brown WJ. Do physical activity, sitting time and body mass index affect fertility over a 15-year period in women? Data from a large population-based cohort study. Hum Reprod 2020;35(3): 676−83.

[37] Orio F, et al. Effects of physical exercise on the female reproductive system. Minerva Endocrinol 2013; 38(3):305−19.

[38] Kermack AJ, et al. Amino acid composition of human uterine fluid: association with age, lifestyle and gynaecological pathology. Hum Reprod 2015;30(4):917−24.

[39] Vujkovic M, et al. The preconception Mediterranean dietary pattern in couples undergoing in vitro fertilization/intracytoplasmic sperm injection treatment increases the chance of pregnancy. Fertil Steril 2010; 94(6):2096−101.

[40] Chavarro JE, et al. Diet and lifestyle in the prevention of ovulatory disorder infertility. Obstet Gynecol 2007;110(5):1050−8.

[41] Kermack AJ, et al. Effect of a 6-week "Mediterranean" dietary intervention on in vitro human embryo development: the Preconception Dietary Supplements in Assisted Reproduction double-blinded randomized controlled trial. Fertil Steril 2020;113(2):260−9.

[42] Ozbakir B, Tulay P. Does cigarette smoking really have a clinical effect on folliculogenesis and oocyte maturation? Zygote 2020:1−4.

[43] de Angelis C, et al. Smoke, alcohol and drug addiction and female fertility. Reprod Biol Endocrinol 2020; 18(1). 21−21.

[44] Jensen TK, et al. Does moderate alcohol consumption affect fertility? Follow up study among couples planning first pregnancy. Bmj 1998;317(7157):505−10.

[45] Gaskins AJ, Chavarro JE. Diet and fertility: a review. Am J Obstet Gynecol 2018;218(4):379−89.

[46] Schliep KC, et al. Preconception perceived stress is associated with reproductive hormone levels and longer time to pregnancy. Epidemiology 2019;30(Suppl. 2):S76−s84.

[47] Magnus MC, et al. Role of maternal age and pregnancy history in risk of miscarriage: prospective register based study. BMJ 2019;364:l869.

[48] Maconochie N, et al. Risk factors for first trimester miscarriage—results from a UK-population-based case−control study. BJOG 2007;114(2):170−86.

[49] Metwally M, et al. Does high body mass index increase the risk of miscarriage after spontaneous and assisted conception? A meta-analysis of the evidence. Fertil Steril 2008;90(3):714−26.

[50] Russo LM, et al. A prospective study of physical activity and fecundability in women with a history of pregnancy loss. Hum Reprod 2018;33(7):1291−8.

[51] Farioli A, et al. Smoking and miscarriage risk. Epidemiology 2010;21(6):918−9.

[52] Lassi ZS, et al. Preconception care: caffeine, smoking, alcohol, drugs and other environmental chemical/radiation exposure. Reprod Health 2014;11(Suppl. 3). S6−S6.

[53] Lashen H, Fear K, Sturdee DW. Obesity is associated with increased risk of first trimester and recurrent miscarriage: matched case-control study. Hum Reprod 2004;19(7):1644−6.

[54] Alecsandru D, et al. Exploring undiagnosed celiac disease in women with recurrent reproductive failure: the gluten-free diet could improve reproductive outcomes. Am J Reprod Immunol 2020;83(2). e13209.

[55] Mishra GD, Dobson AJ, Schofield MJ. Cigarette smoking, menstrual symptoms and miscarriage among young women. Aust N Z J Public Health 2000;24(4):413−20.

[56] George L, et al. Risks of repeated miscarriage. Paediatr Perinat Epidemiol 2006;20(2):119—26.

[57] Stefanidou EM, et al. Maternal caffeine consumption and sine causa recurrent miscarriage. Eur J Obstet Gynecol Reprod Biol 2011;158(2):220—4.

[58] Saraiya M, et al. Cigarette smoking as a risk factor for ectopic pregnancy. Am J Obstet Gynecol 1998; 178(3):493—8.

[59] Xia Q, et al. Relation of *Chlamydia trachomatis* infections to ectopic pregnancy: a meta-analysis and systematic review. Medicine 2020;99(1). e18489—e18489.

[60] Lengen C, Jäger S, Kistemann T. The knowledge, education and behaviour of young people with regard to *Chlamydia trachomatis* in Aarhus, Denmark and Bonn, Germany: do prevention concepts matter? Soc Sci Med 2010;70(11):1789—98.

[61] LaMontagne DS, et al. Establishing the national chlamydia screening programme in England: results from the first full year of screening. Sex Transm Infect 2004;80(5):335—41.

[62] Freda MC, Moos M-K, Curtis M. The history of preconception care: evolving guidelines and standards. Matern Child Health J 2006;10(5 Suppl.l):S43—52.

[63] Heslehurst N, et al. A nationally representative study of maternal obesity in England, UK: trends in incidence and demographic inequalities in 619 323 births, 1989—2007. Int J Obes 2010;34(3):420—8.

[64] Devlieger R, et al. Maternal obesity in Europe: where do we stand and how to move forward?: a scientific paper commissioned by the European Board and College of Obstetrics and Gynaecology (EBCOG). Eur J Obstet Gynecol Reprod Biol 2016;201:203—8.

[65] Heslehurst N, et al. Obesity in pregnancy: a study of the impact of maternal obesity on NHS maternity services. BJOG Int J Obstet Gynaecol 2007;114(3):334—42.

[66] Mantakas A, Farrell T. The influence of increasing BMI in nulliparous women on pregnancy outcome. Eur J Obstet Gynecol Reprod Biol 2010;153(1):43—6.

[67] Sattar N, Greer IA. Pregnancy complications and maternal cardiovascular risk: opportunities for intervention and screening? BMJ 2002;325(7356):157—60.

[68] Baird J, et al. Developmental origins of health and disease: a lifecourse approach to the prevention of noncommunicable diseases. Healthcare (Basel) 2017;5(1).

[69] McKerracher L, et al. Knowledge about the developmental origins of health and disease is independently associated with variation in diet quality during pregnancy. Matern Child Nutr 2020;16(2):e12891.

[70] Crozier SR, et al. Women's dietary patterns change little from before to during pregnancy. J Nutr 2009; 139(10):1956—63.

[71] Haakstad LA, et al. Why do pregnant women stop exercising in the third trimester? Acta Obstet Gynecol Scand 2009;88(11):1267—75.

[72] Ciao AC, Latner JD, Durso LE. Treatment seeking and barriers to weight loss treatments of different intensity levels among obese and overweight individuals. Eat Weight Disord 2012;17(1):e9—16.

[73] Colberg SR, Castorino K, Jovanovič L. Prescribing physical activity to prevent and manage gestational diabetes. World J Diabetes 2013;4(6):256—62.

[74] Aune D, et al. Physical activity and the risk of preeclampsia: a systematic review and meta-analysis. Epidemiology 2014;25(3):331—43.

[75] Poyatos-Leon R, et al. Effects of exercise during pregnancy on mode of delivery: a meta-analysis. Acta Obstet Gynecol Scand 2015;94(10):1039—47.

[76] Bell R. The effects of vigorous exercise during pregnancy on birth weight. J Sci Med Sport 2002;5(1):32—6.

[77] Petrov Fieril K, Glantz A, Fagevik Olsen M. The efficacy of moderate-to-vigorous resistance exercise during pregnancy: a randomized controlled trial. Acta Obstet Gynecol Scand 2015;94(1):35—42.

[78] Jukic AMZ, et al. A prospective study of the association between vigorous physical activity during pregnancy and length of gestation and birthweight. Matern Child Health J 2012;16(5):1031—44.

[79] Ruchat S-M, et al. Effect of exercise intensity and duration on capillary glucose responses in pregnant women at low and high risk for gestational diabetes. Diabetes Metabol Res Rev 2012;28(8):669–78.

[80] Penney DS. The effect of vigorous exercise during pregnancy. J Midwifery Women's Health 2008;53(2):155–9.

[81] Omanwa K, et al. [Is low pre-pregnancy body mass index a risk factor for preterm birth and low neonatal birth weight? Ginekol Pol 2006;77(8):618–23.

[82] Bhattacharya S, et al. Effect of body mass index on pregnancy outcomes in nulliparous women delivering singleton babies. BMC Public Health 2007;7(1):168.

[83] Sebire N, et al. Is maternal underweight really a risk factor for adverse pregnancy outcome? A population-based study in London. BJOG 2001;108(1):61–6.

[84] Rasmussen LB, et al. Folate and neural tube defects. Recommendations from a Danish working group. Dan Med Bull 1998;45(2):213–7.

[85] Picciano MF. Pregnancy and lactation: physiological adjustments, nutritional requirements and the role of dietary supplements. J Nutr 2003;133(6):1997S–2002S.

[86] Scholl TO, et al. Use of multivitamin/mineral prenatal supplements: influence on the outcome of pregnancy. Am J Epidemiol 1997;146(2):134–41.

[87] De-Regil LM, et al. Vitamin D supplementation for women during pregnancy. Cochrane Database Syst Rev 2016;(1).

[88] Javaid MK, et al. Maternal vitamin D status during pregnancy and childhood bone mass at age 9 years: a longitudinal study. Lancet 2006;367(9504):36–43.

[89] Jensen RG. Handbook of milk composition. Academic Press; 1995.

[90] Manson JE, et al. Menopausal hormone therapy and long-term all-cause and cause-specific mortality: the women's health initiative randomized trials. Jama 2017;318(10):927–38.

[91] Type and timing of menopausal hormone therapy and breast cancer risk: individual participant meta-analysis of the worldwide epidemiological evidence. Lancet 2019;394(10204):1159–68.

[92] Liu Y, et al. Menopausal hormone replacement therapy and the risk of ovarian cancer: a meta-analysis. Front Endocrinol 2019;10:801.

[93] Grossman DC, et al. Hormone therapy for the primary prevention of chronic conditions in postmenopausal women: US preventive services task force recommendation statement. Jama 2017;318(22):2224–33.

[94] Marlatt KL, Beyl RA, Redman LM. A qualitative assessment of health behaviors and experiences during menopause: a cross-sectional, observational study. Maturitas 2018;116:36–42.

[95] Gold EB, et al. Longitudinal analysis of the association between vasomotor symptoms and race/ethnicity across the menopausal transition: study of women's health across the nation. Am J Publ Health 2006;96(7):1226–35.

[96] Kroenke CH, et al. Effects of a dietary intervention and weight change on vasomotor symptoms in the Women's Health Initiative. Menopause 2012;19(9):980–8.

[97] Moilanen J, et al. Prevalence of menopause symptoms and their association with lifestyle among Finnish middle-aged women. Maturitas 2010;67(4):368–74.

[98] Lucas M, et al. Inflammatory dietary pattern and risk of depression among women. Brain Behav Immun 2014;36:46–53.

[99] Lee Y, Kim H. Relationships between menopausal symptoms, depression, and exercise in middle-aged women: a cross-sectional survey. Int J Nurs Stud 2008;45(12):1816–22.

[100] Nygaard I, et al. Prevalence of symptomatic pelvic floor disorders in US women. Jama 2008;300(11):1311–6.

[101] Jelovsek JE, Barber MD. Women seeking treatment for advanced pelvic organ prolapse have decreased body image and quality of life. Am J Obstet Gynecol 2006;194(5):1455–61.

[102] Radziminska A, et al. The impact of pelvic floor muscle training on the quality of life of women with urinary incontinence: a systematic literature review. Clin Interv Aging 2018;13:957—65.

[103] Goodridge SD, et al. Association of knowledge and presence of pelvic floor disorders and participation in pelvic floor exercises: a cross-sectional study. Female Pelvic Med Reconstr Surg 2020. https://doi.org/10.1097/SPV.0000000000000813. Online ahead of print.

[104] Ferreira M, Santos P. Impact of exercise programs in woman's quality of life with stress urinary incontinence. Rev Port Saude Publica 2012;30(1):3—10.

[105] Kilpatrick KA, et al. Non-pharmacological, non-surgical interventions for urinary incontinence in older persons: a systematic review of systematic reviews. The SENATOR project ONTOP series. Maturitas 2020; 133:42—8.

[106] Felicissimo MF, et al. Intensive supervised versus unsupervised pelvic floor muscle training for the treatment of stress urinary incontinence: a randomized comparative trial. Int Urogynecol J 2010;21(7): 835—40.

[107] Lamerton TJ, Torquati L, Brown WJ. Overweight and obesity as major, modifiable risk factors for urinary incontinence in young to mid-aged women: a systematic review and meta-analysis. Obes Rev 2018;19(12): 1735—45.

[108] Lee UJ, et al. Obesity and pelvic organ prolapse. Curr Opin Urol 2017;27(5):428—34.

[109] Fjerbaek A, et al. Treatment of urinary incontinence in overweight women by a multidisciplinary lifestyle intervention. Arch Gynecol Obstet 2020;301(2):525—32.

[110] Imamura M, et al. Lifestyle interventions for the treatment of urinary incontinence in adults. Cochrane Database Syst Rev 2015;(12).

[111] Hu JS, Pierre EF. Urinary incontinence in women: evaluation and management. Am Fam Physician 2019; 100(6):339—48.

[112] Wieslander CK, et al. Paper 34: smoking is a risk factor for pelvic organ prolapse. Female Pelvic Med Reconstr Surg 2005;11:S16—7.

[113] Myers DL. Female mixed urinary incontinence: a clinical review. Jama 2014;311(19):2007—14.

[114] Le Berre M, et al. What do we really know about the role of caffeine on urinary tract symptoms? A scoping review on caffeine consumption and lower urinary tract symptoms in adults. Neurourol Urodyn 2020;39(5): 1217—33.

[115] Stewart E. Overactive bladder syndrome in the older woman: conservative treatment. Br J Community Nurs 2009;14(11):466—73.

[116] Ferlay J, et al. Cancer incidence and mortality worldwide: sources, methods and major patterns in GLO-BOCAN 2012. Int J Cancer 2015;136(5):E359—86.

[117] Renehan AG, et al. Body-mass index and incidence of cancer: a systematic review and meta-analysis of prospective observational studies. Lancet 2008;371(9612):569—78.

[118] Sheikh MA, et al. USA endometrial cancer projections to 2030: should we be concerned? Future Oncol 2014;10(16):2561—8.

[119] Koutoukidis DA, et al. Diet and exercise in uterine cancer survivors (DEUS pilot)—piloting a healthy eating and physical activity program: study protocol for a randomized controlled trial. Trials 2016;17(1). 130—130.

[120] Harrison R, et al. Body mass index and attitudes towards health behaviors among women with endometrial cancer before and after treatment. Int J Gynecol Cancer 2020;30(2):187—92.

[121] Shepherd J, et al. Cervical cancer and sexual lifestyle: a systematic review of health education interventions targeted at women. Health Educ Res 2000;15(6):681—94.

[122] Lowy DR, Schiller JT. Reducing HPV-associated cancer globally. Cancer Prev Res 2012;5(1):18—23.

Skin health: what damages and ages skin? Evidence-based interventions to maintain healthy skin

20

Lorna Jeng[1] and Anjaly Mirchandani[2]

[1]*Radiologist, Whittington Hospital, London, United Kingdom;* [2]*General Practitioner, Northfields Surgery, London, United Kingdom*

Introduction

Skin is the largest organ in the body and is the main defence against the environment. Skin is a fundamental component of the innate immune system and acts as a physical barrier. Skin has many other important roles including controlling water loss from the body, preventing infection, thermoregulation, the sense of touch and the production of vitamin D.

Skin, like all organs, will age and is prone to certain diseases, including malignancies. Skin cancers are broadly divided into melanoma and non-melanoma cancers, with the latter group comprising basal cell carcinoma, squamous cell carcinoma and the rarer adnexal tumours. The incidence of skin cancer is increasing annually, but it is reported that up to 86% of melanomas are preventable [1].

This chapter will review basic skin anatomy and physiology and shall explore the pathophysiology of aging and tumourigenesis. Lifestyle factors which can be modified to reduce the incidence of skin cancers shall be discussed.

Anatomy and physiology of the skin

Skin is composed of three main layers, with each layer having its own important physiological functions (Fig. 20.1). The epidermis acts as a physical barrier, regulates water loss, produces keratin which makes the skin waterproof and the melanocytes present melanin synthesis, which helps to protect from the effects of UV radiation and gives skin its colour. The dermis mainly comprises connective tissue, which includes collagen to provide the skin with its tensile strength, and also elastic fibres, which aid with stretch and recoil. Also contained within the dermis are the skin appendages — the sebaceous glands, sweat glands and hair follicles, along with cutaneous nerves, blood vessels and lymphatics. The subcutaneous fat helps to anchor the skin to the underlying fascia and also aids insulating the body.

A Prescription for Healthy Living. https://doi.org/10.1016/B978-0-12-821573-9.00020-5

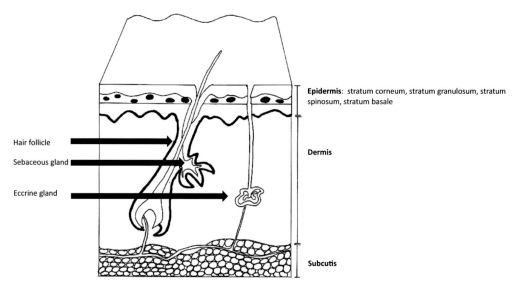

Epidermis: stratum corneum, stratum granulosum, stratum spinosum, stratum basale

Hair follicle

Sebaceous gland

Eccrine gland

Dermis

Subcutis

FIGURE 20.1

Structure of skin.

The specific cell types in each layer are held together by a complex matrix of molecules, the extracellular matrix (ECM), that act as scaffolding to support the cells, give them structure and hold in moisture. Important molecules in the ECM include collagen, elastin and oligosaccharides, which help retain water to provide hydration and 'plumpness' to the skin [2].

Intrinsic and extrinsic factors associated with ageing

Biological ageing refers to the progressive degeneration of tissues over time, due to molecular and cellular damage. Skin ageing is the result of intrinsic and extrinsic factors.

Intrinsic factors are the physiological aging processes which are largely determined by an individual's genetic make-up [3]. Intrinsic aging tends to result in skin thinning, fine wrinkles and dry skin, and occurs due to atrophy, decreased elasticity, reduced metabolic activity and reduced cell turnover [4]. As skin ages its protective barrier function is impaired and it becomes more fragile and prone to mechanical damage.

Extrinsic aging is an acceleration of normal intrinsic aging by environmental factors such as ultraviolet (UV) radiation, pollution, smoking and nutrition. UV radiation makes the greatest contribution, accounting for up to 80% of aging [5].

The intracellular and tissue-level pathways involved in intrinsic and extrinsic ageing are described:

Telomere shortening

The ends of chromosomes are capped with extra nucleotides, telomeres, which function to protect chromosomes from degradation during cell division. Cell replication inevitably results in telomere shortening. When the telomere becomes too short, the cell stops dividing. UV radiation can damage telomeres, resulting in premature cell senescence. This is one mechanism through which the

UV-mediated aging process results in changes to the appearance of the skin: with time, cell turnover reduces so skin becomes thinner.

Oxidative stress

Reactive oxygen species (ROS) are oxygen-containing molecules that are unstable and react with other molecules in a phenomenon described as oxidative stress. ROS are produced as a by-product in many normal cell processes and are required for normal cell functioning. However, pollution, UV radiation and smoking can dramatically increase the levels of ROS within the cell.

ROSs have a range of pathogenic effects, including DNA damage and the triggering of chronic inflammation, which leads to the degradation of the ECM and a reduction in the structural integrity of collagen [5]. Furthermore, the activation of collagenases also contributes to ECM lysis. These effects result in the cosmetic appearance of wrinkles along with deposition of abnormal ECM molecules, classic features of photoaging [3,6].

Hormones

Sex hormones decline over time, particularly in postmenopausal women. This has a major role in the intrinsic aging process of skin. Oestrogen is important in skin hydration and in the production of collagen and other components of the ECM. Reduced levels of oestrogen are therefore associated with dry, thin and wrinkled skin. Oestrogen also has important anti-inflammatory properties, and low levels can result in an exacerbation of any pre-existing inflammatory skin conditions such as acne rosacea [7].

Lifestyle factors to maintain healthy skin

The major lifestyle behaviours to maintain healthy skin are detailed. The role of cosmetic products shall not be discussed.

UV protection

UV damage is by far the greatest contributor to extrinsic aging and the development of skin cancer. Therefore, limiting exposure to UV radiation is one of the best ways to protect the skin. This can be done by reducing time spent in bright sunlight, avoiding exposure to artificial UV radiation, for example sun beds, covering the skin with protective sun wear or applying a protective sunscreen barrier to skin to absorb harmful rays. Studies have shown that the use of sunscreen prevents DNA damage and can protect against malignant melanoma and squamous cell carcinoma [8].

UV light is part of the electromagnetic spectrum emitted by the sun that is not visible to the human eye. UVA has a longer wavelength and results in skin aging, UVB has a shorter wavelength and causes skin burning. Both UVA and UVB radiation can cause DNA damage and increase the risk of skin cancer. Sunscreen can protect against both UVA and UVB radiation. The degree of protection given against UVB radiation is given as the sunscreen's SPF factor and UVA protection is given as the star rating. Sunscreens that include reflective substances such as titanium dioxide provide additional protection [9].

As skin is exposed to UV radiation on a daily basis, even in the colder Winter months, some individuals advocate the use of daily UV protection in face moisturiser or foundation to help slow down signs of aging. A study performed in 2016 reported that using a daily SPF 30 sunscreen throughout the year reduced the speed of the ageing process as well as reversing the signs of ageing [10]. However, other studies have had conflicting results. One potential reason for inconsistent findings is that sunscreen is often only applied once in the morning, therefore the SPF protection declines throughout the day if the sunscreen is not reapplied. Other studies have shown that some make-up products with advertised SPF protection may not be as protective as dedicated SPF sunscreen [11].

Despite the benefits of limiting exposure to UVA and UVB with regards to skin aging and protection against skin cancer, a certain amount of sunlight is essential for vitamin D production. Vitamin D is essential in regulating the amount of calcium and phosphate in the body and is vital for bone metabolism, with vitamin D deficiency causing rickets in children and osteomalacia in adults. In addition, recent evidence suggests that vitamin D is also important in protecting against cardiovascular disease, diabetes, hypertension and some autoimmune conditions [12]. Although some vitamin D is obtained from food, most of the body's vitamin D is obtained from dermal synthesis following skin exposure to UV containing sunlight. Currently around a quarter of the United Kingdom (UK) population and up to 37% of the United States (US) population are vitamin D deficient [12]; therefore, there is a fine balance to be achieved when considering sun protection versus risk of vitamin D deficiency.

Smoking avoidance

As described in Chapter 16 smoking is a significant risk factor for a variety of adverse health outcomes. Smoking also creates high levels of ROS in the skin of the face, particularly around the mouth. This accelerates the signs of aging, particularly wrinkles [13].

Hydration

Skin hydration is regulated by the stratum corneum, the most superficial skin layer. Hydration is maintained by a combination of hygroscopic molecules within the corneocytes attracting water into stratum corneum and intercellular lipid molecules preventing transepidermal loss of water molecules. Inadequate hydration leads to impairment in desquamation and thus dry, flaky skin [14] It is generally understood that humans need a certain daily quantity of water and this is often recommended at 6—8 glasses of water [15]. However, there is limited evidence to support a specific value and whether increasing this has any beneficial effect on the body. A recent literature review addressing whether an increase in dietary fluid affected skin hydration reported that enhanced water consumption was associated with an increase in hydration within the deeper skin layers and a reduction in the appearance of dry skin, especially in those with prior reduced water intake [16].

Sleep

Sleep is important for skin health because it is during sleep that the skin has its greatest turnover. Levels of cortisol are also reduced at night: excess stress hormones can have adverse effects on the skin, increasing its sensitivity, worsening any pre-existing inflammatory conditions such as eczema and causing the breakdown of collagen and other ECM molecules. During sleep skin rebalances its hydration. This has the effect of reducing the appearance of bags under the eyes, reducing dryness, helping to keep skin 'plump' and reducing the appearance of fine wrinkles [17].

Exercise

Exercise has many health benefits, potentially including attenuating skin ageing. It has been reported that endurance exercise may reduce age-associated changes to skin through modulated levels of IL-15 [18].

Exercise also helps to maintain a healthy weight, reducing the risk of developing obesity and diabetes mellitus, two conditions that can affect the integrity and condition of skin.

Antioxidants

Antioxidants can help to buffer the pathogenic effects of ROS and can, therefore, help reduce the extrinsic aging effects caused by oxidative stress.

Antioxidants include vitamins C and E, and certain molecules found in green tea and aloe vera. Eating a balanced, healthy diet, rather than using supplements, is the best way to obtain all the antioxidants necessary [6].

Skin cancer

Non-melanoma skin cancer is the fifth most common cancer worldwide, and melanoma is the 19th most common. Around 147,000 cases of non-melanoma skin cancer are diagnosed in the UK each year [18], although the incidence could be even higher than this due to under reporting. The age-standardised rate for melanoma per 100,000 people is 12.7 in the US and approximately 13,000 cases of melanoma are diagnosed in the UK each year [19]. The pathogenesis and risk factors for skin cancers are multifactorial but UV radiation is the major contributing factor.

Although melanoma accounts for only 4% of skin cancers it is responsible for around 75% of skin cancer deaths, with most cases occurring in Caucasian adults over the age of 30 [20] (Fig. 20.2). The

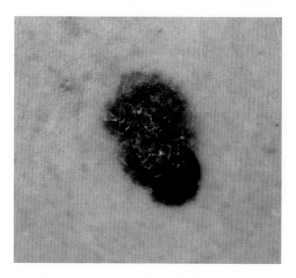

FIGURE 20.2

Malignant melanoma.

incidence of melanoma has tripled over the last 20 years and the highest rates are reported in Australia, New Zealand and in other countries near the equator where there is exposure to high-intensity UV light throughout the year [20]. In addition to UV radiation, risk factors for melanoma include episodes of sunburn, family history, light skin tone and the presence of large numbers of naevi. The most frequent sites for malignant melanoma are the trunk in males and the legs in women, that is sites of high levels of sun exposure.

There are several subtypes of malignant melanoma, including superficial spreading, nodular and acral. All pigmented lesions should be assessed using a checklist, for example NICE [21] recommends the use of a weighted seven-point checklist. Urgent referral following a fast track suspected cancer pathway should be actioned if a score of three or more is reached:

- Major features of the lesions (scoring two points each):
  - Change in size
  - Irregular shape
  - Irregular colour
- Minor features of the lesions (scoring one point each):
  - Largest diameter 7 mm or more
  - Inflammation
  - Oozing or crusting
  - Change in sensation or symptoms, for example itching

A useful acronym to aid with evaluating lesions is as follows:

- Asymmetry: one half of the mole is different to the other
- Border: irregular, uneven, scalloped or poorly defined edge
- Colour: variation in colour or uneven colour within a mole
- Diameter: any lesion growing in size
- Evolving: the mole is changing over time in its size, shape or colour

As with all cancers, histological assessment is essential for diagnosis. Treatment options will depend on the stage of the disease and include surgery, lymph node sampling, chemotherapy, radiotherapy and biologic therapies. In the US, the overall 5-year and 10-year survival rates for individuals with melanoma are 92% and 89%, respectively [22].

Non-melanoma skin cancers

Although the non-melanoma skin cancers usually run a benign course, early recognition and treatment are essential to avoid extensive treatment at a later stage, which can cause significant adverse physiological and psychological consequences for patients [23].

Basal cell carcinomas (BCC) comprise 80%−85% of the non-melanoma carcinomas and account for significant morbidity to patients and also financial burden to healthcare services (Fig. 20.3) [24]. They largely occur in sun-exposed areas such as the head and neck. Risk factors for developing BCC include fair skin type, increasing age, UV exposure, previous history of BCC and certain congenital syndromes such as Gorlin syndrome. Following histological confirmation of the diagnosis, treatment choices are guided by the size and site of the lesion, and include excision, curettage and cautery, cryotherapy, topical and photodynamic therapy.

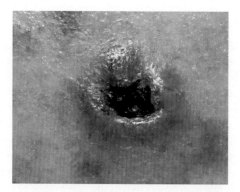

FIGURE 20.3

Basal cell carcinoma.

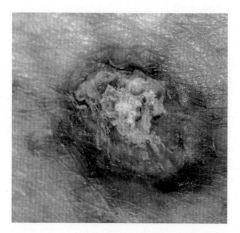

FIGURE 20.4

Squamous cell carcinoma.

Squamous cell carcinomas (SCC) comprise 15%−25% of non-melanoma carcinomas and have been increasing in incidence in the UK over the last two decades (Fig. 20.4). They have the potential to metastasise and occur twice as commonly in men than women. Risk factors for the development of SCC are similar to those for BCC, but SCC can also develop in a chronic wound. Immunosuppressed patients are also at increased risk of developing SCC.

SCCs commonly occur on the head and neck and dorsum of the hands. The lesions are often rapidly growing, indurated and can be hyperkeratotic. They may also be painful. Treatment options include surgical excision and radiotherapy.

Clinical advice for patients

It is important to educate patients so that they are aware how to prevent the development of skin cancers. It is also vital to equip them with knowledge about skin malignancy, so they know when to seek medical help. It has been reported that only a minority of patients ever receive appropriate education relating to skin cancer prevention [25].

The crucial message to give patients is to avoid excessive sun exposure and sunburn to aid skin cancer prevention. Protection of the skin with appropriate clothing, including a hat and sunglasses, as well as frequently and liberally applying appropriate sunscreen is imperative, along with advising to spend time in the shade especially between 11 a.m. and 3 p.m. Patients can be educated to help identify any changes in skin lesions to promptly alert their medical practitioner using tools such as the 'ABCDE' acronym. Patients should also be advised of the importance of early diagnosis and treatment, as this will significantly impact and improve outcome [26].

Conclusion

Skin is the largest organ in the body and has many vital roles including defence against the environment and immune defence. Skin ages and is damaged as a result of both intrinsic and extrinsic factors. The major source of extrinsic skin damage is UV radiation: this increases the risk of skin cancer developing as well as accelerating signs of ageing.

To maintain healthy skin, there are several lifestyle modifications an individual can make. These

Summary

- Skin is the largest organ in the human body and is the body's main defence against the environment.
- Skin ageing is caused by intrinsic and extrinsic factors.
- Lifestyle behaviours are important in slowing the signs of ageing and more importantly in reducing the risk of developing malignancies.
- Skin cancers are common.
- UV exposure is a significant risk factor for the development of skin cancers and appropriate UV protection is crucial.
- 'ABCDE' is a useful acronym to evaluate melanocytic skin lesions.
- Early detection of skin cancers is fundamental in improving outcomes.

include reducing exposure to UV light, avoiding smoking, taking part in regular exercise, eating a healthy diet and getting sufficient sleep. These behaviours will not only have a positive impact on skin health, they will impact the physical and psychological wellbeing of the individual as a whole.

References

[1] Cancer Research UK. Risks and causes of melanoma; n.d. Available at:: http://www.cancerresearchuk.org/about-cancer/melanoma/risks-causes [Accessed November 2019]).

[2] Naylor EC, Watson RE, Sherratt MJ. Molecular aspects of skin ageing. Maturitas 1 July 2011;69(3):249–56.

[3] Rinnerthaler M, Bischof J, Streubel MK, et al. Oxidative stress in aging human skin. Biomolecules June 2015;5(2):545–89.

[4] Sjerobabski-Masnec I, Situm M. Skin aging. Acta Clin Croat 1 December 2010;49(4):515–8.

[5] Nursing Times. Anatomy and physiology of ageing 11: the skin; n.d. https://www.nursingtimes.net/roles/older-people-nurses-roles/anatomy-and-physiology-of-ageing-11-the-skin-27-11-2017/([Accessed December 2019]).

[6] Zhang S, Duan E. Fighting against skin aging: the way from bench to bedside. Cell Transpl May 2018;27(5): 729—38.

[7] Claudia A. Hormones and your skin; n.d. http://www.dermalinstitute.com/uk/library/76_article_Hormones_and_Your_Skin.html.

[8] Olsen CM, Wilson LF, Green AC, et al. Prevention of DNA damage in human skin by topical sunscreens. Photodermatol Photoimmunol Photomed May 2017;33(3):135—42.

[9] Sunscreen preparations; n.d. https://bnf.nice.org.uk/treatment-summary/sunscreen.html ([Accessed November 2019]).

[10] Randhawa M, Wang S, Leyden JJ, et al. Daily use of a facial broad-spectrum sunscreen over one-year significantly improves clinical evaluation of photoaging. Dermatol Surg 1 December 2016;42(12):1354—61.

[11] Séhédic D, Hardy-Boismartel A, Couteau C, Coiffard LJ. Are cosmetic products which include an SPF appropriate for daily use? Arch Dermatol Res 1 September 2009;301(8):603—8.

[12] Liu X, Baylin A, Levy PD. Vitamin D deficiency and insufficiency among US adults: prevalence, predictors and clinical implications. Br J Nutr April 2018;119(8):928—36.

[13] Nicita-Mauro V, Basile G, Maltese G, Nicita-Mauro C, Gangemi S, Caruso C. Smoking, health and ageing. Immun Ageing 2008;5:10.

[14] Verdier-Sévrain S, Bonté F. Skin hydration: a review on its molecular mechanisms. J Cosmet Dermatol June 2007;6(2):75—82.

[15] McCartney M. Waterlogged? BMJ 12 July 2011;343:d4280.

[16] Akdeniz M, Tomova-Simitchieva T, Dobos G, et al. Does dietary fluid intake affect skin hydration in healthy humans? A systematic literature review. Skin Res Technol August 2018;24(3):459—65.

[17] Appold K. 6 Amazing reasons to sleep for skin health; n.d. https://www.everydayhealth.com/skin-and-beauty/amazing-reasons-to-sleep-for-skin-health.aspx.

[18] Crane JD, MacNeil LG, Lally JS, et al. Exercise-stimulated interleukin-15 is controlled by AMPK and regulates skin metabolism and aging. Aging Cell August 2015;14(4):625—34.

[19] Rhee JS, Matthews BA, Neuburg M, et al. The skin cancer index: clinical responsiveness and predictors of quality of life. Laryngoscope March 2007;117(3):399—405.

[20] Skin cancer statistics; n.d. https://www.wcrf.org/dietandcancer/cancer-trends/skin-cancer-statistics ([Accessed January 2020]).

[21] National Institute for Health and Care Excellence (NICE). Suspected cancer: recognition and referral. NICE Guideline [NG12]. 2017. Available from: https://www.nice.org.uk/guidance/ng12. Accessed 2November 2019.

[22] Miller KD, Siegel RL, Lin CC, Mariotto AB, Kramer JL, Rowland JH, Stein KD, Alteri R, Jemal A. Cancer treatment and survivorship statistics. CA Cancer J Clin July 2016;66(4):271—89.

[23] National Collaborating Centre for Cancer Suspected cancer: recognition and referral. Full guideline June 2015.National Institute for Health and Care Excellence; n.d. Available from: https://www.nice.org.uk/guidance/ng12/evidence [Accessed November 2019].

[24] Suárez B, López-Abente G, Martínez C, et al. Occupation and skin cancer: the results of the HELIOS-I multicenter case-control study. BMC Public Health December 2007;7(1):180.

[25] O'Driscoll L, McMorrow J, Doolan P, et al. Investigation of the molecular profile of basal cell carcinoma using whole genome microarrays. Mol Canc December 2006;5(1):74.

[26] Freiman A, Yu J, Loutfi A, et al. Impact of melanoma diagnosis on sun-awareness and protection: efficacy of education campaigns in a high-risk population. J Cutan Med Surg 2004;8:303—9.

Western medical acupuncture

21

Carolyn Rubens

General Practitioner and Medical Acupuncturist, Lighthouse Medical Practice, Eastbourne, United Kingdom

What is acupuncture?

Acupuncture has its origins in traditional Chinese medicine. It has been practiced for 2000 years, although its conception may have dated even further back in time in Persia and other areas of the world where acupuncture-like instruments have been discovered. The word *acupuncture* means to penetrate with a needle, from the Latin: *acu* − needle, *punctura* − prick. Acupuncture can help with a number of symptoms, including pain, nausea and vomiting, and it can also reduce fear and anxiety and improve energy levels.

Traditional Chinese acupuncture is based on the theory that vital energy or 'Qi' circulates around the body along channels called meridians. Blockages in this flow are thought to be the cause of ill health. In Chinese medicine, acupuncture is used to correct the flow of Qi. Western medical acupuncture is based on scientific rationale. Increasing numbers of healthcare professionals are being trained to use it therapeutically alongside conventional management approaches. The remainder of the chapter will consider western medical Acupuncture.

The physiology of acupuncture

Acupuncture is known to stimulate fast nerve fibres called A delta fibres as they enter the spinal cord. These then inhibit pain impulses carried in slower nerve fibres knows as C fibres and through connections with other levels of the spinal cord also inhibit pain impulses there. This helps explain why acupuncture needles in one part of the body can affect pain sensation in another region. Acupuncture is also known to stimulate release of endogenous opioids such as endorphins and dynorphins as well as other neurotransmitters such as serotonin [1].

Acupuncture and symptom control

One of the key reasons people use acupuncture is to help relieve pain, often when they are intolerant to other pain medications. Types of pain which acupuncture can help to relieve include knee osteoarthritis, shoulder pain, back pain, wrist pain, whiplash and dysmenorrhoea. It is helpful in treating

A Prescription for Healthy Living. https://doi.org/10.1016/B978-0-12-821573-9.00021-7

235

headaches and migraines and is recommended by the National Institute of Clinical Excellence (NICE) for the prevention of migraine [2]. In the UK acupuncture is recommended in the latest draft NICE guidelines for chronic pain over and above opioids to treat this often complex and difficult symptom.

Acupuncture can be used for a number of non pain symptoms including irritable bladder, constipation and stress incontinence [3–5]. Acupuncture is also used for individuals with cancer to alleviate nausea caused by chemotherapy, where it can be as effective as standard antiemetics [6]. It can be used to manage other side effects of cancer treatments, such as chemotherapy induced peripheral neuropathy (CIPN), hot flushes, breathlessness and xerostomia [7]. Some people with cancer like to use acupuncture because of its additional beneficial side effects such as improving sleep and reducing levels of anxiety, which in turn can reduce the perception of pain [7].

Evidence to support the use of acupuncture

Numerous scientific trials have been carried out looking at the use of acupuncture for different conditions. It has been difficult to determine the effects of acupuncture in some of these trials, as it has been difficult to design a reliable placebo.

To try to account for the placebo effect, some studies have compared true acupuncture with sham acupuncture. Sham acupuncture uses either off-point needling or a special needle that does not actually penetrate the skin. Some studies have shown no difference between sham and real acupuncture for symptoms such as migraine, but it is important to note that in some of those trials, those receiving either form of acupuncture did better than receiving no treatment at all [8].

There is, however, a growing body of evidence which has found acupuncture to be more effective than sham acupuncture/no acupuncture for treating back pain [9], chronic pain [10], preventing chronic headache [11,12], knee osteoarthritis [13] and shoulder pain [14]. Acupuncture was in the NICE guidelines for back pain in the United Kingdom until recently, and the American College of Physicians recommends acupuncture for back pain [15].

There is also growing evidence for the use of acupuncture in treating non pain symptoms. For example, nausea was the first area to have positive trial results, and the number needed to treat is between four and five for postoperative nausea and vomiting. The latest Cochrane review reported that individuals who receive P6 acupoint stimulation have no difference in postoperative nausea or vomiting compared with individuals who receive antiemetic drugs [16]. There is accumulating evidence that acupuncture may be successful in alleviating symptoms such as for hot flushes, joint pains and other symptoms related to cancer treatment [7]. A recent systematic review showed significant reduction in menopausal symptoms for at least 3 months in menopausal symptoms related to breast cancer [17]. Menopausal hot flushes in women without breast cancer also appear to respond to a similar degree as low dose hormone replacement therapy and better than phytoestrogens [18].

What acupuncture involves

Acupuncture needles are inserted just below the skin at particular points to stimulate nerves. These may be left in place from 5 to 45 min, but usually for 15–20 min. The needles should not hurt when they are in place although there may be a slight prick as they enter the skin. Most acupuncturists use between 4 and 20 points during a session.

The number of treatments will depend on the condition being treated and response to the treatment. A typical schedule might be several weekly treatments followed by top-up treatments as needed every few weeks for a chronic condition. It is also possible to learn how to do 'do it yourself' needling at specific points after an initial course of treatment. If acupuncture is going to help symptoms, an improvement is likely to seen by the fifth or sixth session, but often much sooner. Many acupuncturists now use electroacupuncture to enhance the effects of needling. This involves attaching a very weak electrical current to the needles once they are in the skin. There is also a particular type of acupuncture called auricular acupuncture, where the needles may be left in place for some days.

Possible side effects of acupuncture

When given by professionally qualified therapists, acupuncture is generally very safe. Side effects are rare and are usually preventable with good knowledge of anatomy and safe needling techniques. Due to the use of sterile single-use disposable needles, serious cases of infection are rare.

The most common side effects of acupuncture are failure to remove the correct number of needles, dizziness, possibly from a short-term drop in blood pressure and bruising or slight bleeding at the needle sites. In one large study, it was reported that only 63 out of more than 55,000 individuals had any adverse effects from acupuncture [19]. These were all minor adverse events including needling pain, bruising and bleeding. While serious adverse events such as pneumothorax and acute hypotensive events have been previously described, none were seen in this study. The incidence of serious adverse events has been reported to be 11 serious adverse events over 4,441,103 treatments [19].

There are also positive side effects of acupuncture reported by many patients including feelings of relaxation, improved mood and improved sleep.

Who should not have acupuncture?

Individuals should only have acupuncture if they want to. People with valvular heart disease are not recommended to have acupuncture with indwelling acupuncture needles. Other factors which may affect suitability for acupuncture include an abnormal blood clotting profile, leukopenia and demand pacemakers.

The future of acupuncture

Acupuncture is used increasingly by doctors and other healthcare professionals for pain relief instead of medication. It has pleasant side effects so is liked by patients and better tolerated than a lot of medication. It can also treat the cause of the symptoms, for example headaches can be caused by a trigger point in the trapezius muscle. Trigger points are taut bands of skeletal muscle with a characteristic twitch response which can be quickly treated with acupuncture. They have characteristic referral patterns of pain as mapped by Travell and Simons [20]. This knowledge seems to have been lost in medicine over the years as we tend to palpate muscles less.

There is an increasing role for acupuncture in healthcare systems to help reduce the amount of medication people are prescribed, as well as in supporting those coming off medication or unable to take it. In many cases of musculoskeletal problems, having easy access to acupuncture can prevent the need to be referred for physiotherapy, rheumatology or orthopaedics, thus saving money for healthcare systems. Barriers to implementing acupuncture within primary care include time and scheduling constraints, lack of clinic space and inadequate resources [21].

Finding an acupuncturist

Acupuncture is widely used in many pain clinics, cancer centres and alongside general practice or family medicine. It is very important that the practitioner is properly trained and qualified to use acupuncture for people with cancer. It is not advisable to access treatment from the high street without checking the qualifications of the practitioner first. Many practitioners of traditional Chinese traditional acupuncture also encourage the use of oral herbs, but it is important to understand that these herbs may interact with conventional treatments.

The best way to find a reliable acupuncturist is to identify a reputable organisation in the country that you are in. They will be able to suggest practitioners in your area who have trained with them and so you can be sure they are adequately qualified. While there is no statutory regulation of acupuncture in England, Western medical acupuncturists are listed on the webpage of the British Medical Acupuncture Society, and there is an accredited register of traditional acupuncture practitioners listed on the Professional Standards Authority for Health and Social Care website. In America, most physicians are permitted to practice acupuncture within the scope of their medical practice, although in some states a special license is compulsory, which necessitates additional education and training.

To maintain the excellent profile of safety of acupuncture within modern medicine, it must be performed by suitably qualified acupuncturists who have demonstrable knowledge of anatomy and physiology and clean needling technique.

Conclusion

Acupuncture may have an increasing role in the future of healthcare, as we try to reduce opiate usage and reduce the drug burden in multimorbidity. It is a cheap, safe and effective therapeutic approach for a wide range of conditions, such as musculoskeletal pain, and should be considered first-line therapy in many of these where an acupuncturist is locally available.

Summary

Acupuncture is a safe and effective therapy that is used widely within modern medicine to treat a variety of ailments. It is a safe alternative to a number of pain killers and treats the cause of symptoms in many cases. It has a low side effect profile making it attractive to patients. The potential for acupuncture to reduce referral rates to physiotherapy and orthopaedics and reduce medication burden requires further research.

References

[1] Pomeranz B, Berman B. Scientific basis of acupuncture in basics of acupuncture, pub Springer 5th ed. pp 7–86.
[2] NICE guideline on headaches (2012): diagnosis and management of headaches in young people and adults. 2012. http://guidance.nice.org.uk/CG150.
[3] Peters KM, MacDiarmid SA, Wooldridge LS, et al. Randomised trial of percutaneous tibial nerve stimulation versus extended-release tolterodine: results from the overactive bladder innovative therpay trial. J Urol 2009; 182:1055–61.

[4] Liu Z, Yan S, Wu J, et al. Acupuncture for chronic severe functional constipation: a randomised trial. Ann Intern Med 2016;165:761—9.

[5] Liu Z, Liu Y, Xu H, et al. Effect of electroacupuncture on urinary leakage among women with stress urinary incontinence: a randomised clinical trial. J Am Med Assoc 2017;317:2493—501. https://doi.org/10.1001/jama.2017.7220.

[6] Lee A, Fan LT. Stimulation of the wrist acupuncture point P for preventing postoperative nausea and vomiting. Cochrane Database Syst Rev 2009:CD003281. https://doi.org/10.1002/14651858.CD003281.pub3.

[7] Rubens C, Filshie J. Acupuncture in cancer and palliative care. In: Filshie J, White A, Cummings M, editors. Medical acupuncture — a Western scientific approach. London: Elsevier; 2016. 566—85.

[8] White A. A critical approach to systematic reviews of treatment. In: Filshie J, White A, Cummings M, editors. Medical acupuncture — a Western scientific approach. London: Elsevier; 2016. p. 298—314.

[9] Furlan AD, Yazdi F, Tsertsvadze A, et al. Acupuncture for (sub)acute non-specific low-back pain. Cochrane Database Syst Rev 2011:CD009265.

[10] Vickers AJ, Cronin AM, Maschino AC, et al. Acupuncture for chronic pain: individual patient data meta-analysis. Arch Intern Med 2012;172:1444—53.

[11] Linde K, Allais G, Brinkhaus B, et al. Acupuncture for tension-type headache. Cochrane Database Syst Rev 2009:CD007587.

[12] Linde K, Allais G, Brinkhaus B, et al. Acupuncture for migraine prophylaxis. Cochrane Database Syst Rev: CD001218.

[13] White A, Foster NE, Cummings M, et al. Acupuncture treatment for chronic knee pain: a systematic review. Rheumatology 2007;46:384—90.

[14] Green S, Buchbinder R, Hetrick S. Acupuncture for shoulder pain. Cochrane Database Syst Rev 2005: CD005319. https://doi.org/10.1136/aim.3.1.28.

[15] Qaseem A, et al. Noninvasive treatments for acute, subacute, and chronic low back pain: a clinical practice guideline from the American College of physicians. Ann Intern Med 2017;166(7):514—30.

[16] Lee A, Fan LT. Stimulation of the wrist 69 acupuncture point P6 for preventingpostoperative nausea and vomiting. Cochrane Database Syst Rev 2009:CD003281.

[17] Chien T-J, Liu C-Y, Fang C-J, et al. 2020. The maintenance effect of acupuncture on breast cancer-related menopause symptoms: a systematic review. Climacteric 2020;23(2):130—9.

[18] Palma F, Fontanesi F, Facchinetti F, et al. Acupuncture or phy(F)itoestrogens vs. (E)strogen plus progestin on menopausal symptoms. A randomized study. Gynae Eondcrinol 2019;35(11):995—8. PMID: 31142156. https://doi.org/10.1080/09513590.2019.1621835.

[19] White A. The safety of acupuncture — evidence from the UK. Acupunct Med 2006;24(Suppl. l):53—7.

[20] Simons D, Travell J, Simons L. Myofascial pain and dysfunction: the trigger point manual: two volume set. 2nd ed., vol. 1; 1998. First Edition/vol. 2.

[21] Ledford CJ, Fisher CL, Moss DA, et al. Critical factors to practicing medical acupuncture in family medicine: patient and physician perspectives. J Am Board Fam Med 2018;31(2):236—42.

Nutrition and healthy eating habits

Fruit and vegetables: prevention and cure?

22

Gemma Newman[1,2]

[1]*General Practitioner, National Health Service, Ashford, United Kingdom;* [2]*Advisory Board Member, Plant-Based Health Professionals UK, United Kingdom*

Introduction

Around the world, most nutrition guidelines encourage the consumption of fruit and vegetables. The health benefits of fruits and vegetables are long established — they provide a multitude of vitamins and minerals along with large amounts of fibre. However, there is still a lot to be learnt about the roles that fruit and vegetables can play in disease prevention and management.

Fruit and vegetable consumption in different populations

Fruit and vegetable consumption varies across the globe and according to socioeconomic status. Research has shown that low-income families are hit the hardest with nutritional inadequacy and caloric excess [1,2]. There is a general pattern of increasing fruit and vegetable consumption with increasing educational attainment. Professor Michie, University College London, has highlighted that the social determinants of health related to fruit and vegetable consumption are associated with capability, opportunity and motivation [3].

In the United Kingdom (UK), only 29% of adults eat five portions of fruit and vegetables a day, and only 18% of children [4]. In the United States, the majority of Americans fail to consume the two-and-a-half cups of vegetables and two cups of fruits per day recommended by The Dietary Guidelines for Americans [5]. Barriers to eating more include inadequacy in the food supply, poor availability of produce in rural and underserved urban areas, food cost, time 'poverty' and poor health literacy.

European statistics reveal wide varieties in daily consumption of fruits and vegetables — a quarter or more of the population in the UK, the Netherlands, Denmark and Ireland eat a minimum of five portions daily, compared to less than 8% of the populations of Greece, Slovenia, Austria, Croatia, Bulgaria and Romania. More than half of the population of Bulgaria (58.6%) and Romania (65.1%) stated that fruit and vegetables are not included in their daily diet at all [6].

Fruit, vegetables and fibre

Fruits are the only naturally occurring sugars the human population historically had access to. Humans are drawn to fruits because of their health benefits, but the inherent attraction to sugar-rich, bright products is likely to be the reason we are so drawn to colours and sweet tastes in foods that are manufactured for palatability but contain 'empty calories' today.

Fruits, along with vegetables, are rich in antioxidants, vitamins, minerals, phytonutrients and fibre. Phytonutrients is an umbrella term for the polyphenols contained in fruits and vegetables such as anthocyanins, lignans, flavonoids, sulphorophane, resveratrol, lycopene and pyrroloquinoline quinone. These compounds protect plants from ultraviolet rays and insects, for example, but can also provide benefits for human health. The fibre in fruit allows for slower release of the naturally occurring sugars also present and has a vital role in regulating gut microbiota. The importance of fibre in health is described in Chapter 23.

In brief, fibre helps to prevent constipation, haemorrhoids and diverticular disease, and is also important in providing short-chain fatty acids such as butyrate, proprionate and acetate to aid propagation of beneficial gut microbes [7]. Such microbes have numerous vital functions in the body including vitamin production, immune regulation, the maintenance of the gut mucosa and the synthesis of neurotransmitters such as serotonin [8]. Although the serotonin produced in the gut cannot pass the blood–brain barrier, it has important immunoregulatory functions.

Fruit, vegetables and cancer

The World Cancer Research Fund (WCRF) has fruit and vegetables as the foods at the top of their list at preventing the second largest cause of death in the western world – cancer [9]. Every long-lived population on earth has fresh seasonal locally grown fruits and vegetables as the staple of their diet. Costa Rica, Italy, Japan, California and Greece all contain areas referred to as 'Blue Zones' – small pockets of the world with the highest proportion of centenarians. Although they eat very different cuisines, their diets are 95% plant-based, in which fruits and vegetables play a key role, as well as legumes, soy and whole grains. These populations may seem to have nothing in common on the surface, but from the plant-based Seventh Day Adventists in California, to the elderly sweet potato loving Okinawans off the coast of Japan – they all eat foods grown locally, and mostly plants.

Fruit consumption may reduce breast cancer risk later in life. Epidemiology researchers followed the Nurses' Health Study II cohort of 90,476 premenopausal women for 22 years. It was reported that those who ate the most fruit during adolescence, around three servings a day, compared with those who ate the least, half a serving daily, had a 25% lower incidence of breast cancer. Interestingly, no protection was found from drinking fruit juices at younger ages [10]. Similar patterns were also seen when analysing overall fibre intake from fruits and vegetables in relation to cancer risk.

For patients who already have a cancer diagnosis, it has been shown that a high intake of fruit and vegetables may improve prognosis. There are many potential mechanisms for this. Researchers at the University of California have demonstrated that a dietary flavonoid, apigenin, completely restores p53 nuclear localisation in neuroblastoma cells [11]. Apigenin is found in fruits and vegetables and has been shown to have antitumour effects in several types of cancer [12]. P53 is a tumour suppressor. In the normal state, intracellular levels of p53 are low. When DNA is damaged, p53 levels increase and it is involved in DNA repair process and in initiating apoptosis. In many cancers, p53 is inactivated by cytoplasmic sequestration. Apigenin is able to activate p53 and transport it into the nucleus, so its antitumourigenic effects can be mediated. Apigenin is found in fruit, including apples, cherries and grapes, vegetables such as parsley, artichoke, basil and celery, nuts and in some teas. Apigenin has been shown to have growth inhibitory effects in several cancer cell lines, including breast, colon, skin, thyroid and leukaemia. It may also have beneficial effects in lung cancer [13] and pancreatic cancer [14].

Fruit and vegetables may also have anticarcinogenic effects mediated through an ability to inhibit vascular endothelial growth factor (VEGF), therefore reducing the ability of tumours to obtain a blood supply. Oxygen can diffuse from capillaries for only 150–200 μm. When distances exceed this, cell death follows. Thus, the expansion of tumour masses depends on angiogenesis, which plays a major role in the progression of most solid tumours, including those of the lung, colon, bladder, breast, cervix and prostate. The compound in fruit responsible for this is thought to be Resveratrol, which is found in grape skin and berries. Berries also contain high concentrations of proanthocyanidin, which also inhibits VEGF expression [15]. One study showed that grape seed extract may also *upregulate* oxidant-induced VEGF expression [16], suggesting that it can also induce angiogenesis as part of normal tissue healing. This means the biological effects of fruit and vegetables to modulate cell growth could be selective.

There are many thousands of plant compounds, including other polyphenols such as lignans, flavonoids, quercetin, lycopene and resveratrol. The current understanding of how they might work in synergy with each other is incomplete. More research is needed to address how such foods could be utilised as a way of modulating disease risk or as a potential medical therapy.

Fruit, vegetables and vascular disease

A 2019 American College of Cardiology/American Heart Association guideline on the primary prevention of Cardiovascular Disease stipulated that:

> all adults should consume a healthy diet that emphasises the intake of vegetables, fruits, nuts, whole grains, lean vegetable or animal protein and fish, and minimises red meats and processed red meats, refined carbohydrates and sweetened beverages [17].

Fruit and vegetables are the first two food groups mentioned and have a key role in the prevention of cardiovascular disease. A meta-analysis of cohort studies following 469,551 participants found that a higher intake of fruits and vegetables is associated with a reduced risk of death from cardiovascular disease, with an average reduction in risk of 4% for each additional serving per day of fruit and vegetables [17]. The longest study to date, done as part of the Harvard-based Nurses' Health Study and Health Professionals Follow-up Study, included almost 110,000 people who were tracked for 14 years. It was found that a higher average daily intake of fruits and vegetables was associated with a lower risk of developing cardiovascular disease. Compared with those in the lowest category of fruit and vegetable intake, less than one-and-a-half servings daily, those who averaged eight or more servings a day were 30% less likely to have had a heart attack [18] or stroke [19]. When researchers combined findings from the Harvard studies with several other long-term studies, they reported a similar pattern emerging: individuals who ate more than five servings of fruits and vegetables per day had roughly a 20% lower risk of coronary heart disease and stroke [20], compared with those who ate less than three servings per day. One hypothesis is that the effects might be mediated through the modulation of blood pressure.

The Dietary Approaches to Stop Hypertension Study [21] examined the effect on blood pressure of a diet that was rich in fruits, vegetables and low-fat dairy products with restricted amounts of saturated and total fat. It was found that individuals with hypertension who followed this diet reduced their systolic blood pressure by around 11 mm Hg and their diastolic blood pressure by almost 6 mm Hg — as much as medications can achieve. A randomised trial, the Optimal Macronutrient Intake Trial for Heart Health (OmniHeart), showed that a fruit and vegetable-rich diet lowered blood pressure even more when some of the carbohydrate was replaced with healthy unsaturated fat or protein [22].

Fruit, vegetables and diabetes

Whole food plant-based diets contain an abundance of fruit and vegetables. This eating pattern has been shown to make the development of diabetes less likely [23–25]. Furthermore, research suggests that whole food plant-based diets can improve or even reverse diabetes that has already been diagnosed [26–29]. The American Association of Clinical Endocrinologists Guidelines (2018) suggests a plant-based diet as the preferred eating pattern for patients with known type 2 diabetes.

A study of over 2300 Finnish men showed that vegetables and fruits, especially berries, may reduce the risk of type 2 diabetes [30]. Additionally, a study of over 70,000 female nurses aged 38–63 years, who were free from cardiovascular disease, cancer and diabetes, showed that consumption of green leafy vegetables and fruit was associated with a lower risk of diabetes over time. Whilst not conclusive, this research also indicated that the consumption of fruit juices may be associated with an increased risk amongst women [31].

Dietary guidelines in the UK regarding the consumption of starchy vegetables and fruits have been conflicting in the management of diabetes. To understand this, it is helpful to review the three main techniques that are commonly used to induce clinical remission of type 2 diabetes. Along with other lifestyle factors, dietary approaches include caloric restriction, limiting carbohydrate-rich foods and adopting a whole-foods plant-based approach.

The Direct Trial showed beneficial results for individuals undertaking caloric restriction. Participants were provided with soups and meal replacement shakes and maintained a very low-calorie diet: 24% of the intervention group achieved a weight loss of at least 15 kg [32]. Diabetes remission was achieved in almost half of the intervention group compared to 4% of the control group. Nine serious adverse events were reported in the intervention group, with two events, biliary colic and abdominal pain, occurring in the same participant. Although this study suggests caloric restriction can be a highly effective strategy, such diets can be challenging to maintain in the long term.

Low carbohydrate high-fat diets (LCHF) are another strategy effective in achieving weight loss. However, a study using data from the UK National Diet and Nutrition Survey (2008–16) showed that an increased consumption of protein and fat, accompanied by a reduction in carbohydrates, was correlated with increased rates of diabetes [33]. In patients with established diabetes, LCHF diets may result in poor disease control [34–36].

A study comparing dietary strategies for weight loss and health suggested that the *quality* rather than the *ratio* of macronutrients is key — individuals who consume healthy high carbohydrate or healthy low carbohydrate diets, consuming whole grains, non-starchy vegetables, whole fruits and nuts, have a lower risk of premature death than those who eat low-quality carbohydrates, animal protein and saturated fat [37]. It has also been shown in metabolic ward studies that ketogenic approaches to weight loss results in loss of lean body mass and that 2 months of a tightly controlled ketogenic diet saw slowing of body fat loss compared to higher carbohydrate diet [38].

Why have whole food plant-based diets been recommended in the American College of Clinical Endocrinologists Guidelines? Research suggests that having a diet rich in fruits and vegetables, as well as other healthy carbohydrates, can lead to significantly reduced intra-myocellular fat [39], higher insulin sensitivity [40], better blood glucose and insulin levels and improved β-cell function [41]. Ultimately, approaches to diabetes have to suit the patient sitting in front of you — the most effective weight loss and disease management diet strategy is the one the patient is willing to do. But being familiar with what works and what is sustainable helps patients to make an informed choice they can adhere to.

Fruit and vegetable consumption guidelines

The National Diet and Nutrition Survey estimated that, in 2018, the fruit and vegetable intake in the UK amounted to only 8% of total food consumption [42]. This is concerning. It has been shown that for every 10% increase in processed foods in the diet, there is a corresponding 10% increase in the risk of developing a malignancy [43].

The official dietary advice in the UK is to eat five portions of fruit and vegetable daily. However, in one study, it was found that increasing this intake to seven portions a day had the potential to reduce the risk of premature death by 42% [44]. Furthermore, it has been suggested that aiming for 10 portions confers an even greater benefit. It has been reported that individuals who consume 10 portions of fruit and vegetables daily have nearly a third lower risk of death than those who eat none [45]. Most of the benefits were linked to reductions in heart disease and cancer.

Simple tips to encourage your patients to consume more fruit and vegetables are listed in Box 22.1.

Box 22.1 Tips to help encourage your patients to eat more fruit and vegetables

- Encourage your patients to 'eat the rainbow'. This helps them to understand that the greater the variety of colours they can get into their fruit and vegetable choices the better. You can explain the colours represent different beneficial phytonutrients they need for good health. Red for lycopene, orange for carotenoids, blue for resveratrol and green for antioxidants. This will hopefully remind them to pick up colourful vegetables that they might not usually buy.
- Suggest that your patients keep a rainbow chart on the fridge to remind them of all the colours of fruits and vegetables to enjoy each week — children especially enjoy ticking the colours they have consumed.
- If finances are tight, frozen fruit and vegetables can be a convenient and cost-effective way of increasing consumption.
- Some individuals might consider getting a regular organic fruit and vegetable box if budget and location allows. This encourages buying more local and organic produce to increase the variety of their fruit and vegetable consumption and fosters increasing creativity.

Organic produce

It is frequently asked whether it is healthy to eat fruit and vegetables if they have been grown conventionally, with the use of pesticides? Pesticides were originally patented as antibiotics and antifungals, but they also have a nonselective ability to kill plants [46]. The use of pesticides worldwide has risen almost 15-fold since 1996 [47] when crops genetically engineered to resist pesticides were introduced [48]. Reassuringly, the European Food Safety Authority has found no link to cancer in humans, yet a study published in the Journal of the American Medical Association suggested that consuming organic produce did reduce the risk of certain cancers. The research followed nearly 70,000 people for 7 years. After accounting for potential confounding factors such as lifestyle and family history, it was found that those who ate the most organic food had a 25% reduced relative risk of developing a malignancy (hazard ratio for quartile 4 vs quartile 1, 0.75; 95% CI, 0.63–0.88; P for trend = .001; absolute risk reduction, 0.6%; hazard ratio for a 5-point increase, 0.92; 95% CI, 0.88–0.96) [49].

The International Agency for Research on Cancer classifies a small number of pesticides as known carcinogens and a few others as possible carcinogens [50]. In animal studies, some pesticides have been reported to be carcinogenic, for example organochlorines, creosote and sulfallate, whilst others, including organochlorines DDT, chlordane and lindane, have been designated as tumour promoters [51]. Human data are limited by the small number of studies that evaluate individual pesticides. To add to confusion, epidemiological studies are sometimes contradictory. Some have linked phenoxy acid herbicides, or contaminants within them, with sarcoma and lymphoma [52]. Organochlorine insecticides have also been associated with non-Hodgkin's lymphoma (NHL), leukaemia and, less consistently, with cancers of the lung and breast [53].

Organophosphorus compounds are linked with NHL and leukaemia, and triazine herbicides with ovarian cancer [52].

However, none of these associations can be considered causal; therefore, further epidemiologic studies are necessary and important. There is also little doubt that even if there is negligible effect on human cells when traces of pesticides are consumed, they will have a more marked effect on our microbial populations, given their mechanism of action. Furthermore, pesticides may also assist in propagating antibiotic resistant strains of bacteria [54].

It is interesting to note that some fruit and vegetables are less concentrated with pesticide residues than others. Pesticide residues are tested annually by the United States Department of Agriculture [55]. The products with the least concentrations of residues include avocado, broccoli, cauliflower, asparagus, melons, kiwi, papaya, onions, peas and cabbages [55]. In the UK, data from 2018 indicate that only 0.1% of wheat crops grown in the UK remained untreated with pesticides, and wheat received on average three fungicides, three herbicides, two growth regulators and one insecticide [56]. The Expert Committee on Pesticide Residues in Food released a report in September 2019 which surveyed 498 samples of 20 different foods — 28 of the samples contained residues above those permitted by law, which included beans, cabbage, peppers, meat, cheese, lemons and rice — however, the report found that the residue levels isolated were unlikely to cause harm when consumed [57].

The bottom line is that, based on current evidence, the overall importance of consuming fruit and vegetables outweighs the potential adverse effects of the consumption of commercially grown produce. Widespread adoption of regenerative agricultural techniques and changes to how industrial subsidies are allocated are likely to be beneficial for food production long term, but until more evidence emerges, the data suggest that it is important to consume as many fruits and vegetables as possible regardless of their mode of production. The future of sustainable agricultural practices may necessitate far more judicial use of pesticides alongside no-till restorative soil practices.

Conclusions

Although more research will be beneficial in understanding the many ways in which fruits and vegetables exert their effects on the body, it is already clear that they are a vital part of a healthy diet.

Guidelines on heart disease, diabetes and cancer prevention all emphasise the importance of fruits and vegetables in the diet, as do the healthy eating guidelines used to aid patients to make health decisions.

Fibre, vitamins, minerals and polyphenols are all constituents of these foods through which human health benefits may be mediated, and evidence seems to be pointing to increasing the consumption of fruits and vegetables from the minimum of five a day that is stipulated in the current official nutritional guidelines.

Although evidence is mixed, emerging data point towards the benefits of consumption of organic produce over conventionally grown. For many, inadequacy in the food supply, poor availability of produce in rural and underserved urban areas, food cost, time 'poverty' and poor health literacy mean that increasing fresh produce consumption will be a challenge. This is why public health campaigns as well as changes to food industry practices and support of no-till regenerative agricultural practices are as important to population health as basic education on the benefits of these healthy foods.

Summary

- Fruit and vegetable intake is related to socioeconomic status.
- Fruits and vegetables contain fibre, vitamins and minerals, which are all important for optimum health.
- A diet rich in fruits and vegetables lowers the risk of cardiovascular disease, can reduce blood pressure, reduces the risk of developing certain types of cancer and can have a positive effect upon blood sugar levels, therefore reducing the risk of diabetes.
- Organic produce may be a superior choice for cancer risk reduction, but the evidence to support this is currently inconclusive.
- All adults should strive to consume at least five portions of fruit and vegetables every day.

References

[1] Patience S. Supporting low income families with nutrition. Journal of health visiting 2013;1(6).

[2] Casey P, Szeto K, et al. Arch Pediatr Adolesc Med 2001;155(4):508−14 [Prevalence, Health and Nutrition Status].

[3] Atkins L, Miche S. Changing eating behaviour: what can we learn from behavioural science? Nutr Bull 2013;38(1):pp30−35.

[4] NHS digital. https://digital.nhs.uk/data-and-information/publications/statistical/statistics-on-obesity-physical-activity-and-diet/statistics-on-obesity-physical-activity-and-diet-england-2019/part-6-diet.

[5] Yeh M-C, et al. Chapter 19 − fruit and vegetable consumption in the United States: patterns, barriers and federal nutrition assistance programs fruits. Vegetables and Herbs; 2016. p. 411−22.

[6] EHIS (European health interview survey) ec.europa.eu.

[7] Montagne L, Pluske JR, Hampson DJ. A review of interactions between dietary fibre and the intestinal mucosa, and their consequences on digestive health in young non-ruminant animals. Anim Feed Sci Technol 2003;108(1−4):95−117. https://doi.org/10.1016/S0377-8401(03)00163-9.

[8] Yano1 JM, Yu1 K, Donaldson1 GP, Shastri1 GG, Ann1 P, Liang M2, Nagler3 CR, Ismagilov2 RF, Mazmanian1 SK, Hsiao1 EY. Indigenous bacteria from the gut microbiota regulate host serotonin biosynthesis. Cell April 9, 2015;161(2):264−76. https://doi.org/10.1016/j.cell.2015.02.047.

[9] www.wcrf-uk.org/uk/preventing-cancer/cancer-prevention-recommendations.

[10] Farvid MS, Chen WY, Michels KB, Cho E, Willett WC, Eliassen AH. Fruit and vegetable consumption in adolescence and early adulthood and risk of breast cancer: population based cohort study. BMJ May 11, 2016;353:i2343.

[11] Cai X, Liu X, et al. Inhibition of Thr-55 phosphorylation restores p53 nuclear localization and sensitizes cancer cells to DNA damage. Proc Nat Acad Sci November 2008;105(44):16958−63.

[12] Galati G, O'Brien PJ. Potential toxicity of flavonoids and other dietary phenolics: significance for their chemopreventive and anticancer properties. Free Radic Biol Med 2004;37:287−303.

[13] Liu L-Z, Fang J, Zhou Q, Hu X, Shi X, Jiang B-H. Apigenin inhibits expression of vascular endothelial growth factor and angiogenesis in human lung cancer cells: implication of chemoprevention of lung cancer the American society for pharmacology and experimental therapeutics. Mol Pharmacol 2005;68:635−43.

[14] Ujiki MB, Ding X, Salabat MR, et al. Apigenin inhibits pancreatic cancer cell proliferation through G2/M cell cycle arrest. Mol Canc 2006;5:76. https://doi.org/10.1186/1476-4598-5-76.

[15] Sagar SM, Yance D, Wong RK. Natural health products that inhibit angiogenesis: a potential source for investigational new agents to treat cancer-Part 1. Curr Oncol 2006;13(1):14−26.

[16] Khanna S, Roy S, Bagchi D, Bagchi M, Sen CK. Upregulation of oxidant-induced vegf expression in cultured keratinocytes by a grape seed proanthocyanidin extract. Free Radic Biol Med 2001;31:38−42.

[17] Wang X, Ouyang Y, Liu J, Zhu M, Zhao G, Bao W, Hu FB. Fruit and vegetable consumption and mortality from all causes, cardiovascular disease, and cancer: systematic review and dose-response meta-analysis of prospective cohort studies. BMJ July 29, 2014;349:g4490.

[18] Hung HC, Joshipura KJ, Jiang R, Hu FB, Hunter D, Smith-Warner SA, Colditz GA, Rosner B, Spiegelman D, Willett WC. Fruit and vegetable intake and risk of major chronic disease. J Natl Cancer Inst November 3, 2004;96(21):1577–84.

[19] He FJ, Nowson CA, Lucas M, MacGregor GA. Increased consumption of fruit and vegetables is related to a reduced risk of coronary heart disease: meta-analysis of cohort studies. J Hum Hypertens September 2007; 21(9):717.

[20] He FJ, Nowson CA, MacGregor GA. Fruit and vegetable consumption and stroke: meta-analysis of cohort studies. Lancet January 28, 2006;367(9507):320–6.

[21] Appel LJ, Moore TJ, Obarzanek E, Vollmer WM, Svetkey LP, Sacks FM, Bray GA, Vogt TM, Cutler JA, Windhauser MM, Lin PH. A clinical trial of the effects of dietary patterns on blood pressure. N Engl J Med April 17, 1997;336(16):1117–24.

[22] Appel LJ, Sacks FM, Carey VJ, Obarzanek E, Swain JF, Miller ER, Conlin PR, Erlinger TP, Rosner BA, Laranjo NM, Charleston J. Effects of protein, monounsaturated fat, and carbohydrate intake on blood pressure and serum lipids: results of the OmniHeart randomized trial. J Am Med Assoc November 16, 2005; 294(19):2455–64.

[23] Chiu THT, Huang H-Y. Yen-Feng Chiu, Wen-Harn Pan et al Taiwanese Vegetarians and Omnivores: dietary Composition, Prevalence of Diabetes and IFG. PloS One 2014;11 9(2):e88547.

[24] Vang A, Singh PN, Lee J, Haddad E, Brinegar C. Meats, processed meats, obesity, weight gain and occurrence of diabetes among adults: findings from Adventist Health Studies. Ann Nutr Metab 2008;52: 96–104. https://doi.org/10.1159/000121365.

[25] Ye EQ, Chacko SA, Chou EL, et al. Greater whole-grain intake is associated with lower risk of type 2 diabetes, cardiovascular disease, and weight gain. J Nutr 2012;142:1304–13.

[26] Barnard ND, Cohen J, Jenkins DJ, et al. A low-fat vegan diet and a conventional diabetes diet in the treatment of type 2 diabetes: a randomized, controlled, 74-wk clinical trial. Am J Clin Nutr 2009;89:1588s–96s.

[27] Lim EL, Hollingsworth KG, Aribisala BS, et al. Reversal of type 2 diabetes: normalisation of beta cell function in association with decreased pancreas and liver triacylglycerol. Diabetologia 2011;54:2506–14.

[28] Barnard ND, Katcher HI, Jenkins DJ, et al. Vegetarian and vegan diets in type 2 diabetes management. Nutr Rev 2009;67:255–63.

[29] Lee YM, Kim SA, Lee IK, et al. Effect of a brown rice based vegan diet and conventional diabetic diet on glycemic control of patients with type 2 diabetes: a 12-week randomized clinical trial. PloS One 2016;11: e0155918.

[30] Mursu J, Virtanen JK, Tuomainen TP, Nurmi T, Voutilainen S. Intake of fruit, berries, and vegetables and risk of type 2 diabetes in Finnish men: the kuopio ischaemic heart disease risk factor study–. Am J Clin Nutr November 20, 2013;99(2):328–33.

[31] Bazzano LA, Li TY, Joshipura KJ, Hu FB. Intake of fruit, vegetables, and fruit juices and risk of diabetes in women. Diabetes Care 2008;31(7):1311–7. https://doi.org/10.2337/dc08-0080.

[32] Lean ME, Leslie WS, Barnes AC, Brosnahan N, Thom G, McCombie L, Peters C, Zhyzhneuskaya S, Al-Mrabeh A, Hollingsworth KG, et al. Primary care weight-management for type 2 diabetes: the cluster-randomised diabetes remission clinical trial (DiRECT). Lancet February 10, 2018;391(10120):541–51.

[33] C. Chaitong, Lean MEJ, Combet E. Lower carbohydrate and higher fat intakes are associated with higher haemoglobin A1c: findings from the UK national diet and nutrition Survey 2008-2016.

[34] Fretts AM, Follis JL, Nettleton JA, et al. Consumption of meat is associated with higher fasting glucose and insulin concentrations regardless of glucose and insulin genetic risk scores: a meta-analysis of 50,345 Caucasians. Am J Clin Nutr 2015;102:1266–78.

[35] Feskens EJ, Sluik D, van Woudenbergh GJ. Meat consumption, diabetes, and its complications. Curr Diabetes Rep 2013;13:298–306.

[36] Kim Y, Keogh J, Clifton P. A review of potential metabolic etiologies of the observed association between red meat consumption and development of type 2 diabetes mellitus. Metabolism 2015;64:768–79.

[37] Shan Z, Guo Y, Hu FB, Liu L, Qi Q. Association of low-carbohydrate and low-fat diets with mortality among US adults. JAMA Intern Med 2020;180(4):513–23. https://doi.org/10.1001/jamainternmed.2019.6980. [published online ahead of print, 2020 Jan 21].

[38] Kevin DH, Kong YC, Juen G, Yan Y LAJCN. Energy expenditure and body composition changes after an isocaloric ketogenic diet in overweight and obese men. Am J Clin Nutr August 2016;104(2):324–33.

[39] Goff LM, Bell JD, So PW, Dornhorst A, Frost GS. Veganism and its relationship with insulin resistance and intramyocellular lipid. Eur J Clin Nutr February 2005;59(2):291–8 [c].

[40] Gojda J, Patkova J, Jacek M, Potockova J, Trnka J, Kraml P, Andel M. Higher insulin sensitivity in vegans is not associated with higher mitochondrial density. Eur J Clin Nutr December 2013;67(12):1310–5.

[41] Oh YS, et al. Fatty acid-induced lipotoxicity in pancreatic beta-cells during development of type 2 diabetes. Front Endocrinol 16 July, 2018;9 384. https://doi.org/10.3389/fendo.2018.00384.

[42] Roberts C, Steer T, Maplethorpe N, Cox L, et al. National diet and nutrition Survey. Results from years 7-8 (combined) of the rolling programme (2014/15 to 2015/16). PHE publications; March 2018. Published.

[43] Fiolet T, Srour B, Selle L, Kesse-Guyot E, Allès B, Méjean C, et al. Consumption of ultra-processed foods and cancer risk: results from NutriNet-Santé prospective cohort. BMJ 2018;360:k322.

[44] Oyebode O, Gordon-Dseagu V, Walker A, et al. Fruit and vegetable consumption and all-cause, cancer and CVD mortality: analysis of Health Survey for England data. J Epidemiol Commun Health 2014;68:856–62.

[45] Aune D, Giovannucci E, Boffetta P, Fadnes LT, Keum NN, Norat T, Greenwood DC, Riboli E, Vatten LJ, Tonstad S. Fruit and vegetable intake and the risk of cardiovascular disease, total cancer and all-cause mortality—a systematic review and dose-response meta-analysis of prospective studies. Int J Epidemiol June 2017;46(Issue 3):1029–56.

[46] Aktar MW, Sengupta D, Chowdhury A. Impact of pesticides use in agriculture: their benefits and hazards. Interdiscipl Toxicol 2009;2(1):1–12. https://doi.org/10.2478/v10102-009-0001-7.

[47] Garthwaite D, Ridley L, Mace A., Parrish G et al reportPesticide usage Survey report 284, arable crops in the United Kingdom 2018, national statistics.

[48] Brookes G, Barfoot PGM. Crops: the global economic and environmental impact. The First Nine Years1996 − 2004 AgBioForum 2005;8(2&3):187–96 [AgBioForum].

[49] Baudry J, Assmann KE, Touvier M, et al. Association of frequency of organic food consumption with cancer risk: findings from the NutriNet-santé prospective cohort study. JAMA Intern Med. 2018;178(12): 1597–606. https://doi.org/10.1001/jamainternmed.2018.4357.

[50] Vainio H, Wilbourn J. Identification of carcinogens within the IARC monograph program. Scand J Work Environ Health 1992;18(1):64–73.

[51] Dich J, Zahm SH, Hanberg A, et al. Pesticides and cancer. Cancer Causes Control 1997;8:420–43. https://doi.org/10.1023/A:1018413522959.

[52] Jayakody N, Harris EC, Coggon D. Phenoxy herbicides, soft-tissue sarcoma and non-Hodgkin lymphoma: a systematic review of evidence from cohort and case-control studies. Br Med Bull 2015;114(1):75–94. https://doi.org/10.1093/bmb/ldv008.

[53] Alavanja MC, Bonner MR. Occupational pesticide exposures and cancer risk: a review. J Toxicol Environ Health B Crit Rev 2012;15(4):238–63. https://doi.org/10.1080/10937404.2012.632358.

[54] Kurenbach B, Marjoshi D, Carlos F, Amábile-Cuevas, Gayle C, Ferguson, Godsoe W, Gibson P, Jack A. Heinemann sublethal exposure to commercial formulations of the herbicides Dicamba, 2,4-Dichlorophenoxyacetic acid, and Glyphosate cause changes in antibiotic susceptibility in *Escherichia coli* and *Salmonella enterica* serovar typhimurium mBio. Mar 2015;6(2):e00009–15.

[55] John S.P, Martha L, Diana H, Robert LE, US department of agriculture, USDA pesticide data program: pesticide residues on fresh and processed fruit and vegetables, grains, meats, milk, and drinking water.

[56] Pesticide residues in food report published by the department for environment, food and rural affairs last updated 22nd January 2020 helena cooke pesticides policy health & safety executive chemicals regulation division, Mallard House, Kings Pool, 3 Peasholme Green, York YO1 7PX.

[57] Department for environment, food and rural affairs, the expert committee on pesticide residues in food (PRIF) report on the pesticide residues monitoring programme: quarter 1 2019. September 2019.

Macronutrients and micronutrients 23

Laura Gush[1], Sonal Shah[2] and Farah Gilani[3]

[1]*General Practitioner, National Health Service, Bridgend, United Kingdom;* [2]*General Practitioner, National Health Service, London, United Kingdom;* [3]*General Practitioner, Ayrshire Medical Group, National Health Service, Scotland, United Kingdom*

A healthy diet is a vital component of a healthy lifestyle. There are a myriad of different diets and eating plans which people follow for a variety of reasons, including religious and ethical beliefs, medical requirements or for proposed health benefits. Some of the popular diets include vegetarianism, veganism, ketogenic, paleo, macrocounting and intermittent fasting.

While there is evidence both for and against the health benefits of each of these, for most people it is advisable to eat a balanced diet. This will include the three main macronutrients, proteins, carbohydrates and fats, along with fibre and a range of micronutrients, which are vitamins and minerals.

This chapter will explain what macronutrients and micronutrients are, why they are important and how to counsel patients to find an eating plan which suits them as an individual.

Protein

Proteins are complex molecules composed of amino acids. They have numerous roles in the body, as well as acting as a source of energy. One gram of protein contains four kilocalories, which is the same amount of energy as carbohydrates but less than fat or alcohol.

Proteins act as enzymatic catalysts in numerous intracellular reactions, play a role in the structural scaffolding of cells and are a key component of muscles and connective tissues. They are important in cell signaling, give sperm their motility, have a role in the immune system as antibodies, carry oxygen around the body as haemoglobin and are also components of hormones.

Protein is the second most abundant compound in the human body after water, with approximately 43% contained in muscle, 15% in skin and 16% in the blood.

How much protein is required in the diet?

An individual's daily protein requirements depend on factors such as their age, sex, weight and activity level. It is also important to consider whether they are aiming to lose, gain or maintain weight and muscle mass. Recommendations vary between 0.6 and 0.8 g per kilogram of body weight per day for adults [1–3]. As an approximation, this equates to around 45 g for average sedentary females or 56 g for males. These values will be higher for those looking to build muscle or for people who are very active [4].

Older adults have increased protein requirements [5], with recommendations for those over 50 years being around 50% higher than the younger adult to prevent osteoporosis and sarcopenia. Patients recovering from injury also have increased protein requirements to allow tissue repair.

Various studies have shown that a high-protein diet can help aid weight loss. The mechanisms for this include increased satiety compared with high-carbohydrate or high-fat diets and increased thermogenesis which augments energy expenditure and has additional influences on satiety. In the long term, protein consumption is involved in muscle anabolism and, with appropriate overall energy balance, can favour the retention of lean muscle mass, which in turn raises metabolic rate [6–8].

Is it possible to consume too much protein?

There have been controversial claims in the popular media that excess protein intake can cause osteoporosis. It has been reported that the metabolism of protein, particularly from meat sources, can lead to an 'acidic environment' which causes calcium to leach from bones to mediate a buffering effect. This is said to be illustrated by protein intake leading to a spike in urinary calcium excretion [9]. However, these claims have been refuted. Experiments have been performed in which study participants ingested radio-labelled calcium isotopes before commencing a high-protein diet. Although urinary calcium excretion increased, the hypercalcuria was shown to be the radio-labelled isotope, showing that the excess calcium in the urine came from the diet and not from the bones [10]. Furthermore, research has also shown that a diet high in protein seems to cause upregulation of intestinal absorption of calcium, which would compensate for any increased calcium excretion, leading to no overall net loss or gain [11].

It has also been suggested that the processing of metabolites generated as a result of protein consumption places a high burden on the kidneys. One study showed that patients with chronic kidney disease on very low protein diets had slower deterioration of their eGFR and a slower progression to renal failure [12,13]. It was concluded that patients with a glomerular filtration rate of less than 25 mL/min/1.73 m^2 should consume no more than 0.6 g/kg/day of protein.

While patients with renal impairment may need to moderate their protein intake, several studies have demonstrated no negative effects on renal function in the healthy population [14–16]. Indeed, the main risk factors for renal disease are hypertension and diabetes, and it has been suggested that high-protein diets may improve blood glucose control and control of blood pressure [17].

It is evident that there is a limit to how much protein the body can use to build muscle at a given time. Protein intake above a certain level does not increase muscle protein synthesis [18,19]. Therefore, excess protein intake will lead to increased storage of fat.

Protein sources

Protein sources are described as 'complete' if they contain all nine essential amino acids which cannot be synthesised in the body. These include red meat, poultry, eggs, fish, milk, cheese and yoghurt. Vegetarian complete protein sources include soy, amaranth, buckwheat, hemp, seafood, seaweed, spirulina and quinoa. Examples of the protein contents of common foods are illustrated in Table 23.1.

Table 23.1 The left column includes animal sources and the right column includes plant sources of protein.

Protein content per 100 g of common protein sources [20,21]			
Food	**Protein content (g)**	**Food**	**Protein content (g)**
Chicken (grilled, no skin)	32.0	Red lentils	7.6
Beef steak (lean, grilled)	31.0	Chickpeas	8.4
Lamb chop (lean, grilled)	29.2	Kidney beans	6.9
Pork chop (lean, grilled)	31.6	Baked beans	5.2
Tuna (canned, in brine)	23.5	Tofu	8.1
Salmon (grilled)	24.2	Wheat flour	12.6
Prawns	22.6	Bread (white or brown)	7.9
Cod (uncooked weight)	18.0	Rice (boiled)	2.6
Eggs	12.5	Pasta (cooked)	6.6
Cow's milk (skimmed/ semi/full fat)	3.4/3.4/3.3	Oatmeal	11.2
Cottage cheese	12.6	Almonds	21.1
Cheddar cheese	25.4	Walnuts	14.7
Whole-milk yoghurt	5.7	Quinoa (cooked)	4.4
Low-fat yoghurt	4.8	Peanut butter	22.5

A typical daily portion of protein could include the following:

- Grilled chicken with two eggs and a glass of milk
- A salmon fillet with a matchbox-sized piece of cheddar cheese

Carbohydrates

Carbohydrates are a diverse group of substances, involved in a range of functions in the body. Principally, carbohydrates are substrates for energy metabolism. They also play a role in satiety, blood glucose control and lipid metabolism. When they are fermented in the gut, they are important in regulating the metabolism and balance of commensal flora. Furthermore, they are implicated in the health of the large bowel mucosa [22].

The carbohydrate family is diverse and includes everything from grains in bread and pasta, to table sugar, to the fibre found in fruit and vegetables. International recommendations are that diets should contain at least 400 g or five fruit and vegetables a day, excluding potato, sweet potatoes, cassava and other starchy roots [23].

At a chemical level, carbohydrates are molecules composed of carbon, hydrogen and oxygen, which bond to form 'sugar' or 'saccharide' units. The number of these units and how they are joined determines the properties of the carbohydrate. Broadly speaking, they are divided into three groups, as seen in Table 23.2.

Table 23.2 Classification of carbohydrates based on their chemical structure.

Class	Chain length	Subgroup	Examples
Sugars	1	Monosaccharides	Glucose Fructose Galactose
	2	Disaccharides	Sucrose (glucose and fructose) Lactose (glucose and galactose) Maltose (glucose and glucose)
Oligosaccharides (short-chain carbohydrates)	3−9	Malto-oligosaccharides (α-glucans)	Maltodextrins
		Non-α-glucans oligosaccharides	Raffinose Stachyose Verbascose
Polysaccharides	>10	Starch	Amylose Amylopectin Modified starches
		Fibres	Cellulose Hemicellulose Pectin Plant gums

Modified from Cummings J, Carbohydrate terminology and classification. Eur J Clin Nutr, 2007; 61; S5−S18.

Simple carbohydrates

The simplest sugar unit is glucose; it is the most abundant carbohydrate and is utilised as a source of energy by every cell in the body. Glucose is a monosaccharide; others include fructose and galactose. Most simple sugars occurring in nature are disaccharides, made of two units of sugar held together by a glycosidic bond.

Oligosaccharides

Oligosaccharides are composed of chains of three to eight basic sugar units. These occur in a limited number of plant foods such as beans and lentils, onions, garlic and artichokes. The tight bonding between sugar units means they cannot be digested in the small intestine and are instead fermented by resident colonic bacteria. They are often thought of as indigestible carbohydrates and are an example of prebiotics [24], which can modulate the composition and/or activity of the gut microbiota and confer a beneficial physiological effect on the host [25].

Naturally rich sources of prebiotics include the chicory root, leeks, onions, garlic, asparagus, bananas and some nuts. There is an emerging interest in the role of prebiotics and their effects on health, including modulating lipid metabolism [24] and satiety [25]. Prebiotics intake should be encouraged and included as part of a healthy, balanced diet.

Complex starchy carbohydrates

Starch, the principal carbohydrate in many diets, is the storage carbohydrate of plants such as cereals, root vegetables and legumes. Starch consists of long chains of glucose molecules loosely joined together [22]. During digestion, these chains are easily digested into single glucose units by enzymes found along the small intestine.

Food which are high in starch, such as potatoes, corn, oats and rice, are easily digested. Glucose enters the blood stream rapidly, causing glucose levels to rise very quickly. Foods such as these are described as having a high glycaemic index, which means they cause a rapid spike in blood sugar levels.

Complex fibrous carbohydrates/fibre

Complex carbohydrates, such as fibre, are made of long complex chains tightly bound together by glycosidic bonds. In nature, their role is to maintain a plant's structure; hence they are principally found in the plant cell wall. As the bonding between molecules is so tight, it is difficult for the body to break it down. Fibre, like resistant starch, cannot be digested and absorbed in the small intestine. Instead it moves to the colon, where it is digested by resident bacteria.

Not all fibres are the same. Broadly speaking, there are two types of fibre, soluble and insoluble. Soluble, or rapidly fermentable fibre, includes beans, oat bran, avocado and berries. The bacteria in the upper part of the large intestine can break these down relatively quickly. Insoluble, or poorly fermentable fibre, includes whole grains, legumes, wheat germ, beans, flaxseed, leafy vegetables and nuts.

Fibre, the new superfood?

Fibre used to be thought of as an inert component of food, related to the elimination of waste in the gut, but it has now been found to have a vast range of properties important in health and disease.

The European Food Safety Authority (EFSA) suggests that including fibre-rich foods in a healthy balanced diet can improve weight maintenance [26]. Foods that contain fibre tend to be fibrous in nature; mastication therefore requires time and effort. Prolonged exposure to food in the mouth allows time for the activation of signaling mechanisms that mediate satiety. This has been shown to reduce oral intake [27]. Satiety is also driven by gastric distension and delayed gastric emptying, which can be observed when individuals consume soluble fibre. Such fibre absorbs water, which increases its volume and viscosity [28].

Insoluble fibre that reaches the colon is fermented by commensal bacteria such as *Bifidobacteria* and *Lactobacilli*, which results in the production of gases and short-chain fatty acids (SCFAs). SCFAs provide a direct energy source for colonic epithelial cells and influence hepatic insulin sensitivity [29]. They also have antiinflammatory properties and may play a key role in the prevention of certain cancers, in particular colorectal cancer [30].

A diet rich in fibre is associated with delayed absorption of glucose, which helps to keep blood glucose levels stable [28,31]. Fibre is also known to play a role in reducing the risk of heart disease, type-2 diabetes, hypertension and high cholesterol [32].

Table 23.3 Recommended reference intake of fibre by age group.

Age (years)	Recommended intake of fibre
2–4	10 g per day
4–6	14 g per day
7–10	16 g per day
11–14	19 g per day
15–17	21 g per day
≥18	25 g per day

Table 23.4 Fibre content of different food sources.

Food	Serving size	Fibre content
Pear	One medium	5.5 g
Strawberry	One cup	3 g
Avocado	One avocado	6.7 g
Raspberries	One cup	8 g
Carrots (skin on)	One cup	6.9 g
Wholegrain bread	One slice	2.4 g
Apple	One medium	2.2 g
Artichoke	One artichoke	10.3 g
Broccoli	One cup	2.4 g
Popcorn (popped)	One cup	1.2 g
Lentils	100 g	3.7 g
Red kidney beans	100 g	6.5 g
Rolled oats	Half cup	4.5 g
Brown rice	One cup	2.7 g

Although there is variation across the world, most national guidelines stipulate that dietary fibre intake for both adult men and women should be above 25–30 g daily [33–35].

The recommended reference intake for children, based on guidance from the EFSA [36], is shown in Table 23.3, and foods that are good sources of fibre are described in Table 23.4.

What is the harm of eating processed carbohydrates?

Modern western diets are high in processed carbohydrates and saturated fats. This refers to whole food that has been refined in some way, often leading to the loss of fibre, vitamins and minerals [28]. Manufacturers often add sugars, artificial sweeteners and fats to improve taste and texture and to increase the shelf life. Examples of refined carbohydrates include pasta, especially white pasta, white rice, cakes, cookies, muffins and most baked desserts. These foods are often very calorific but provide no nutritional value and so are referred to as 'empty' calories. They are low in fibre and extremely palatable. They are digested quickly, leading to major swings in blood sugar levels, which can contribute to overeating [37].

Sugar

Many of the concerns about the overconsumption of carbohydrates refer to eating sugar.

'Free sugars' is an emerging term defined by the World Health Organization (WHO) as sugars that are added to foods, as well as those naturally present in honey, syrups and fruit juices [38]. It is evident that the consumption of free sugars can increase overall energy intake and may reduce the intake of foods containing more nutritionally adequate calories, leading to an unhealthy diet, weight gain and increased risk of noncommunicable diseases [39–41].

Diets that are persistently high in sugar disrupt the body's glucose metabolism, and a state of insulin resistance can develop, leading to diseases such as diabetes, obesity, dementia and cancer [42].

Over the past decade, global sugar consumption has grown from about 130 to 178 million tonnes [41]. The WHO's sugar guideline [38], issued in March 2015, recommends that adults and children should restrict their sugar intake to less than 10% of total energy intake per day, which is the equivalent of around 12.5 teaspoons of sugar for adults.

Fat

Fat, the final macronutrient, is a vital part of any diet. In addition to being a rich source of energy, it provides essential fatty acids, it is important in the absorption of vitamins A, D, E and K and it has a role in the production of hormones.

Fats are classified as saturated, unsaturated, polyunsaturated and trans-fats, depending on their chemical compositions.

Fat is composed of fatty acids and glycerol. Fatty acids are carboxylic acids and can either be saturated or unsaturated with hydrogen atoms. If there is just one double bond between carbon atoms, it is *unsaturated*; if there are multiple double bonds, it is described as *polyunsaturated*. *Trans-fats* are unsaturated fats which contain *trans* rather than *cis* double bonds as a result of the manufacturing process.

Saturated fats

Natural saturated fats are largely found in animal products such as fatty meats, eggs and dairy products including cheese, butter, ghee, lard and cream. Some vegetable fats, such as cocoa butter, palm oil and coconut oil, also contain saturated fat. These tend to be chemically very stable and solid at room temperature.

A recent review of high-fat dairy products such as whole milk and cheese showed no association with obesity or cardiovascular risk [43]. In fact, it has even been suggested that diets high in dairy fat may protect against obesity and improve blood pressure and the risk of diabetes in adults [44] although the mechanism for this is not clear.

Monounsaturated fats

Monounsaturated fats are found in olive oil, rapeseed oil, avocados, almonds, hazelnuts, peanuts and olives. In recent years, the consumption of a Mediterranean-style diet has been widely promoted. The backbone of this diet is the liberal use of olive oil. Olive oil contains polyphenols which have anti-oxidant properties. A large study looking at over 800,000 patients found that diets with a high intake of olive oil reduced the risk of death by 11%, cardiovascular events by 12% and stroke by 17% [45].

Polyunsaturated fats

Linoleic acid (LA) and α-linolenic acid (ALA) are essential fatty acids and are intrinsic to cell membranes and the structure of the central nervous system. They are precursors of eicosanoids, which are involved in inflammation, cardiac rhythm, vascular function and many other processes [46]. LA is the parent fatty acid of the omega-6 family, and ALA is the parent acid of the omega-3 family. Both LA and ALA can be saturated to form other fatty acids required by the body.

ALA is found in small amounts in meat, eggs, fish and aquatic plants. ALA can be converted to docosahexaenoic acid (DHA) and eicosapentaenoic Acid (EPA), which are considered to be the most important omega-3 fatty acids in nutrition. Good sources of foods which are rich in these include oily fish, seeds, walnuts and seaweed. Nonanimal sources of DHA and EPA include chia seeds, linseeds, hemp seeds, nuts and products made from soybeans.

Omega-3 fatty acids have a number of proven health benefits. Both DHA and EPA have antiin-flammatory properties [47] and may protect against cardiovascular disease [48], fatty liver, arthritis and cancer, especially colorectal cancer and hepatic tumours [49].

Currently, it is unclear over the exact quantities and in which forms, omega-3 should be consumed. Research is being conducted to address this.

Omega-6-containing foods include cooking oils, some nuts and sunflower seeds. Linoleic acid is converted to arachidonic acid, which is a precursor of eicosanoids. Eicosanoids are associated with promoting inflammation and cell damage, as well as increasing vascular thrombosis, vascular stenosis and the development of heart disease [48]. Despite evidence suggesting LA to be proinflammatory, in large study of 128,000 adults followed up over 32 years, it was found that those with higher intakes of LA had a lower risk of coronary heart disease, cancer and overall mortality [50].

It has been suggested that healthy diets should have a balance between the consumption of omega-6 and omega-3 fatty acids. Currently, there is no guidance on the exact ratio, but some studies suggest a ratio of 4:1 omega-6:omega-3. This is significantly different to the ratio that is seen in modern western diets, which tends to be between 15:1−30:1 [51].

Diets that involve eating oily fish once or twice a week, regular consumption of nuts and avoiding excessive use of vegetable oils may help to achieve this.

There are small amounts of trans-fats found naturally in dairy products, beef and lamb. These are not thought to be harmful to health. However, industrially produced trans-fats are a different story. Trans-fats are very cheap to produce, improve the taste and texture of food and are chemically ideal to be used repeatedly for deep-frying.

In the 1990s, concerns developed about the effect of trans-fats on the levels of unhealthy cholesterol in the blood. Research suggested that even small amounts of daily trans-fat intake had a significant effect on the blood lipid profile and were associated with a threefold increase in the risk of heart disease and sudden death [52].

In 1992, a landmark study [53] found those with the highest intake of trans-fats had a 50% increase in coronary heart disease. This prompted Denmark's Nutrition Council to monitor trans-fat in their food supply [54], and they estimated that 50,000 Danes were at high risk for developing cardiovascular disease. Given this, they introduced an outright ban on trans-fats in food.

The WHO held an expert consultation in 2008, which concluded that there was significant evidence to conclude that partially hydrogenated vegetable oils increase the risk of coronary heart disease, sudden cardiac death, metabolic syndrome components and diabetes [55]. The WHO subsequently asked all governments to take action to ensure that diets should have a trans-fatty acid intake of less than 1% of total energy intake [56,57].

Current recommendations

In the past half century, the dietary advice from the scientific community was to adopt a low-fat diet due to concerns related to cardiovascular disease. However, in the past few years, the understanding of fat has changed. There has been a shift in guidance, which now advocates the intake of polyunsaturated fats and suggests the avoidance of trans-fats.

Current expert opinion from the Food and Agriculture Association of the United Nations is that total fat should not exceed 30% of total energy intake and saturated fat intake should be less than 10% of total energy intake [58]. Fat is high in energy. A gram of fat provides 9 kcal (37 kJ) of energy compared with 4 kcal (17 kJ) for carbohydrate and protein. If energy intake exceeds energy expenditure, fat is stored in the body, and there is no limit to how much fat the body can store.

Advice regarding the safe consumption of fats is a hotly debated issue. There are ongoing controversies about the types of fats which should be included in the diet and in what quantities. Government guidelines around the world maintain the advice that reducing saturated fat consumption reduces the risk of heart disease and hypercholesterolaemia [59–61]. There is no clear evidence for its effect on blood pressure, stroke, type 2 diabetes, cancer or cognition [59].

One of the major controversies has been the development of a hypothesis which states that diets high in saturated fat do not cause heart disease. This has encouraged many to adopt high-fat diets or ketogenic diets [62,63]. Rather than the traditional view that cardiovascular disease is due to atherosclerosis, some believe it results from inflammation damaging myocardial cells [64]. It has been argued that inflammation results from a diet high in processed carbohydrates and sugars, insulin resistance and limited exercise, and it is these areas that need addressing rather than fat consumption.

There is, however, a clear and robust body of evidence linking high levels of low-density lipoprotein cholesterol with atherosclerosis and adverse outcomes. Therefore, it is prudent to reduce consumption of saturated fats as well as processed carbohydrates in conjunction with promoting physical activity and a healthy weight.

Although there is no agreed viewpoint among professionals, there are sensible approaches which can be adopted. Eating healthy fats from a wide variety of sources will ensure adequate quantities of the essential fatty acids which are needed for the body to function. Good sources include oily fish, whole milk, moderate amounts of cheese, eggs, avocados, olive oil, nuts, seeds and occasional red meat.

There is clear advice that trans-fats are harmful and only limited amounts should be consumed. This means avoiding foods such as solid margarines and food that is deep-fried or heavily processed and limiting the use of food made with partially hydrogenated vegetable oil.

Micronutrients

Vitamins and minerals are micronutrients essential for the body to function. They are substances which are required in tiny amounts compared with macronutrients, but without which we could not survive. Humans have been aware of the importance of micronutrients from as early as the 15th century, when sailors discovered that consuming citrus fruits containing vitamin C could cure scurvy. Vitamins and minerals are not synthesised in the body, meaning that they have to be obtained from external sources. The vast majority come from food, but some are attained from elsewhere, such as vitamin D which is made in the skin following exposure to sunlight [65].

Deficiencies in micronutrients are rare in modern Western society [66,67], but sales of vitamins and mineral supplements are higher than ever and are projected to rise over the coming years [68,69]. It is thought that most people take these supplements either because they believe that they do not consume enough through their diets [68] or as a form of 'insurance policy' ensuring that any potential dietary deficiencies are addressed through the use of a supplement [69]. Supplements are usually taken to achieve good health and well-being rather than because there is a need to address specific deficiencies. Multivitamin supplements, along with fish oils, are the most popular dietary supplements [68,69].

The functions of micronutrients and recommended daily intake

The roles of the various vitamins and minerals are explained in Table 23.5, along with their recommended daily amounts [65,66,70]. While these are general daily requirements, the specific recommended intake can depend on age, gender, growth and, for women, whether they are pregnant or breastfeeding [71].

Headlines often tout the health benefits of specific vitamins or minerals. For example, there has been great interest in the antioxidant properties of vitamins A, C, E and selenium and their potential to prevent cancer and reduce the incidence of chronic disease such as cardiovascular disease [65,67]. Moreover, multiple vitamins and minerals have been reported to improve the body's resistance to infection [72].

However, large-scale studies consistently fail to show tangible health benefits when diets are supplemented with these micronutrients. Studies have found no reduction in death or chronic disease such as heart disease or cancer relating to specific multivitamin or multimineral supplements [66,67,73–76]. Researchers do acknowledge that it is difficult to assess the benefits and harms of nutritional supplements, as, unlike drugs, the intake of vitamins and minerals cannot be controlled strictly in a trial as they occur naturally in foods [77].

There are, however, numerous studies showing that a varied, balanced diet, rich in fruits and vegetables, has multiple health benefits. In particular, the WHO has endorsed the Mediterranean diet as an effective means to prevent and control chronic diseases such as heart disease and cancer [78]. This is a largely plant-based diet, consisting of whole, natural, unprocessed foods and containing plenty of fresh fruit and vegetables, nuts and cereals [79]. It is suggested that vitamins and minerals work in synergy as part of a balanced diet, to create an effect that is more than the sum of their parts [80]. For example, iron is absorbed more efficiently when paired with vitamin C.

Table 23.5 Micronutrient function and recommended intake.

Vitamin/Mineral	Recommended daily amount	Major functions	Sources
Calcium	800 mg	Bone, muscle and teeth health, blood coagulation	Dairy, green leafy vegetables, nuts, fortified foods
Chromium	40 µg	Facilitating insulin function	Meat, wholegrains, lentils, broccoli
Copper	1 mg	Blood cell production, immune system and bone health, brain development	Nuts, shellfish, offal
Iodine	150 µg	Thyroid hormone synthesis	Fish, shellfish, cereals, grains
Iron	14 mg	Red blood cell production, immune system health	Red meat, lean poultry, beans, broccoli, kale, dried fruits, fortified foods
Magnesium	375 mg	Releasing energy from food, bone and muscle health	Green leafy vegetables, nuts, brown rice, fish, meat, dairy, bread
Manganese	2 mg	Breaking down food	Tea, bread, nuts, cereals, green vegetables
Phosphorus	700 mg	Bone and teeth health, releasing energy from food	Red meat, dairy, fish, poultry, brown rice, oats
Potassium	2000 mg	Cardiovascular health, fluid balance	Bananas, broccoli, brussel sprouts, pulses, nuts, seeds, fish
Selenium	55 µg	Immune-system health, prevents cell damage, antioxidant	Button mushrooms, broccoli, garlic, sardines, tuna, nuts, barley
Vitamin A	800 µg	Immune system, vision, skin health	Sweet potatoes, pumpkin, carrots, cantaloupe, squash, liver, eggs, dairy, oily fish
Thiamin (Vitamin B1)	1.1 mg	Nervous system, digestive health	Peas, eggs, wholegrains, fortified breakfast cereals
Riboflavin (Vitamin B2)	1.4 mg	Skin, vision, nervous system, digestive health	Milk, eggs, rice, fortified breakfast cereals
Niacin (Vitamin B3)	16 mg	Skin, nervous system, digestive health	Meat, fish, wheat flour, eggs, milk
Pantothenic acid (Vitamin B5)	6 mg	Digestive health	Meats, vegetables, wholegrains, oats
Pyridoxine (vitamin B6)	1.4 mg	Oxygen transport, releasing energy from food	Lean poultry, coldwater fish, potatoes, chickpeas, bananas
Folic acid (Vitamin B9)	200 µg	Red blood cell production	Fortified foods (breads, cereals, pastas, whole grains), beans, peas, green leafy vegetables
Vitamin B12	2.5 µg	Red blood cell production, using folic acid, releasing energy from food	Meat, fish, dairy, fortified breakfast cereals

Continued

Table 23.5 Micronutrient function and recommended intake.—cont'd

Vitamin/Mineral	Recommended daily amount	Major functions	Sources
Vitamin C	80 µg	Support the immune system, antioxidative, wound healing	Citrus fruits, green leafy vegetables, bell peppers, brussel sprouts, strawberries, papaya
Vitamin D	5 µg	Muscle, teeth and bone health	Sunlight (smaller amounts in oily fish, eggs, fortified foods)
Vitamin E	12 mg	Skin, eyes and immune system health	Nuts, seeds, plant oils, broccoli, spinach
Vitamin K	75 µg	Blood coagulation, wound healing	Green leafy vegetables, plant oils, cereal grains
Zinc	10 mg	New cell formation, releasing energy from food	Oysters, wheat germ, crab, lean meats, beans, yoghurt, chickpeas

The role of vitamin and mineral supplementation

In the majority of cases, supplements are only beneficial if an individual is not obtaining sufficient quantities of a micronutrient from their diet [65]. In the absence of deficiency, excesses of vitamins or minerals are usually excreted, rather than benefiting the body. In addition, focussing on replenishing vitamins and minerals via supplementation rather than focusing on a varied, whole-food diet, means the individual will miss out on the synergistic benefits of these micronutrients, when taken through foods which contain multiple micronutrients, fibre and macronutrients [65]. Although most people do not need to take supplements, there are some circumstances in which they are recommended [65]:

Vitamin D: There has been a growing spotlight on vitamin D in recent years with studies showing that many individuals in the northern hemisphere have low levels of vitamin D, which can impact on bone and muscle health. As a result, in some countries such as the United Kingdom (UK), it is now recommended that everyone considers taking a 10 µg vitamin D supplement between October and March, when there is insufficient sunlight for the production of vitamin D in the skin [65]. In other countries such as the United States (US), only high-risk groups are advised to consider taking a supplement. Such groups include individuals with limited skin exposure to sun, people with darker skin, those over the age of 65 years, pregnant and breastfeeding women and children up to the age of 5 years [81].

Folic acid: For women trying to conceive, or who are in the first 12 weeks of pregnancy, folic acid is recommended due to its role in reducing the incidence of neural tube defects in the developing baby. Supplementing with 400 µg/day is the general recommendation [65,66].

Country-specific recommendations: In the United Kingdom, there is thought to be an increased risk of deficiencies in vitamins A, C and D in young children, leading to the recommendation that children should take a daily supplement containing these vitamins [65]. In the United States, people over the age of 50 years are advised to consider B12 supplements, and pregnant women are advised to take iron supplements [66]. Individuals who are vegan may also benefit from a B12 supplement, as it can be difficult to obtain sufficient quantities of vitamin B12 from a wholly plant-based diet [65,66].

As recommended by a medical professional: Individuals may be advised by their doctor to take a specific supplement, such as iron or vitamin B12, if a deficiency is diagnosed.

In addition to these groups, studies have shown benefits with some specific supplements [65]. For example, a regular zinc supplement can reduce the likelihood of catching a cold, and taking it within 24 h of the onset of symptoms can reduce the length of illness by a day. However, the studies showing these benefits tend to be small, and the potential benefits have to be balanced against the costs of buying the supplement and possible side effects such as diarrhoea, vomiting and abdominal pain.

Potential harms of supplementation

As well as the cost implications of supplementing with vitamins and minerals, consideration must also be given to potential harms [65,66]. Vitamin C and the B vitamins are water soluble, which means they can be excreted in the urine if taken in excess. However, vitamins A, D, E and K are fat soluble, so are stored in the body in liver and adipose cells. This means that their levels can build up, leading to toxicity. For example, an excess of vitamin A can lead to gastrointestinal upset, headaches and blurred vision. Excess vitamin A can also cause birth defects if taken during pregnancy [65]. Some supplements can also interact with prescribed medications, with evidence showing that taking high levels of vitamins C and E can modulate some chemotherapy regimes [72]. Finally, there is evidence that excess amounts of certain vitamins and minerals, for example Vitamin A, can impair the immune system [72]. Thus, while it can be tempting to take vitamin and mineral supplements 'in case of' benefit, it is equally important to be aware of the potential harms.

Nutrition labels

There has been increasing global interest in nutrition labelling as a policy tool through which governments can guide consumers to make informed food purchases and healthier eating choices [82]. Food labels display the ingredients and composition of a food item, providing individuals with basic nutritional information. However, research has indicated that consumers generally regard standard nutrition labelling as complex and confusing, which is largely due to the use of technical terms and numerical data [83]. There is also a lack of awareness about what roles the nutrients featured on the labels play in the diet and in health. It is thought that older people and those with lower levels of education or income are the least likely to understand nutrition labels [84].

Each country has its own regulations regarding nutrition labelling. For some countries such as the United States, Australia, Canada and New Zealand, nutrition labelling on all prepackaged food products is mandatory. In other countries, these regulations are voluntary but become mandatory if a health claim is made or if there is food with special dietary uses. 'Front-of-pack' repeat nutritional information, and labelling for nonprepacked food, is also on a voluntary basis.

The exact information provided, and its format, varies from country to country. Table 23.6 lists the mandatory nutritional information that is most commonly required.

Table 23.6 Mandatory nutritional information on food labels.

	Per serving (g)	Per 100 g/100 mL	% daily value/reference intakes (RIs)
	Nutrient information is based on a specific amount of food, usually labelled as 'per serving'; this may allow individuals to compare it with their own intake, for example half a bar, or whole packet.	Most labels will give nutrient information for servings of 100 g or ml. This ensures that different foods can be compared directly. However, for many, this may limit its usefulness; for some people, the quantity of a food item they may eat is more than the suggested serving size or more than 100 g/mL, and so to calculate the nutritional content requires some basic mathematic skills.	This gives an amount of the nutrient in a specified amount of food and how these correlates to recommendations for healthy eating. RI based on an 'average' adult, with an estimated energy requirement of 8400 KJ/2000 kcal [19,85].
Energy Usually listed as kilocalories (kcal) or kilojoules (KJ)			
Total fat (g)		**Low fat means** 3 g or less per 100 g. **High fat means** 17.5g or more per 100 g.	
Saturated fat (g)	Note other names for ingredients high in saturated fat: animal fat/oil, beef fat, butter, chocolate, milk solids, coconut, coconut oil/milk/cream, copha, cream, ghee, dripping, lard, suet, palm oil, sour cream, vegetable shortening.	**Low saturated fat means** 1.5 g or less per 100 g. **High saturated fat means** 5 g or more per 100 g.	
Carbohydrate (g)			
Sugars (g)	Note for sugars: It is important to advise that added sugar may be described as dextrose, fructose, glucose, golden syrup, honey, maple syrup, sucrose, brown sugar, lactose, sucrose, raw sugar.	Low sugar means 5 g of total sugars or less per 100 g. High sugar means more than 22.5 g of total sugars per 100 g.	
Protein (g)			
Sodium/salt Note most labels include the term salt, as it felt it is more readily comprehensible to consumers than sodium. But if listed as sodium, then this can be multiplied by 2.5 to calculate the salt content.		**Low salt means** 0.3 g or less per 100 g (or 0.1 g sodium or 100 mg sodium). **High salt means** 1.5 g or more per 100 g (or 0.6 g sodium or 600 mg sodium).	

Prepackaged food should also contain an ingredient list, which states all the individual components used to make a product. The order of ingredients equates to their amount in the product, with the most abundant ingredient featuring first on the list.

In addition to mandatory labelling, some countries encourage the voluntary use of health rating systems or traffic light labelling on the front of packages. This provides readily understood information and assists consumers to easily compare similar items, for example breakfast cereals.

Food labelling is a key area in helping to support consumer choice, but the decision-making process of consumers also requires individuals to be able to decode such labels. This requires basic literacy and numeracy skills and may be influenced by socioeconomic status, the cultural norms of the individual and self-perception of psychophysical well-being. These factors must therefore be addressed by policymakers and those championing improvements in health and well-being.

Conclusions

Eating a combination of protein, carbohydrates, fats and micronutrients is essential for a balanced, healthy, sustainable diet. International consensus is that adults and children should be eating around 400 g of fruits and vegetables a day and that adults should eat at least 30 g of fibre daily. When discussing carbohydrate choices with patients, they should be encouraged to aim for more vegetables, fruits, legumes such as lentils and beans, nuts, seeds and whole grains including oats, quinoa and brown rice. Diets should limit the amount of processed foods that are consumed, and sugar intake should be restricted to less than 10% of energy intake. It is desirable to include a protein source with most meals and to opt for healthy fats such as those in avocados or olive oil.

Supplementation with vitamin pills is not necessary for most individuals, providing they consume a varied and balanced diet, which is predominantly plant based and which includes all the colours of the rainbow. Consideration should be given to taking a vitamin D supplement in the northern hemisphere, particularly in the winter months. And of course, if individuals are concerned about a specific deficiency of a vitamin or mineral, it is important to discuss further with a doctor.

Summary

- Macronutrients and micronutrients are the key elements of a healthy diet - they comprise protein, carbohydrates and fats, along with vitamins and minerals.
- Specific daily requirements depend on a range of variables including age, gender, weight, activity level and general health.
- Proteins have multiple functions, including acting as an energy source, and aiding in tissue growth and repair.
- Carbohydrates primarily provide energy but also have other roles including the regulation of satiety.
- A balanced diet should include healthy fats from a variety of natural sources.
- Vitamins and minerals have multiple functions and often work synergistically.
- Micronutrient deficiencies are rare in the developed world, with supplementation usually only beneficial in specific circumstances.
- To meet all of the body's nutritional requirements, it is important to maintain a good balance of macronutrients and micronutrients in the diet.

References

[1] British Nutrition Foundation. Nutrients, food and ingredients. [Online] https://www.nutrition.org.uk/nutritionscience/nutrients-food-and-ingredients.html.

[2] British Nutrition Foundation. Nutrient requirements. [Online] https://www.nutrition.org.uk/nutritionscience/nutrients-food-and-ingredients/nutrient-requirements.html?start=1.

[3] US department of agriculture. Dietary Guidance. National Agricultural library. [Online] https://www.nal.usda.gov/fnic/dietary-guidance-0.

[4] Lemon PW. Do athletes need more dietary protein and amino acids. Int J Sports Nutr June 1995;5:39−61.

[5] Gaffney S. Dietary protein requirements in elderly people for optimal muscle and bone health. J Am Ger Soc June 6, 2009;57:1073−9.

[6] Veldhorst MA. Gluconeogenesis and energy expenditure after a high-protein, carbohydrate-free diet. Am J Clin Nutr September 3, 2009;90:519−26.

[7] Johnston CS. Postprandial thermogenesis is increased 100% on a high-protein, low-fat diet versus a high-carbohydrate, low-fat diet in healthy, young women. J Am Coll Nutr February 1, 2002;21:55−61.

[8] Johnstone AM. Effect of overfeeding macronutrients on day to day food intake in man. Eur J Clin Nutr July 7, 1996;50:418−30.

[9] Ausman LM. Estimated net acid excretion inversely correlates with urine pH in vegans, lacto-ovo vegetarians, and omnivores. J Ren Nutr September 5, 2008;18:456−65.

[10] Kerstetter JE. The impact of dietary protein on calcium absorption and kinetic measures of bone turnover in women. J Clin Endo Metab 2005;90:26−31.

[11] Calvez J. Protein intake, calcium balance and health consequences. Eur J Clin Nutr 2012;66:281−95.

[12] Levey AS. Effects of dietary protein restriction on the progression of advanced renal disease in the Modification of Diet in Renal Disease Study. Am J Kid Dis 1996;27:652−63.

[13] Effects of dietary protein restriction on the progression of moderate renal disease in the Modification of Diet in Renal Disease Study. J Am Soc Nephrol 1997;8.

[14] Martin WF. Dietary protein intake and renal function. Nutr Metab (London) 2005;2.

[15] Manninen AH. High protein weight loss diets and purported adverse effects: where is the evidence? J Int Soc Sports Nutr 2004;1:45−51.

[16] Elswyck. A systematic review of renal health in healthy individuals associated with protein intake above the US recommended daily allowance in randomized controlled trials and observational studies. Adv Nutrit July 4, 2019;9:404−18.

[17] Gannon MC. An increase in dietary protein improves the blood glucose response in persons with type 2 diabetes. Am J Clin Nutri October 4, 2003;78:734−41.

[18] Deutz NEP, Wolfe RR. Is there a maximal anabolic response to protein intake with a meal? Clin Nutr 2013;32:309−13.

[19] Symons TB. A moderate serving of high-quality protein maximally stimulates skeletal muscle protein synthesis in young and elderly subjects. J Am Diet Assoc 2009;109:1582−6.

[20] The association of UK dietitians. Portion sizes. [Online] https://www.bda.uk.com/foodfacts/portion_sizes.

[21] British nutrition foundation. Protein. [Online] https://www.nutrition.org.uk/nutritionscience/nutrients-food-and-ingredients/protein.html?start=4.

[22] Cummings J. Carbohydrate terminology and classification. Eur J Clin Nutr December 2007;61:S5−18. Supplement.

[23] World Health organization. Diet, nutrition and the prevention of chronic diseases: report of a Joint WHO/FAO Expert Consultation. Geneva: WHO technical report series; 2003. No 916.

[24] Gibson R. Dietary modulation of the human colonic microbiota: introducing the concept of prebiotics. J Nutri 1995;125:1401−12. 6, s.l.

[25] Thomas L. Probiotics: a proactive approach to health. A symposium report. Br J Nutr December 1, 2015: S1−15.

[26] Scientific opinion on dietary reference value for carbohydrates and dietary fibre. Europ Food Safety Auth 2010;8:1462. Parma, Italy.

[27] Sakata T. A very-low-calorie conventional Japanese diet: its implications for prevention of obesity. Obes Res September 1995;3:233S−9S.

[28] Rebello C. Dietary fiber and satiety: the effects of oats on satiety. Nutri Rev February 2, 2016;74:131−47.

[29] Bach K. Microbial degradation of whole grain complex carbohydrates and impact on short chain fatty acids and health. Adv Nutr 2015;6:206−13.

[30] McNabney S, Henagan T. Short chain fatty acids in the colon and peripheral tissues: a focus on butyrate, colon cancer, obesity and insulin resistance. Nutrients 2017;9. 12, 12.

[31] Spector T. The diet myth. Orion books; 2015. p. 184. 9781780229003.

[32] Reynolds A, Mann J. Carbohydrate quality and human health: a series of systematic reviews and meta-analyses. The Lancet 2019. https://www.ncbi.nlm.nih.gov/pubmed/30638909.

[33] Nutrient Reference Values. Ministry of health; 2019. https://www.nrv.gov.au/nutrients/dietary-fibre.

[34] Promotion, office of disease prevention and health. Dietary guidelines for Americans 2015−2020. 8th ed. 2015.

[35] England, public health. Government dietary recommendations. 2016.

[36] (EFSA), European Food Safety Authority. Dietary reference values for nutrients. 201714. e15121.

[37] Ludwig D. High glycemic index foods, overeating, and obesity. Paedatrics March 3, 1999;103.

[38] World Health Organisation. Guideline: sugars intake for adults and children. 2015. Geneva : s.n.

[39] Hauner H, Bechthold A. Evidence-based guideline of the German Nutrition Society: carbohydrate intake and prevention of nutrition related disease. Ann Nutr Metab 2012;60:1−58.

[40] Malik V, Pan A. Sugar-sweetened beverages and weight gain in children and adults: a systematic review and meta-analysis. Am J Clin Nutr 2013;98:1084−102.

[41] International, World Cancer Research Fund. Curbing global sugar consumption: effective food policy actions to help promote healthy diets and tackle obesity'. 2015.

[42] Orgel E. The links between insulin resistance, diabetes, and cancer. Curr Diab Rep 2014. https://doi.org/10.1007/s11892-012-0356-6. Aorik.

[43] Kratz M. The relationship between high fat dairy consumption and obesity, cardiovascular and metabolic disease. Eur J Nutr 2013;52:1−24.

[44] Yu E. Dairy products, dairy fatty acid and the prevention of cardiometabolic disease: a review of recent evidence. Curr Atherosclerosis Rep 2018;Vol. 20:s.l.

[45] Schwingshack HL. Monounsaturated fatty acids, olive oil and health status: a systematic review and meta-analysis of cohort studies. Lipids in health and disease. Lipis Health Dis 2014;Vol. 1:154.

[46] Forouhi N, Krauss R. Dietary fat and cardiometabolic health: evidence, controversies and consensus for guidance. BMJ 2018;361.

[47] Calder P. Omega-3 fatty acids and inflammatory processes: from molecules to man. Biochem Soc Trans October 2017;15. 45.

[48] Patterson E, Wall RA. Health implications of high dietary omega-6 polyunsaturated fatty acids. J Nutr Metab 2012;2012:539426. https://doi.org/10.1155/2012/539426.

[49] Weylandt K, Serini S. Omega-3 polyunsaturated fatty acids: the way forward in times of mixed evidence. Biomed Res Int August 2015;(2).

[50] Wang DD, Li Y. Association of specific dietary fats with total and cause specific mortality. JAMA Int Med 2016;176:1134−45.

[51] Simopoulos A. The importance of the omega-6/omega-3 fatty acid ratio in cardiovascular disease and other chronic diseases. Exp Biol Med June 6, 2008;233.

[52] Spector T. The diet myth. Orion books; 2015. p. 87. 9781780229003.

[53] Willett W. Intake of trans fatty acids and risk of coronary heart disease among women. Lancet 1993;341: 581−5.

[54] The World Health organisation. who.int. [Online] 14 5 2018. https://www.who.int/news-room/feature-stories/detail/denmark-trans-fat-ban-pioneer-lessons-for-other-countries.

[55] Food and agriculture organisation of the United Nations. Fats and fatty acids in human nutrition: report of an expert consultation. 2008.

[56] Nishida C, Uauy R. WHO Scientific Update on health consequences of trans fatty acids: introduction. Eur J Clin Nutr 2009;63:S1−4.

[57] Allen K. Potential of trans fats policies to reduce socioeconomic inequalities in mortality from coronary heart disease in England: cost effectiveness modelling study. The BMJ September 2015;351.

[58] FAO Food and nutrition paper 91. Fats and fatty acids in human nutrition: report of an expert consultation. Food and Agriculture Organization of the United Nations; 2010.

[59] British Nutrition foundation. British Nutrition foundation. 2012 [Online] https://www.nutrition.org.uk/nutritionscience/nutrients-food-and-ingredients/fat.html?start=3.

[60] American Heart Association. The American Heart Association Diet and Lifestyle Recommendations. 2017 [Online], https://www.heart.org/en/healthy-living/healthy-eating/eat-smart/nutrition-basics/aha-diet-and-lifestyle-recommendations.

[61] National Health and Medical Research Council. Australian dietary guidelines. Eat for health. 2013. p. 1864965789.

[62] Oh K. Dietary fat intake and the risk of coronary heart disease in women: 20 years of follow up of the nurses health. Am J Epidemiol 2005;1. 161.

[63] Ieosdottir M. Cardiovascular event risk in relation to dietary fat intake in middle aged individuals: data from the Malmo diet and cancer study. Eur J Cardiovasc Prev Rehabil 2007;14.

[64] Malhotra A. Saturated fat does not clog up arteries: CHD is a chronic inflammatory condition. Bri J Sports Med 2018;51. 15.

[65] Supplements Who needs them? NHS Choices. A behind the headlines report. June 2011. https://www.nhs.uk/news/2011/05May/Documents/BtH_supplements.pdf.

[66] Multivitamin/mineral supplements. Fact sheet for health professionals. National Institutes of Health. Office of Dietary Supplements. Updated October 17, 2019. https://ods.od.nih.gov/factsheets/MVMS-HealthProfessional/.

[67] Schwingshackl L, et al. Dietary supplements and risk of cause-specific death, cardiovascular disease, and cancer: A systematic review and meta-analysis of primary prevention trials. Adv Nutr 2017;8:27−39.

[68] Mintel - "Vitamins and supplements" report. September 2016. UK.

[69] Dietary supplements market size analysis report by ingredient (botanicals, vitamins). May 2019. by form, by application (immunity, cardiac health), by end user, by distribution channel, and segment forecasts, 2019-2025. Grand view research.

[70] NHS Choices. Vitamins and minerals https://www.nhs.uk/conditions/vitamins-and-minerals/.

[71] British Nutrition Foundation. Nutrition requirements. 2017. https://www.nutrition.org.uk/attachments/article/261/Nutrition%20Requirements_Revised%20Oct%202017.pdf.

[72] Alpert PT. The role of vitamins and minerals on the immune system. Home Health Care Manag Pract 2017; 29(3):199−202.

[73] Jenkins, et al. Supplemental vitamins and minerals for CVD prevention and treatment. J Am Coll Cardiol 2018;71(22):2570−84.

[74] Bjelakovix G, et al. Antioxidant supplements for prevention of mortality in healthy participants and patients with various diseases. Cochrane Database Syst Rev 2012;3.

[75] Macpherson H, et al. Multivitamin-multimineral supplementation and mortality: a meta-analysis of randomized controlled trials. Am J Clin Nutr 2013;97:437—44.

[76] Mochamat, et al. A systematic review on the role of vitamins, minerals, proteins, and other supplements for the treatment of cachexia in cancer: a European Palliative Care Research Centre cachexia project. J Cachexia Sarcopenia Muscle 2016;8:25—39.

[77] Chang SM. Should meta-analyses trump observational studies? Am J Clin Nutr 2013;97(2):237—8.

[78] Renzella J, et al. What national and subnational interventions and policies based on Mediterranean and Nordic diets are recommended or implemented in the WHO European Region, and is there evidence of e ectiveness in reducing noncommunicable diseases? World Health Organisation; Health evidence network synthesis report 58. 2018.

[79] Sofi F, et al. Adherence to Mediterranean diet and health status meta-analysis. BMJ 2008;337:a1344.

[80] Larsson SC. Dietary approaches for stroke prevention. Stroke 2017;48:2905—11.

[81] Vitamin D. Fact sheet for health professionals. National Institutes for Health. Office of dietary supplements; August 2019. https://ods.od.nih.gov/factsheets/VitaminD-HealthProfessional/Updated.

[82] World health organisation. Guiding principles and framework manual for front-of-pack labelling for promoting healthy diet. Geneva : s.n.: Department of nutrition for health and development; 2019.

[83] Cowburn G. Consumer understanding and use of nutrition labelling: a systematic review. Public Health Nutr 2004;vol. 8:21—8.

[84] World Health Organisation. Nutrition labels and health claims: the global regulatory environment. 2004.

[85] Council of the European Union. European Commission. Regulation (EU) No 1169/2011 of the European Parliament and of the Council on the provision of food information to consumers. 2011. Brussels : s.n.,

Caffeine in health and disease: a brief overview

24

Emmajane Down

General Practitioner, National Health Service, London, United Kingdom

Introduction

Caffeine is the most commonly used drug in the world. It has been implicated as a cause of, or contributing factor to, a range of diseases, and conversely has been utilised as a prescribed therapy. This chapter shall explore the roles of caffeine in health and disease and will describe the current global recommendations for safe caffeine consumption.

A brief history of caffeine

The human population is thought to have been enjoying coffee since the 1400s when coffee beans were first roasted and brewed in Yemen. Tea originated in Southwest China, and both coffee and tea were imported to Europe in the 16th century. Although we have been drinking these beverages for hundreds of years, it was not until 1819 that the German scientist, Friederich Runge, identified the drug, caffeine, which is now the world's most commonly used psychoactive drug.

Coffee is the most frequently consumed beverage around the world. Nearly 90% of the American population are reported to enjoy it [1]. As a consequence, its pharmacology, physiological effects and safety have long been topics of discussion and research.

What is caffeine and what effects does it have on the body?

Caffeine is a stimulant, which enhances mood and promotes wakefulness [2]. In nonhabitualised subjects, caffeine also causes diuresis, bronchodilation and a rise in systolic blood pressure [3].

The precise mechanism of action of caffeine is not entirely clear: its major effects are mediated through the inhibition of adenosine receptors, and it also acts to inhibit phosphodiesterase enzymes, leading to an increase in intracellular levels of cyclic adenosine monophosphate (cAMP). Furthermore, caffeine has additional effects in mobilising intracellular calcium and has activity at benzodiazepine and dopamine receptors [4].

Caffeine levels peak in the blood within 1 h after ingestion and its effects last for around 3 h. It is broken down in the liver and excreted in urine.

Studies have reported that caffeine can be useful in improving the performance of sleep-deprived individuals, with some evidence showing that shift workers who use caffeine make fewer mistakes due to drowsiness [5]. During periods of sleep deprivation, caffeine can act to enhance alertness and vigilance, which is an effective aid for special operations military personnel, as well as athletes during times of exhaustive exercise that requires sustained focus [6].

In the sporting arena, caffeine has been reported to be an effective ergogenic aid for sustained maximal endurance activity and to enhance time trial performance. Furthermore, it has been demonstrated that caffeine can enhance glycogen synthesis during the recovery phase of exercise. Caffeine may also play a role in maintaining the performance of athletes during prolonged high-intensity exercise such as soccer, field hockey and rowing, but the effects of caffeine on strength and power activities are inconsistent [7].

The amount of caffeine required to boost performance in athletes is that regarded as being within a 'normal' daily intake, which equates to less than four cups a day — higher doses are more likely to be detrimental to performance [6].

Caffeine intake was limited in 1984 at the Olympic games, but the restriction was lifted in 2004.

In the clinical setting, caffeine has been used as an adjuvant to analgesia in the treatment of headaches [8], migraines [9] and in the management of premature neonatal apnoea [10]. It has also been used therapeutically in post-prandial hypotension in the elderly [11], to enhance seizure duration during electroconvulsive therapy and to treat post-lumbar puncture cerebrospinal fluid leaks.

Caffeine and genetics

There is growing evidence that an individual's genetic make-up plays a role not only in their responses to caffeine but also in their level of consumption [12]. Variability in the rate of caffeine metabolism and its physiological effects is partly mediated through inherited variants in the enzyme cytochrome P-450 and in caffeine's main target receptors A1R and A2AR [13]. Genetic factors are also important in explaining why some individuals are more sensitive to the anxiogenic effects of caffeine [14] and others are more susceptible to sleep disturbances and insomnia [15].

Adverse effects of caffeine

For most individuals, the consumption of three or four cups of coffee a day is not considered to be harmful. However, some people can experience feelings of anxiety, hyperactivity, nervousness and sleep disturbance due to caffeine intake. For individuals with an underlying anxiety or panic disorder, caffeine can be a symptomatic trigger. The physiological and behavioural effects of caffeine vary with chronic versus acute exposure, the dose consumed and the mode of administration [14].

Tolerance to the acute effects of caffeine develop rapidly [16] but caffeine-naive individuals experience different effects to habitual users, and the effects are more pronounced. At low doses, psychological effects include mild euphoria and alertness [17], whereas high doses can produce nausea, anxiety, trembling and jitteriness in sensitive individuals [18].

Caffeine dependence

Caffeine consumption can lead to a mild physical dependence. As it does not threaten physical, social or economic health, it is not classified as addictive and therefore is not considered to be a major health concern. Habitual users may experience mild withdrawal symptoms if they stop drinking caffeine. These symptoms can include headaches, fatigue, depressed or irritable mood, difficulty concentrating or flu-like symptoms [19]. The effects are short lived, lasting between 2 and 4 days.

There is, however, a more serious caffeine dependency recognised by the World Health Organisation. The diagnosis is reserved for those who are significantly distressed and functionally impaired by their dependence to an extent they require formal treatment. However, it does not yet feature in the Diagnostic and Statistical Manual of Mental Disorders as its clinical significance is unclear [20].

Caffeine, chronic disease and other health effects

There is no robust evidence that moderate caffeine intake increases the risk of developing hypertension, cardiac dysrhythmias, myocardial infarction or cancer [21,22]. In the short term, it can transiently raise blood pressure and heart rate, but these effects are short lived [23].

Caffeine and pregnancy

It is suggested that high levels of caffeine intake could cause low birthweight babies [24]. As a result of this, The World Health Organisation has recommended that the safest approach is for pregnant women to reduce caffeine intake to 200 mg per day, which is equivalent to two cups of instant coffee. There is no strong clear evidence that caffeine is responsible for miscarriages or infertility [25].

Caffeine bone health

Caffeine has been reported to have an adverse effect on bone health and may be associated with osteoporosis. However, a definitive relationship has not been established, as most studies considering this effect have been carried out in populations with suboptimal calcium intake [26] so are complicated by confounding factors.

Caffeine overdose

Although caffeine is freely available, it is important to acknowledge that it is a drug and there have been some reports of death due to overdose [27]. However, this is extremely rare and is likely to result from consuming extremely high quantities of caffeine in pill form. Toxic levels are equal to approximately 10 g of caffeine, which would be virtually impossible to drink as it equates to around 80−100 cups of coffee.

The World Health Organisation advises that caffeine is safe if consumed in low-to-moderate levels. That equates to 400 mg, which is roughly four cups of coffee per day.

Are there any health benefits of caffeine?

In the past, studies investigated a possible link between regular caffeine intake and a reduced risk of some cancers, including brain tumours [28], head and neck malignancy, prostate cancer and hepato-cellular carcinoma [29]. However, results did not provide conclusive evidence of a protective effect [29].

It has also been suggested that caffeine may protect against dementia and/or other neurological disorders. Whilst animal models have indicated that this hypothesis warrants further investigation, a link remains to be confirmed [30].

Which drinks and food contain caffeine?

Caffeine is legal and unregulated. The exact amount in different foods and drinks varies and does not need to be listed on the packaging, leading to potential confusion for the consumer. For an example, the exact amount of caffeine in 'real' coffee varies depending on how finely coffee beans are ground, the darkness of the roast and the brewing method used. Freshly brewed coffee contains between 70 and 140 mg caffeine per cup, instant coffee contains 30–90 mg, and tea between 15 and 70 mg.

Caffeine is also found in kola nuts, some fizzy drinks and cocoa pods used in the production of chocolate. Synthetic caffeine is added to some medicines, such as pain and migraine tablets, as well as energy drinks. Table 24.1 illustrates the caffeine content of some popular drinks and tablets used worldwide:

Table 24.1 Caffeine content of beverages and medications.

Drink	Caffeine content (mg)
Starbucks brewed coffee small	150
Starbucks espresso	75
Starbucks americano small	75
Starbucks green tea	25
Starbucks English breakfast tea	40
Instant coffee	60
Coke fizzy drink (1 can)	34
Diet coke (1 can)	46
Energy drink: Red Bull (1 can)	80
Anacin tablets × 2 (aspirin and caffeine)	64

Conclusion

Caffeine is the most commonly used drug around the world. It may have beneficial effects in enhancing mood and promoting alertness and can enhance athletic performance. Whilst low-to-moderate levels of consumption appear to be safe, it can trigger adverse psychological effects such as anxiety and insomnia, and in the short term it can raise blood pressure and heart rate. It is important to be aware that water is always the best option to quench thirst, and if caffeine is being used to manage tiredness or stress, the individual should be encouraged to address the underlying causes for this, rather than automatically reaching for a caffeine hit.

Summary

- Caffeine is the most commonly used drug in the world.
- Responses to caffeine depend on genetic factors.
- Most of its effects are mediated through the inhibition of adenosine receptors.
- Caffeine is a stimulant, which enhances mood and promotes wakefulness.
- In nonhabitualised subjects, caffeine causes diuresis, bronchodilation and a rise in systolic blood pressure.
- Caffeine may improve performance in athletes and sleep-deprived individuals.
- Caffeine can trigger anxiety, hyperactivity, nervousness and sleep disturbance.
- There is no robust evidence that moderate caffeine intake increases the risk of developing hypertension, cardiac dysrhythmias, myocardial infarction or cancer.
- Caffeine can cause a mild physical dependence.
- Caffeine intake should be restricted during pregnancy.
- It is recommended that the maximum daily consumption of caffeine does not exceed 400 mg, which equates to around four cups of coffee.
- If caffeine is being used to manage tiredness or stress, an individual should be encouraged to address the underlying causes for this, rather than automatically reaching for a caffeine hit

References

[1] Frart CD, et al. Food sources and intakes of caffeine in the diets of persons in the United States. J Am Diet Assoc January 2005;105(1):110.

[2] Nehlig A, et al. Caffeine and the central nervous system: mechanisms of action, biochemical, metabolic and psychostimulant effects. Brain Res Brain Res Rev 1992;17(2):139−70.

[3] Mosqueda-Garcia R, et al. The cardiovascular effects of caffeine. In: Garett, editor. Caffeine. Raven New York: Coffee and Health; 1993. p. 157−76.

[4] A.H. Nall et al. Caffeine promotes wakefulness via dopamine signaling in *Drosophila*. Published online 2016 Feb 12. https://doi.org/10.1038/srep20938.

[5] Ker K, Edwards PJ, et al. Caffeine for the prevention of injuries and errors in shift workers. Cochrane database of systematic reveiws; 2010.

[6] Burke L. Caffeine and sport performance. Appl Physiol Nutr Metabol 2009;33(6):1319−34.

[7] Goldstein ER, et al. International society of sports nutrition position stand: caffeine and performance. J Int Soc Sports Nutr 2010;7:5. https://doi.org/10.1186/1550-2783-7-5.

[8] Migliardi JR, et al. Caffeine as an analgesic adjuvant in tension headaches. Clin Pharmacol Ther November 1994;56(5):576−86.

[9] Richard B, Lipton MD, et al. Efficacy and safety of acetaminophen, aspirin, and caffeine in alleviating migraine headache pain three double-blind, randomized, placebo-controlled trials. Arch Intern Med 2000.

[10] Scmidt B, et al. Long-term effects of caffeine therapy for apena pf prematurity. N Engl J Med November 8, 2007;357(19):1893−902.

[11] Heseltine D, et al. The effects of caffeine on postprandial blood pressure in the frail elderly. Postgrad Med J June 1991;67(788):543−7.

[12] Yang A, et al. Genetics of caffeine comsumption and responses to caffeine. Psychopharmacology(Berl) August 2010;211(3):245−57.

[13] Daly JW, et al. The role of adenosine receptors in the central action of caffeine. Pharmacopsychoecolgia 1994;7(2):201−13.

[14] Silverman K, Griffiths RR. Low-dose caffeine discrimination and self-reported mood effects in normal volunteers. J Exp Anal Behav 1992;57:91−107.

[15] Bchir F, et al. Differences in pharmacokinetic and electroencephalographic responses to caffeine in sleep-sensitive and non-sensitive subjects. CR Biol 2006;329:512−9.

[16] Ferré S. An update on the mechanisms of the psychostimulant effects of caffeine. J Neurochem 2008;105:1067−79.

[17] Lieberman HR, et al. The effects of low doses of caffeine on human performance and mood. Psychopharmacology 1987;92:308−12.

[18] Daly JW, Fredholm BB. Caffeine—an atypical drug of dependence. Drug Alcohol Depend 1998;51:199−206.

[19] Juliano L, et al. Characterization of individuals seeking treatment for caffeine dependence. Psychol Addict Behav 2012;26:948−54.

[20] Meredith SE, et al. Caffeine use disorder: a comprehensive review and research agenda. J Caffeine Res September 2013;3(3):114−30.

[21] Gronros N, et al. Diet and risk of atrial fibrillation epidemiologic and clincal evidence. Circ J 2010;74(10):2029−38.

[22] Loomis, et al. Carcinogenicity of drinking coffee, mate, and very hot beverages. Lancet Oncol 2016.

[23] Green PJ, Kirby R, Suls J. The effects of caffeine on blood pressure and heart rate: a review. Ann Behav Med September 1996;18(3):201−16.

[24] Chen L-W. Maternal caffeine intake during pregnancy is associated with risk of low birth weight: a systematic review and dose−response meta-analysis. BMC Med 2014;12:174.

[25] Ricci E, et al. Coffee and caffeine intake and male infertility: a systematic review. Nutr J June 24, 2017;16(1):37.

[26] Heaney RP, et al. Effects of caffeine on bone and the calcium economy. Food Chem Toxicol September 2002;40(9):1263−70.

[27] Jones AW. Review of caffeine-related fatalities along with postmortem blood concentrations in 51 poisoning deaths. J Anal Toxicol 2017;41:167−72.

[28] Holick CN, Smith SG, Giovannucci E, Michaud DS. Coffee, tea, caffeine intake and risk of adult glioma in 3 prospective cohort studies. Cancer Epidemiol Biomark Prev 2010;19:39−47.

[29] Bravi F, Tavani A, Bosetti C, Boffetta P, La Vecchia C. Coffee and the risk of hepatocellular carcinoma and chronic liver disease: a systematic review and meta-analysis of prospective studies. Eur J Cancer Prev September 2017;26(5):368−77.

[30] Driscoll I, et al. Relationships between caffeine intake and risk for probable dementia or global cognitive impairment: the women's health initiative memory study. J Gerontol A Biol Sci Med Sci 2016;71(12):1596−602.

Water: how much should be consumed and what are its health benefits?

25

Ekua Annobil

General Practitioner, Sydney, NSW, Australia

Water, water, everywhere, Nor any drop to drink!

Unlike the unfortunate woebegone Ancient Mariner whose seafaring exploits surrounded by gallons of undrinkable salt water were chronicled by the English poet Samuel Taylor Coleridge, the majority of inhabitants of the developed world are fortunate enough to have access to unlimited supplies of clean, safe drinking water.

Despite this readily available resource being at their disposal any time of the day or night, many overlook the apparent benefits of consuming this colourless, odourless, tasteless liquid, instead opting for all manner of flavoured, fruit-derived, fizzy or hot beverages.

So, what are the advantages of drinking water, and how much, if any, should be consumed each day?

The percentage of water in the human body varies according to both the age and gender of the individual, ranging from 50% to 75% of the total body composition. The average adult human body is made up of as much as 65% water, with babies' bodies often containing close to 75%, dropping to 65% by 1 year of age [1].

Water in the body

Two-thirds of the body's water is stored located intracellularly, with the remainder comprising extracellular fluid and blood [1]. Water is mainly lost from the body as urine, with variable amounts also being lost as sweat, through breath and in faeces. Losses need to be replaced for optimum physiological functioning and to avoid dehydration.

What are an individual's daily water requirements?

There are many schools of thought regarding daily water requirements. Some advise specific volumes, for example 2 L over a 24-h period, whereas others advocate a 'don't get thirsty' approach to hydration, encouraging frequent sips throughout the day.

According to the results of a study conducted by Michael Farrell at Monash University in Melbourne, Australia, individuals are able to subconsciously self-regulate their water intake. He reported that swallowing feels difficult when in a state of overhydration, thus deterring the individual from drinking more [2].

A Prescription for Healthy Living. https://doi.org/10.1016/B978-0-12-821573-9.00025-4

Farrell suggests that this discovery is evidence that controversial advice to deliberately drink fluids in the absence of thirst is wrong: '*It shows we have several very subtle mechanisms for regulating the amount we drink. If left to your own devices, you will drink the requisite amount of water to maintain balance*'.

Furthermore, Farrell states that even people doing exercise just need to drink according to their levels of thirst rather than consuming excessive quantities of fluids. '*These are well refined mechanisms forged on the anvil of evolution*' [2].

How much water are individuals currently advised to drink?

When questioned, the majority of individuals believe that they do not drink enough water. Around the world, a common recommendation is eight glasses a day, but this is not based on identifiable scientific data. No one really knows where the eight glasses concept originated from, but it has been widely quoted and circulated by doctors, nutritionists and allied health and wellness professionals over several decades.

It is possible that the advice was derived from a 1945 recommendation by the US National Research Council (NRC) that adults should consume 1 mL of water for each calorie of food eaten. This adds up to approximately two and a half litres per day for men and 2 L for women. The recommendation also stated that much of this water could be obtained from foods consumed, though this appears to have been overlooked over time.

According to Barbara Rolls, nutrition researcher and author of the 1984 book *Thirst* [3], this amount is 'appropriate for people in a temperate climate who are not exercising vigorously'.

The United Kingdom's National Health Service (NHS) Choices webpage about Water and Drinks [4] states that while some water can be obtained from food '*Your body needs water or other fluids to work properly and to avoid dehydration. That's why it's important to drink enough fluids. In climates such as the UK's, this would equate to approximately 1.2 litres (six to eight glasses) being consumed every day to prevent dehydration. In hotter climates, the body may require significantly larger volumes than this*'.

Guidelines vary globally, with the National Academies of Sciences, Engineering, and Medicine in the United States (US) proposing that an adequate daily fluid intake is 15.5 cups (3.7 L) of fluids for men and 11.5 cups (2.7 L) of fluids a day for women in the US population [5]. These recommendations include fluids from water, other beverages and food. About 20% of daily fluid intake usually comes from food, with the remainder being consumed as drinks.

Scientific recommendations given by the European Food Safety Authority (EFSA) relating to water consumption appear to loosely support the US guidelines. The EFSA released its Scientific Opinions on Dietary Reference Values for Water and on Water-Related Health Claims in March 2010 and April 2011, respectively. These opinions were based on an extensive review of scientific evidence. They clarified the appropriate reference values for water intake among the population in Europe according to age and physiological status and confirmed the importance of water for the maintenance of basic physiological functions such as thermoregulation, physical and cognitive function [6].

Does water intake vary between different populations?

The National Health and Nutrition Examination Survey (NHANES) is an American cross-sectional survey designed to monitor the health and nutritional status of the civilian noninstitutionalised US population. The survey is conducted by the National Centre for Statistics at the Centre for Disease

Control and Prevention (CDC). Data from survey years 2009−10 and 2011−2012 has been analysed to monitor how water intake varies according to age, race and Hispanic origin and physical activity.

Between 2009 and 12, the daily average total water intake from all foods and liquids among US adults aged 20 years and over was 3.46 L for men, with 30% coming from plain water, and 2.75 L for women, with 34% being plain water (Fig. 25.1) [7]. Total water intake was reportedly lower among individuals aged 60 years and over compared with younger adults. The survey results showed that non-Hispanic black men and women had the lowest average total water intake. For US adults aged 20 years and over, total water intake increased with physical activity levels [7].

The average total daily water intake among male participants was found to be approximately 0.25 L less than the recommended adequate intake, while women's consumption was approximately equivalent to the recommended volumes. On average, men and women aged 60 years and over, non-Hispanic black men and women, Hispanic men and women, men and women with low physical activity, and men with moderate physical activity consumed less than the adequate daily intake.

Previous studies have demonstrated that adults aged 60 years and over are among the most vulnerable to dehydration. This report found that men aged 60 years and over consumed 2.92 L daily, which is approximately 0.8 L less than the recommended intake, and women aged 60 years and over consumed 2.51 L, equating to 0.2 L less than advised [7].

Analysis of water and beverage consumption among children aged 4−13 years in the United States has shown that an average of 0.4 L of water is consumed daily, with plain water intake being lower in younger children and in non-Hispanic Black and Mexican Americans [8].

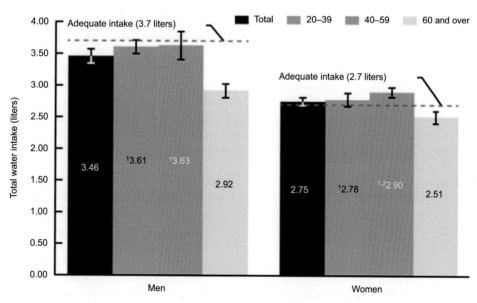

FIGURE 25.1

Mean total water intake per day among adults aged 20 and over, by sex and age group: United States, 2009−12.

CDC/NCHS, National Health and Nutrition Examination Survey, 2009−12.

Amongst US adults, plain water intake was found to be lower in older adults, lower-income adults, and those with lower education [9,10].

Interestingly, in US adolescents, there appeared to be a correlation between drinking less water and drinking less milk, eating less fruits and vegetables, drinking more sugar-sweetened beverages, eating more fast food and getting less physical activity [11].

Does water intake affect energy levels and brain function?

It is often claimed that inadequate hydration can impair energy levels and brain function. There are numerous studies which appear to support this. Research carried out in a small study on the effect of dehydration in 23 women showed that a fluid loss of 1.36% after exercise adversely impacted mood and concentration and increased the frequency of headaches [12].

Other studies have investigated the effects of mild dehydration due to exercise or excessive sweating in hot climates, resulting in the loss of one to 3% of body weight. It has been shown that this degree of fluid loss can impair many aspects of brain functioning, leading to subjective feelings of tiredness, difficulty concentrating and reports of reduced mental alertness [13].

Evidence also suggests that mild dehydration can negatively impact physical performance and can reduce capacity and endurance [14].

Can drinking water help with weight loss?

It has been suggested that increased water intake may reduce body weight by both boosting metabolism and suppressing appetite. It stands to reason as when feelings of hunger and thirst are confused, inadvertent consumption of more calories may result from eating to satisfy the wrong appetite.

According to studies, drinking as little as 500 mL of water can temporarily boost metabolism by 24%−30% [15] with researchers estimating that drinking 2 L in 1 day potentially increases energy expenditure by as much as 96 calories per day. Drinking cold water was thought to be even more beneficial, because the body utilises additional calories to heat the water to body temperature.

Consuming water approximately 30 min before a meal can also reduce the number of calories ingested, particularly in older individuals, with one study demonstrating that dieters who drank 500 mL of water before each meal lost 44% more weight over 12 weeks, compared with those who did not [16].

One study looked at 89 overweight and obese women between the ages of 18−50 years, with a body mass index of 27−40 kg/m^2, who usually consumed diet beverages as part of their daily routine. They were asked to either substitute their diet drinks for water or continue drinking diet beverages five times per week after their lunch for 24 weeks. They also visited a dietitian twice weekly to promote adherence to a hypoenergetic diet. Diet beverages were supplied to the relevant group at their dietician visits, and both groups followed a specific diet protocol, which also included advice to gradually increase activity levels to achieve 60 min of moderate activity on 5 days of the week.

Interestingly, the group who consumed only water had a greater decrease in weight, fasting insulin and 2-h postprandial glucose over the 24 weeks when compared with the group who continued to drink diet beverages [17]. These findings suggest that replacement of diet beverages with water after the main meal may lead to greater weight reduction during a weight loss program. It may also offer clinical benefit by improving insulin resistance.

Overall, it seems that in combination with a healthy lifestyle, drinking adequate amounts of water, particularly before or after meals, may be beneficial for those trying to lose weight, especially if water is being consumed in the place of sugary or diet drinks.

Are there additional health benefits of water?

The positive effects of drinking water are plentiful and varied, ranging from a potential reduction in fatal heart disease by impacting risk factors such as whole blood and plasma viscosity, haematocrit and fibrinogen levels [18] to reduced frequency of urinary tract infections, kidney stones [19] and possibly even colon cancer [20] and constipation [21,22].

It is important that individuals remain adequately hydrated to ensure optimum physiological functioning. The volume required will depend on an individual's age, activity levels, climate and preexisting health conditions. Intake should therefore be adapted for specific circumstances and encouraged as part of a healthy lifestyle, with particular emphasis placed on ensuring education regarding adequate hydration in vulnerable populations.

Summary

- Education, lifestyle, age, gender and ethnicity all influence the volumes of fluid consumed daily.
- The desirable fluid intake depends on many factors including age, gender, health status, activity levels and climate.
- Some guidelines suggest the specific volumes that should be consumed daily; others support drinking in response to thirst clues.
- On average, men and women aged 60 years and over, non-Hispanic black men and women, Hispanic men and women, men and women with low physical activity, and men with moderate physical activity consume less than adequate daily fluid intake.
- Adequate water intake is necessary for the maintenance of basic physiological functions such as thermoregulation, physical functioning and cognitive function.
- Mild dehydration can negatively impact physical performance, including reduced capacity and endurance. It can also impair cognitive functioning.
- Adequate water consumption may reduce the incidence of fatal heart disease, urinary tract infections, kidney stones and constipation.
- It is important to educate individuals that adequate water intake, in addition to other healthy lifestyle modifications, is important to reduce morbidity and mortality.

References

[1] Helmenstine AM. How much of your body is water? 2018. https://www.thoughtco.com/how-much-of-your-body-is-water-609406.
[2] Pascal S, et al. Swallowing inhibition emerges from overdrinking. Proc Natl Acad Sci USA 2016;113(43): 12274—9.
[3] Rolls BJ, et al. Thirst. Cambridge: Cambridge Univ. Press; 1982.
[4] Water, drinks and your health. 2018. https://www.nhs.uk/live-well/eat-well/water-drinks-nutrition/.
[5] Dietary reference intake: electrolytes and water. The National Academies of Science, Engineering, and Medicine; n.d. http://www.nationalacademies.org/hmd/Activities/Nutrition/DRIElectrolytes.aspx.

[6] Le Bellego L, et al. Guidelines for adequate water intake: a public health rationale. In: Proceedings from EFBW symposium IUNS 20th international congress of nutrition granada; 2013 [Spain].

[7] Rosinger A, Herrick K. Daily water intake among U.S. men and women, 2009–2012. NCHS data brief, no 242. Hyattsville, MD: National Center for Health Statistics; 2016.

[8] Drewnowski A, Rehm CD, Constant F. Water and beverage consumption among children age 4-13y in the United States: analyses of 2005–2010 NHANES data. Nutr J 2013;12(1):85.

[9] Drewnowski A, Rehm CD, Constant F. Water and beverage consumption among adults in the United States: cross-sectional study using data from NHANES 2005–2010. BMC Publ Health 2013;13(1):1068.

[10] Kant AK, Graubard BI, Atchison EA. Intakes of plain water, moisture in foods and beverages, and total water in the adult US population-nutritional, meal pattern, and body weight correlates: National Health and Nutrition Examination Surveys 1999–2006. Am J Clin Nutr 2009;90(3):655–63.

[11] Park S, Blacnk HM, Sherry B, Brener N, O'Toole T. Factors associated with low water intake among US high school students – National Youth Physical Activity and Nutrition Study, 2010. J Acad Nutr Diet 2012;112: 1421–7.

[12] Armstrong LE, et al. Mild dehydration affects mood in healthy young women. J Nutr 2012;142(2):382–8.

[13] Maughan RJ. Impact of mild dehydration on wellness and on exercise performance. Eur J Clin Nutr 2003; 57(2):S19–23.

[14] Judelson DA, et al. Hydration and muscular performance: does fluid balance affect strength, power and high-intensity endurance? Sports Med 2007;37:907.

[15] Boschmann M, et al. Water-induced thermogenesis. J Clin Endocrinol Metabol 2003;88(12):6015–9.

[16] Dennis EA, et al. Water consumption increases weight loss during a hypocaloric diet intervention in middle-aged and older adults. Obesity 2009;18(2):300–7.

[17] Madjd A, Taylor MA, Delavari A, Malekzadeh R, Macdonald IA, Farshchi HR. Effects on weight loss in adults of replacing diet beverages with water during a hypoenergetic diet: a randomized, 24-wk clinical trial. Am J Clin Nutr December 2015;102(6):1305–12. https://doi.org/10.3945/ajcn.115.109397.

[18] Chan J, et al. Water, other fluids, and fatal coronary heart disease: the *Adventist Health Study*. Am J Epidemiol 2002;155(9):827–33.

[19] Xu C, et al. Self-fluid management in prevention of kidney stones: a PRISMA-compliant systematic review and dose-response meta-analysis of observational studies. Medicine 2015;94(27):e1042.

[20] Shannon J, et al. Relationship of food groups and water intake to colon cancer risk. Cancer Epidemiol Biomarkers Prev 1996;(5):495–502.

[21] Klauser AG, et al. Low fluid intake lowers stool output in healthy male volunteers. Gastroenterol 1990; 28(11):606–9.

[22] Chung BD1, et al. Effect of increased fluid intake on stool output in normal healthy volunteers. J Clin Gastroenterol 1999;(1):29–32.

Intermittent fasting: a health panacea or just calorie restriction?

26

Ellen Fallows[1] and Hayley S. McKenzie[2]

[1]*General Practitioner, The British Society of Lifestyle Medicine, London, United Kingdom;* [2]*Medical Oncology, University Hospital Southampton NHS Foundation Trust, Southampton, United Kingdom*

Caloric restriction

Caloric restriction (CR) and weight loss have been known to improve health and longevity for decades, with a proverb dating as far back as the 1700s:

> *He that eats till he is sick must fast till he is well* [1].

However, evidence suggests that most people struggle to stick to a CR diet, resulting in a failure to achieve sustained weight loss. For example, it has been reported that only around 20% of overweight people are able to maintain $\geq 10\%$ weight loss after 1 year of a weight loss program [2]. In the Look AHEAD study of overweight individuals, just 26.9% of participants lost $\geq 10\%$ of their body weight after 8 years of participating in the weight loss program [3]. It is thought that CR is not just difficult to adhere to but may be hindered by 'metabolic adaptation', causing a slowing of the resting metabolic rate (RMR) in response to weight loss. This is thought to be a homoeostatic mechanism to reduce energy needs during times of starvation. For example, the 'Biggest Loser' study found that 14 participants lost a mean weight of 58.3 kg during a short-term competition, but their RMR slowed by a mean of 610 calories per day. After 6 years, a mean of 41.0 kg of weight per participant was regained, but their mean RMR remained at 704 cals/day below baseline [4].

Fasting

Given the problems associated with CR, there has been increasing interest in fasting regimes. During evolution, humans had to forage for scarce and low-calorie food requiring significant physical endeavour: we adapted to endure periods of fasting and physical activity. Modern life is a different scenario, where the norm is more often continual grazing on high-calorie, low-nutrient foods whilst exhibiting high levels of sedentary behaviour.

The concept of intermittent fasting (IF) is not new. It has been an established practice in many religions and cultures, for example, fasting during Ramadan. A paper published in the Lancet in 1966 reported successful treatment of obesity through total fasting for up to 249 days. Participants consumed clear fluids and a daily multivitamin tablet only. The patient fasting for the full 249 days lost 34 kg with the researchers being surprised by the 'ease with which the prolonged fast was tolerated' [5].

A Prescription for Healthy Living. https://doi.org/10.1016/B978-0-12-821573-9.00026-6

Prolonged fasting has been shown stimulate specific metabolic pathways in humans that could explain many of the health benefits seen, including reduced risks for many long-term conditions such as type 2 diabetes, cardiovascular (CV) disease, cancer, dementia and Parkinson's disease.

Cellular metabolic pathways stimulated by fasting

A process of 'metabolic switching' occurs during fasting, in order to maintain optimal brain performance and increase resistance to injury or disease. This process ensures that blood glucose is maintained at a normal level. Liver glycogen stores are mobilised, triggered by reduced energy levels in the form of adenosine triphosphate which upregulates cAMP response element binding protein (CREB) [6]. Fat stores are also broken down to free fatty acids to produce ketones, such as β-hydroxybutyrate and acetate. Ketones can be used by the brain as an alternative energy source to glucose.

However, ketones produced in this fasted metabolic state are not just an alternate fuel source but act to stimulate a cascade of pathways involving hormones and cellular processes that alter gene expression. Many of these pathways have been studied in rodent models and could explain some of the health benefits seen during fasting. For example, the hormones leptin, ghrelin and the incretin hormone glucagon-like peptide 1 (GLP-1) are affected by low glucose and the production of ketones during fasting. Levels of leptin and GLP-1 drop whilst ghrelin rises. Ghrelin has been found to enhance hippocampal synaptic plasticity and neurogenesis resulting in improved memory formation in mice [7]. This may be one of the pathways which explain the improvement of memory observed following fasting in human studies [8].

Along with appetite hormones, growth hormones (GHs) are also influenced by fasting. The production of insulin-like growth factor (IGF-1) and GH is reduced from the liver during fasting in rodent models. Raised IGF-1 levels have been correlated with increased cancer risk in both humans and rodents [9]. However, IGF-1 levels in humans are not reduced with chronic CR unless protein is also restricted [10]. So drawing parallels from rodents needs to be done with caution. In contrast, insulin levels are reduced during fasting in both rodent and human studies and again lower insulin is correlated with lower risk of many long-term conditions such as obesity, cancer and CV disease.

At a cellular level, myokines produced from muscles are also involved in this metabolic switch. IL-6, one of the myokines produced during fasting, has been found in mouse models to be antiinflammatory and increase insulin sensitivity of muscle cells [11].

The cellular pathways triggered by ketones during fasting also activate the so-called 'longevity genes' [12] such as SIRT1 and SIRT3 and downregulate the mammalian target of rapamycin (mTOR) pathways. mTOR is an enzyme that regulates pathways involved in DNA repair, inflammation and autophagy. mTOR is involved in growth and ageing if activated and longevity if downregulated by fasting [13].

Ketones also stimulate the production of brain-derived neurotrophic factor (BDNF) from neuronal synapses that act to increase the numbers of mitochondria, referred to as mitochondrial biogenesis, to facilitate more efficient energy use [14].

In summary, the low glucose levels during fasting trigger a metabolic switch resulting in the production of ketones and hormones that shift cellular pathways from a 'cell growth mode' to a 'cell preservation mode'.

This is summarised in Table 26.1.

Table 26.1 Metabolic and hormonal changes that are observed during fed and fasted states.

Fasted state = cell preservation and longevity		Fed state = growth, reproduction and ageing	
Increased	**Reduced**	**Increased**	**Reduced**
Ketones	Glucose	Glucose	Ketones
Ghrelin	Leptin	Leptin	Ghrelin
Myokines	Insulin	Insulin	Myokines
GLP-1	IGF-1	GLP-1	IGF-1
BDNF	Proinflammatory cytokines	Proinflammatory cytokines	BDNF
CREB	mTOR	mTOR	CREB
SIRT1/3	Protein synthesis	Protein synthesis	SIRT1/3
Autophagy + DNA repair	Mitochondrial biogenesis	Mitochondrial biogenesis	Autophagy + DNA repair

Intermittent fasting

Despite the benefits of fasting for prolonged periods, it can carry health risks and be difficult to adhere to over the long term. For example, inadequate B1 can cause beriberi, resulting in irreversible brain damage within days [15] and vitamin C deficiency can cause scurvy within weeks [16]. Following reintroduction of food, refeeding syndrome can occur causing a potentially fatal shift of electrolytes and fluids even after just 5 days of fasting [17]. There are also reports of initial irritability, muscle cramps, changes in bowel habit, fatigue, sleep disturbances and dizziness [18]. Therefore, efforts have been made to harness the benefits of prolonged fasting without these side effects and risks, in the form of IF. IF involves shorter periods of fasting or calorie restriction alternating with normal or 'ad libitum' eating.

There may be additional benefits from the repeated switching from fasted to fed state through IF. It is thought that this switching could improve 'metabolic flexibility' by training the body to use alternate fuel stores, a process called 'intermittent metabolic switching'. The additional physiological benefits of metabolic switching during IF have been confirmed in human studies and thought to be over and above the effects of any CR [19]. The reverse situation, often seen with today's eating habits, is continuous feeding. It is hypothesised that continuous feeding could cause 'mitochondrial metabolic gridlock' [20] whereby we become less efficient at switching to use alternate fuels and triggering survival pathways. There are many different IF protocols used in the research literature. Some of the most commonly used are listed in Table 26.2.

Unfortunately, there are as yet, insufficient studies comparing one IF method to another in order to make any robust conclusions about which methods are superior [25]. However, time-restricted feeding (TRF), a form of IF where food is consumed over a limited time period, shows a particular promise[27]. Pilot trials have shown that the results of TRF depend on the time of day of the eating window, with midday TRF reducing body weight, fat, fasting glucose, insulin, hyperlipidaemia and inflammation, whereas late-day TRF either had no effect or worsened postprandial glucose, blood pressure (BP) and lipid profiles [26]. A small proof-of-concept study in 2018 showed that early time-restricted feeding (eTRF; feeding restricted to 6 h with dinner before 3p.m.) in men with prediabetes for 5 weeks showed improved insulin sensitivity, β-cell responsiveness, BP, oxidative stress and reduced appetite. Participants

Table 26.2 Different patterns of IF.

Type of intermittent fasting	Description
Alternate-day fasting	A fast day with no food, alternated with ad libitum eating.[21]
Modified alternate-day fasting	A cyclical feeding pattern with <25% of usual intake (approximately 500 kcal) with fasting for 24 h followed by ad libitum eating for 24 h.[22]
Periodic fasting	A weekly cyclical feeding pattern of fasting (<25% of required calories) for 1−2 days per week with ad libitum eating for the remaining days, e.g., 6:1 or 5:2 regimes.[23]
Time-restricted feeding	Complete fasting for at least 8−12 h per day with ad libitum eating for the rest of the day repeated every day.[24].Variations include early, mid or late time-restricted feeding.

were provided with enough food to maintain their weight so these improvements were unrelated to weight loss [27]. It is thought that the benefits of eTRF relate to the circadian system that upregulates energy metabolism in the morning. This is in keeping with evidence that shift workers who have to eat at night are more likely to suffer with metabolic diseases [28] and evidence that those who consume most calories at dinner are at increased risk of developing obesity and metabolic syndrome [29].

Calorie restriction and intermittent fasting for longevity

CR has been purported to extend life span in every species studied. For example, the average life span of rats can be increased by up to 80% with an alternate-day fasting regime [30]. Rhesus monkeys, who share approximately 93% genome sequence identity with the humans and have many more similarities to the human life span than rats, show similar improved life span. When on a CR diet over 20 years, the monkeys had delayed onset of age-related diseases and 50% of control-fed animals survived compared with 80% of the CR animals [31]. These effects on life span have been extrapolated to humans, given the pathways affected by fasting in animals such as downregulation of mTOR pathways and IGF-1 reduction are key pathways in human ageing. In addition, there are epidemiological studies such as the association between lower IGF-1 in Italian centenarians [32].

The first human clinical trial of CR, 'CALERIE', was a 2-year randomised controlled trial and achieved an average 11.7% CR over 2 years [33]. Participants' RMR and thyroid hormones reduced over the 2 years leading the authors to suggest that CR is likely to improve longevity, given that a high metabolic rate is a risk factor for mortality [34].

It is likely that IF will show similar benefits to life span to those seen with CR, as the cellular pathways involved in IF are similar to those activated by CR; however, it is thought that additional benefits from metabolic switching may be seen over and above the effects of CR [35]. Adding to the strength of this hypothesis, IF has been found to reduce the risk of many age-related conditions (see below), and correlations have been seen, for example, between the fasting behaviour of Greek Orthodox Christians and longevity in the so-called 'Blue Zone' of Ikaria, Greece, where one-third of the population lives past the age of 90 [36].

However, there are as yet no human clinical trials showing that IF prolongs human life span, and this is likely to remain the case due to the difficulties in running such a long trial.

IF for weight loss

The failure of CR to sustain weight loss has led to an interest in IF for weight loss — the hypothesis being that this may be an easier regime to adhere to and may have additional health benefits. To determine whether IF is superior to CR for weight loss, these were compared in a 2013 study considering two low carbohydrate IF regimes to an isocaloric 25% CR Mediterranean diet. This study showed that there was no significant difference in weight loss between the two groups, but a greater loss of body fat with the IF regimes [37]. A metanalysis of IF for weight loss in 2018 also suggested that there was no significant difference in weight loss achieved by CR versus IF, with both achieving approximately 7 kg loss over the intervention period [38]. A systematic review of IF and weight loss in 2020 concluded similarly that overall IF is an effective weight loss tool with all 27 trials of IF showing weight loss of 0.8% to 13%, regardless of changes in overall caloric intake [39]. However, the studies were small and of short duration. Importantly, this review notes that there were no serious adverse events, the weight loss was predominantly fat loss, rather than lean mass, and hunger remained stable or decreased with IF. This is in keeping with evidence that ketones produced during fasting can reduce appetite. Despite the reported reduction in hunger, dropout rates from many IF trials were >25%, which is equivalent to that seen with CR, but greater than 12—14% with other long-term diet patterns.[45] Similarly, the amount of weight regained after IF and CR was found to be equivalent.

The type of fasting pattern and the dietary pattern consumed when not fasting may influence the success of IF as a weight loss strategy. For example, recent trials of eTRF have shown that levels of the satiety hormone peptide YY and subjective 'stomach fullness' were higher, while ghrelin, 'perceived huger' and the 'desire to eat' were all significantly reduced [40]. Metabolic flexibility was also increased in the eTRF group, who more effectively burned fat during their fasting period.

Intermittent fasting for diabetes and cardiovascular disease

Eating more frequently throughout the day has been found to be associated with increased CV risk in a prospective cohort study, which showed that the hazard ratios for chronic heart disease were 1.10 for men who ate one to two times a day versus 1.26 for those who ate $\geq$ six times a day [41].

Ramadan fasting has also been shown to improve the lipid profile in healthy, obese and those with dyslipidaemia [42]. The 2017 American Heart Association (AHA) consensus statement on meal timings and frequency for the primary prevention of CV disease concluded that trials of IF resulted in a reduction of between 6%—21% of total cholesterol, 7%—32% of low density lipoprotein cholesterol and 14%—42% of triglycerides [43]. These IF trials also showed a decrease in systolic and diastolic BP in trials if a 6—7% weight loss was achieved.

A review of IF versus CR found 11 RCTs showing overall no significant benefit of glucose, HbA1c and triglyceride concentrations [23]. However, fasting insulin levels were significantly lower in the IF group suggesting that IF may be of particular use in diabetes. This has been seen clinically in a 2018 case report showing that, independent of weight loss, IF reversed insulin resistance in prediabetes or type 2 diabetes [44]. In addition, an eTRF study showed improvements in glycaemic control, BP and oxidative stress again, independent of weight loss [45]. Rodent studies suggest that the mechanism behind this improvement may be due to IF reducing inflammation [46], which is a key process in the development of diabetes [47].

Intermittent fasting for neurological conditions

Epidemiological evidence suggests that excess calories, particularly in midlife, increase the risk of stroke, Alzheimer's and Parkinson's disease [48], and animal models suggest that IF can delay progression of both these diseases [14]. In humans, the evidence is suggestive, with CR for 3—4 months improving cognitive function in overweight women [49].

Parallels have been made between pathways triggered by exercise and those involved in the metabolic switch during fasting. It is well established that these exercise-induced pathways improve brain health and reduce the risks of mood disorders and neurodegenerative conditions [50]. Similarly, ketones produced following the fasting metabolic switch may also improve brain health given small pilot studies suggesting that ketogenic diets improve symptoms in children with autism spectrum disorder and epilepsy [51]. Therefore, it has been hypothesised that fasting would give similar brain health benefits as those seen from exercise [52]. However, randomised controlled trials in humans showing reduced risk of neurological conditions as a result of IF are lacking.

Intermittent fasting for cancer

Many cancers are associated with dietary habits, and there is strong evidence according to the World Cancer Research Foundation that obesity increases the risk of at least 12 types of malignancy [53].

Many rodent studies have shown that IF reduces the occurrence of sporadic tumours during normal rodent ageing [54]. IF also suppresses the growth of many types of induced rodent tumours and increases their sensitivity to chemotherapy and irradiation treatment [55]. It is thought that IF impairs energy metabolism in cancer cells during the metabolic switch to survival over growth, with downstream effects to reduce known cancer promoters such as insulin, leptin and GH.

Interventions to restrict eating in cancer patients appear to be feasible with a human trial of CR in men with prostate cancer showing 95% adherence and no adverse events [56].

Preliminary clinical trials suggest that fasting for at least 48 h may prevent chemotherapy-induced DNA damage to healthy tissues and help improve quality of life during chemotherapy [57]. However, there are no data on the effects of IF on cancer rates or recurrence in humans, but many trials are currently in the pipeline.

Risks of intermittent fasting

Much of the evidence of benefits of IF has been extrapolated from animal studies, although early data in humans are encouraging. No studies of any fasting regimes have been performed in children, pregnant or lactating women, the very old or underweight individuals, and it is possible that IF could be harmful to these groups.

Additionally, deprescribing or medication titration may be required to prevent overtreatment in particular with antihypertensives, diuretics and hypoglycaemic agents in particular to avoid potentially fatal hypotension, electrolyte disturbances, renal damage and hypoglycaemia [58]. This is particularly the case for patients on medications for type 2 diabetes where fasting increases the risks of hypoglycaemia, but these risks have been found shown in one study to be reduced with education and medication reduction [59].

There have been concerns that episodes of fasting may trigger or worsen binge-eating behaviours, but the evidence suggests that this is unlikely to be the case [60]. In fact, participants of IF trials report improved body image and less depression [61]. This is likely due to metabolic switching triggering a rise in hormones that suppress appetite.

Applications in clinical practice

IF as a treatment modality has yet to be integrated into many widely used clinical pathways, despite being the most popular diet in 2018, according the International Food Information Council [62], with a recent expansion of books and websites on how to fast and the purported benefits for health and diabetes in particular.

A successful program in Greater Manchester, United Kingdom, has information to support the choice of an IF eating pattern for those who wish to do so as part of their 'Diabetes My Way' Programme [63]. However, the National Health Service Choices website provides information on fasting regimes but suggests that these are 'extreme diets' and that the evidence is limited [64]. The British Dietetics Association lists fasting as a 'fad diet' that is not to be recommended [65]. 'DiabetesUK', an educational charity, only gives advice on how to fast safely with diabetes if this is being done for cultural or religious reasons rather than promoting this as a health intervention [66]. Similarly, the American Diabetes Association 'found limited evidence about the safety and/or effects of intermittent fasting on type 1 diabetes' and only 'preliminary results of weight loss for type 2 diabetes' so does not recommend any specific dietary pattern for the management of diabetes until more research is done [67].

The AHA states that IF may 'produce weight loss, reduce insulin resistance, and lower the risk of cardiometabolic diseases, although its long-term sustainability is unknown' [68]. The AHA suggests that IF can be used as an 'intentional approach to eating that focuses on the timing and frequency of meals and snacks as the basis of a healthier lifestyle and improved risk factor management' and that people should work to 'promote consistent overnight fast periods'.

Access to good quality and up-to-date information for patients and physicians about fasting and the risks, benefits and long-term outcomes is lacking. Most of the better quality and regulated information sources sign post to seeking the advice of medical practitioners, many of whom likely lack training and knowledge of this field. As the evidence base increases, this is an area in need of improved patient accessible information and physician training.

Summary

- IF is as effective as CR for weight loss with similar adherence and rates of weight regain
- IF may have additional benefits for health due to metabolic switching
- Metabolic switching involves changing from growth and reproduction to cell survival pathways
- IF may improve glycaemic control in diabetics and reduce CV risk and the risk of neurological disorders and cancer
- IF may be safer than prolonged periods of fasting with reduced risks of nutritional deficits or refeeding syndrome
- IF has not been associated with worsening or triggering of disordered eating
- There are many types of IF including alternate-day fasting, modified alternate-day fasting, periodic fasting and TRF
- Health systems have yet to integrate IF as a treatment modality, citing lack of long-term data
- Patients can easily access information on IF, but this is neither always up-to-date nor safe
- Patients need to be supervised with IF due to risks, in particular those on medications

References

[1] Fuller T. Gnomologica: adages and proverbs. 1732.

[2] Wing RR, Phelan S. Long-term weight loss maintenance. Am J Clin Nutr 2005;82:222S–5S.

[3] Look Ahead Research Group. Eight-year weight losses with an intensive lifestyle intervention: the look AHEAD study. Obesity 2014;22:5–13. https://doi.org/10.1002/oby.20662.

[4] Fothergill E, et al. Persistent metabolic adaptation 6 years after "The Biggest Loser" competition. Obesity 2016;24:1612–9.

[5] Thomson TJ, et al. Treatment of obesity by total fasting for up to 249 days. Lancet 1966;2(7471):992–6.

[6] Koo SH, et al. The CREB coactivator TORC2 is a key regulator of fasting glucose metabolism. Nature 2005; 437(7062):1109–11.

[7] Kim Y, et al. Ghrelin is required for dietary restriction-induction enhancement of hippocampal neurogenesis: lessons from ghrelin knockout mice. Endocr J 2015;62:269–75.

[8] Witte, et al. Caloric restriction improves memory in elderly humans. Proc Nat Acad Sci Jan 2009;106(4): 1255–60.

[9] Renehan AG, Zwahlen M, Minder C, O'Dwyer ST, Shalet SM, Egger M. Insulin-like growth factor (IGF)-I, IGF binding protein-3, and cancer risk: systematic review and meta-regression analysis. Lancet 2004;363: 1346–53.

[10] Fontana L, et al. Long-term effects of calorie or protein restriction on serum IGF-1 and IGFBP-3 concentrations in humans. Aging Cell 2008;7:681–7.

[11] Wueest S, et al. IL-6 Contributes to early fasting-induced free fatty acid mobilisation in mice. Am J Physiol Regul Integr Comp Physiol 2014;306:861–7.

[12] Sinclair D, Guarente L. Unlocking the secrets of longevity genes. Sci Am 2006;294(3):48–51.

[13] Axton RA, Sabatini DM. mTOR signalling in growth, metabolism and disease. Cell 2017;168(6):960–76.

[14] Matterson MP, et al. Intermittent metabolic Switching, neuroplasticity and brain health. Nat Rev Neurosci 2018;19:63–80.

[15] Drenick EJ, et al. Occurrence of acute Wernicke's encephalopathy during prolonged starvation for the treatment of obesity. N Engl J Med 1966;274(17):937–9.

[16] Hodges RE, et al. Clinical manifestations of ascorbic acid deficiency in man. Am J Clin Nutr 1971;24(4): 432–43.

[17] Kraaijenbrink BV, et al. Incidence of refeeding syndrome in internal medicine patients. Neth J Med 2016; 74(3):116–21.

[18] de Toledo W, et al. safety, health improvement and well-being during a 4-21 day fasting period in an observational study including 1422 subjects. PloS One 2019;14(1).

[19] Seimon RV, et al. Do intermittent diets provide physiological benefits over continuous diets for weight loss? A systematic review of clinical trials. Mol Cell Endocrinol 2015;418(Pt 2):153–72.

[20] Muoio DM. Metabolic inflexibility: when mitochondrial indecision leads to metabolic gridlock. Cell 2014; 159(6):1253–62.

[21] Varady KA, Hellerstein MK. Alternate-day fasting and chronic disease prevention: a review of human and animal trials. Am J Clin Nutr 2007;86(1):7–13.

[22] Harris L, et al. Intermittent fasting interventions for treatment of overweight and obesity in adults: a systematic review and meta-analysis. JBI Data Sys Rev Implement Rep 2018;16(2):507–47.

[23] Cioffi I, et al. Intermittent versus continuous energy restriction on weight loss and cardiometabolic outcomes: a systematic review and meta-analysis of randomized controlled trials. J Transl Med 2018;16(1):371.

[24] Stote KS, et al. A controlled trial of reduced meal frequency without caloric restriction in healthy, normal weight middle aged adults. Am J Clin Nutr 2007;85:981–8.

[25] Mattson MP, et al. Impact of Intermittent Fasting on health and disease processes. Ageing Res Rev 2017;39: 46−58.

[26] Moro T, et al. Effects of 8 weeks of time-restricted feeding on basal metabolism, maximal strength, body composition, inflammation and cardiovascular risk factors in resistance-trained males. J Transl Med 2016; 14:290.

[27] Sutton EF, et al. Early time-restricted feeding improves insulin sensitivity, blood pressure and oxidative stress even without weight loss in men with pre-diabetes. Cell Metabol 2018;27:1212−21.

[28] Wang F, et al. Meta-analysis on night shift work and risk of metabolic syndrome. Obes Rev 2014;15:709−20.

[29] Bo S, Musso G, et al. Consuming more of daily caloric intake at dinner predisposes to obesity. A 6-year population-based prospective cohort study. PloS One 2014;9:9.

[30] Goodrick CL, Ingram DK, Reynolds MA, Freeman JR, Cider NL. Effects of intermittent feeding upon growth and life span in rats. Gerontology 1982;28(4):233−41.

[31] Colman RJ, et al. Caloric restriction delays disease onsent and mortality in Rhesus monkeys. Science 2009: 201−4.

[32] Bonafe M, et al. Polymorphic variants of insulin-like growth factor I (IGF-I) receptor and phosphoinositide 3-kinase genes affect IGF-I plasma levels and human longevity: cues for an evolutionarily conserved mechanism of life span control. J Clin Endocrinol Metab 2003;88:3299−304.

[33] Ravussin E, et al. A 2-year randomized controlled trial of human caloric restriction: feasibility and effects on predictors of health span and longevity. J Gerontol Biol Sci Med Sci 2015;70(9):1097−104.

[34] Ruggiero C, et al. High basal metabolic rate is a risk factor for mortality: the Baltimore Longitudinal Study of Aging. J Gerontol Biol Sci Med Sci 2008;63:698−706.

[35] DeCabo R, et al. Effects of intermittent fasting on health, aging and disease. NEJM 2019;381:2541−51.

[36] Panagiotakos DB, et al. Sociodemographic and lifestyle statistics of oldest old people (>80yrs) living in Ikaria island: the Ikaria study. Cardiol Res Pract 2011:679187.

[37] Harvie M, et al. The effect of intermittent energy and carbohydrate restriction v. daily energy restriction on weight loss and metabolic disease risk markers in overweight women. Br J Nutr 2013:1−14.

[38] Leanne H, et al. Intermittent Fasting interventions for treatment of overweight and obesity in adults: a systematic review and meta-analysis. JBI Data Sys Rev Implement Rep 2018;16(2):507−47.

[39] Welton S, et al. Intermittent Fasting and weight loss; systematic review. Can Fam Phys 2020;66:117−25.

[40] Ravussin E, et al. Early time-restricted feeding reduces appetite and increases fat oxidation but does not affect energy expenditure in humans. Obesity 2019;27:1244−54.

[41] Cahill LE, et al. Prospective study of breakfast eating and incident coronary heart disease in a cohort of male US health professionals. Circulation 2013;128:337−43.

[42] Santos HO, Macedo RC. Impact of intermittent fasting on the lipid profile: assessment associated with diet and weight loss. Clin nut ESPEN 2018;24:14−21.

[43] St-Onge MP, et al. Meal timing and Frequency: implications for Cardiovascular disease prevention: a scientific statement from the AHA. Circulation 2017;135(9):96−121.

[44] Furmli S, et al. Therapeutic Use of intermittent fasting for people with type-2 diabetes as an alternative to insulin. BMJ Case Rep 2018. https://doi.org/10.1136/bcr-2017-221854.

[45] Sutton EF, et al. Early TRF improves insulin sensitivity, Blood Pressure and oxidative stress even without weight loss in men with prediabetes. Cell Metabol 2018;28(6):1212−21.

[46] Hatori M, et al. Time-restricted feeding without reducing caloric intake prevents metabolic diseases in mice fed a high-fat diet. Cell Metabol 2012;15(6):848−60.

[47] Donath MY, Shoelson SE. Type 2 diabetes as an inflammatory disease. Nat Rev Immunol 2011;11(2): 98−107.

[48] Arnold SE. Brain insulin resistant in type 2 diabetes and Alzheimer's disease: concepts and conundrums. Nat Rev Neurol 2018;14:168−81.

[49] Kretsch MJ, et al. Cognitive effects of a long-term weight reducing diet. Int J Obes Relat Metab Disord 1997; 21:14−21.

[50] Cotman CW. Exercise builds brain health: key roles of growth factor cascades and inflammation. Trends Neurosci 2007;30(10):489.

[51] Kraeuter AK, et al. Ketogenic Therapy in Neurodegenerative and psychiatric disorders: from mice to men. Prog Neuro-psychopharmacol Biol Psychiatr 2020;101:109913.

[52] Mattson MP. Intermittent metabolic switching, neuroplasticity and brain health. Nat Rev Neurosci 2018; 19(2):63−80.

[53] World Cancer Research Fund/American Institute for Cancer Research. Diet, nutrition, physical activity and cancer a global perspective. Continuous update project expert report. 2018. Available at: http://dietandcancerreport.org (accessed May 2020).

[54] Descamps O, Riondel J, Ducros V, Roussel AM. Mitochondrial production of reactive oxygen species and incidence of age-associated lymphoma in OF1 mice: effect of alternate-day fasting. Mech Ageing Dev 2005; 126:1185−91.

[55] Shi Y, et al. Starvation-induced activation of ATM/Chk2/p53 signalling sensitizes cancer cells to cisplatin. BMC Canc 2012;12:571.

[56] Demark-Wahnefried W, et al. Feasibility outcomes of a presurgical randomized controlled trial exploring the impact of caloric restriction and increased physical activity versus a wait-list control on tumour characteristics and circulating biomarkers in men electing prostatectomy for prostate cancer. BMC Canc 2016;16:61.

[57] Bauersfeld SP, et al. The effects of short-term fasting on quality of life and tolerance to chemotherapy in patients with breast and ovarian cancer: a randomized cross-over pilot study. BMC Canc 2018;18:476.

[58] Grajower MM, Horne BD. Clinical management of intermittent fasting in patients with diabetes mellitus. Nutrients 2019;11(4):873.

[59] Corley BT, Carroll RW, Hall RM, Weatherall M, Parry-Strong A, Krebs JD. Intermittent fasting in Type 2 diabetes mellitus and the risk of hypoglycaemia: a randomized controlled trial. Diabet Med 2018;35: 588−94.

[60] Gabel K, et al. Safety of 8 hr time restricted feeding in adults with obesity. Appl Physiol Nutr Metabol 2019; 44(1):107−9.

[61] Hoddy KK, et al. Safety of alternate day fasting and effect on disordered eating behaviours. Nutr J 2015;14:44.

[62] https://foodinsight.org/wp-content/uploads/2018/05/2018-FHS-Report-FINAL.pdf.

[63] https://diabetesmyway.nhs.uk/resources/internal/what-can-i-do-to-lose-weight.

[64] https://www.nhs.uk/news/obesity/alternate-day-fasting-may-help-aid-weight-loss/ (accessed May 2020).

[65] https://www.bda.uk.com/resource/detox-diets.html.

[66] https://www.diabetes.org.uk/guide-to-diabetes/enjoy-food/eating-with-diabetes/fasting (accessed May 2020).

[67] Evert AB, Dennison M, Gardner CD, Garvey WT, Lau KH, MacLeod J, et al. Nutrition therapy for adults with diabetes or prediabetes: a consensus report. Diabetes Care May 2019;42(5):731−54.

[68] St-Onge MP, et al. American heart association meal timing and frequency: implications for cardiovascular disease prevention: a scientific statement from the American heart association. Circulation 2017;135(9): e96−121.

Concept boxes: mindfulness, healthy weight and bone health

Throughout this book, several concepts have been referred to, including mindfulness, the importance of maintaining a healthy weight and bone health. The concept boxes in the following give additional information about each of these.

Mindfulness: mind full or mindful?

Saba Jaleel

Psychiatrist, Change Grow Live, Birmingham, United Kingdom

The quieter you become, the more you can hear.
- For many individuals, life in the 21st century is busy. Many people hurtle through each day, rushing from one task to another, with continually active minds. It is common to focus on tasks that need to be completed without noticing and enjoying what is happening in the present day.
- Mindfulness is a technique used to bring the mind back to the current moment.
- The health benefits of mindfulness are so great that in October 2015, the United Kingdom (UK) government published recommendations that mindfulness should be incorporated across healthcare, education, the workplace and criminal justice systems [1].
- What is mindfulness? Mindfulness is often assumed to be a Buddhist ideology, one which requires education and training. But it is not a religious, cultural or philosophical belief [2]; it merely describes living in the present moment. It is the practice of intentionally paying attention to what is happening around you, be that your actions, thoughts or emotions.
- Why is this so important? Humans have a unique ability to be doing one thing while thinking about another [3], for example planning the day ahead while brushing teeth. The human brain can enter an 'autopilot mode' [4], which allows the completion of simple tasks while the mind is busy planning, assessing and evaluating. The mind does not focus on the task being executed but wanders. Mind-wandering is considered to be the brain's default position [5] with research showing that individuals spend an average of 47% of their waking hours thinking about something other than what they are doing at the time [6].
- Mind-wandering is associated with negative emotions such as unhappiness, boredom and stress, along with an impaired performance of tasks requiring executive attention.
- When the mind is focused and present, it is in a 'mindful state': events around us are acknowledged in a nonjudgemental way [7]. The 'autopilot mode' is switched off, and the wandering mind is allowed to stop and rest.
- The benefits of mindfulness: Mindfulness helps individuals to achieve an emotional equilibrium and better equips society to manage day-to-day life stresses. Specific benefits of mindfulness include the following:
  - Improved short-term memory [8]
  - Improved ability to process information quickly and accurately [9]
  - Decreased levels of stress, anxiety and depression [10]
  - Improved empathy and compassion [11]
  - Increase in creativity [12]

- **How to start the practice of mindfulness:** Mindfulness is a tool that can be used by healthcare practitioners for their own well-being, and it can be recommended to patients to enhance their health. There is no mindfulness manuscript to read, and there are no right or wrong ways to practice it. The key principle is to be present in the here and now and to focus. For individuals with very active minds, or for those who get distracted easily, formal mindfulness/meditation workshops or yoga classes may be beneficial. Furthermore, there are many mindfulness apps and online practices that can be followed at home.

However, mindfulness is a simple activity that can easily be carried out without any formal guidance. The strategies in the following outline methods that a healthcare professional can use when recommending mindfulness to their patients:

- *Choose something to concentrate on*, for example, the water when you are showering. Think about how it feels, the temperature, its strength and the sounds it makes. It is natural for your mind to wander after a few seconds. You may not even notice for a while. At the point you become aware of it, bring yourself back to the present moment and continue practicing as before. Repeating this cycle will focus the mind, and over time, you will find your periods of attention, and awareness will increase and mind-wandering will fade.

Every time you practice mindfulness, think about all your senses — sight, hearing, smell, taste and touch.

- *Breathing.* Focusing on breathing is a form of mindfulness that can be done anywhere at any time. Bring your attention to the depth of your breath and concentrate on the filling of your lungs followed by exhaling. Take five long mindful breaths. You could aim to repeat this routine at least once every day.
- *Day-to-day tasks.* A popular way to practise mindfulness is to concentrate on mundane tasks so that mindfulness becomes an integral part of your daily routine. For example before you brush your teeth take time to pay attention to the toothpaste spreading on to the toothbrush and then the movements of the brush on your teeth and gums. Become aware of the smell and sensation and of how the bubbles look in the sink.
- *Your environment and nature.* Any time spent outside is a fantastic opportunity to practise mindfulness. Every time you go for a walk, spend a few minutes along the way focusing on the sounds around you — birds, traffic, chatter, your footsteps. Hold your attention to these sounds and sights for as long as possible, bringing the mind back into focus when it wanders.

The practice of mindfulness is simple and free and requires minimal effort. If individuals can spend just a few minutes every day devoting their attention to the present moment, they may experience a tremendous and positive impact on their perception of themselves and the world around them.

The importance of maintaining a healthy weight

Emma Short

Department of Cellular Pathology, Division of Cancer and Genetics, Cardiff University, University Hospital of Wales, Cardiff, United Kingdom

- Obesity is a major health concern worldwide, affecting affects one in four adults in the United Kingdom and over four in ten adults in the United States [13,14].
- Obesity can reduce life expectancy by up to ten years and is thought to contribute to at least one in every 13 deaths in Europe [13].

- Obesity is associated with socioeconomic disadvantage.
- Obesity increases the risk of many chronic diseases including type 2 diabetes, cardiovascular disease, asthma, colorectal cancer, breast cancer, endometrial cancer, reflux, gallstones and osteoarthritis [13].
- Obesity is also associated with metabolic abnormalities such as insulin resistance, hyperglycaemia and dyslipidaemia [15].
- Obesity mediates its pathogenic effects through a variety of mechanisms including metabolic disturbances, hormonal changes, chronic inflammation and mechanical factors.
- As well as being a serious risk to health, obesity can affect quality of day-to-day life. Obese individuals can feel tired, find it difficult to perform physical activities and suffer back pain. They can also have feelings of low self-esteem and lack of confidence [13].
- Obesity is defined by an individual's body mass index, or BMI, which takes into consideration height and weight. The BMI equation is BMI = weight in kg/(height in metres x height in metres). According to the National Health Service (NHS) [13], a 'healthy weight' is a BMI of 18.5−24.9, overweight individuals have a BMI of 25−29.9, obesity is classified as BMI 30 to 39.9 and a BMI of 40 plus indicates severe obesity.
- Although BMI is one of the most commonly used tools in assessing health throughout the developed world, it does have its drawbacks. For example, it cannot distinguish between weight due to fat or muscle, and it does not take into account body fat percentage.
- Measuring the waist is also useful in assessing whether someone carries too much fat [13]. As a guide, men with a waist circumference greater than 94 cm and women with a circumference greater than 80 cm are at an increased risk of obesity-related health problems [13].
- An individual's weight is a result of the balance between energy input and energy expenditure. Several complex pathways are involved in regulating energy balance, including the appetite centres in the hypothalamus and brain stem, reward and motivation networks in the limbic regions and cerebral cortex and hormone signals from the gut, adipose tissue and other organs.
- **Can an individual be 'fat, fit and healthy'?**

In 2013, it was reported by the European Society of Cardiology that obese individuals who were metabolically healthy had no increased risk of death compared with metabolically healthy individuals who had normal fat levels [16]. This seemed to suggest that if an obese person did enough exercise, they could reduce their risk of heart disease and cancer to that of a person with a healthy weight.

But more recent evidence has refuted this claim. In 2017, research published by the American College of Cardiology, looking at approximately 3.5 million individuals, stated that obese individuals have a 49% higher risk of heart disease, 7% increased risk of stroke and 96% higher risk of heart failure than normal weight individuals, even if they are metabolically healthy [17].

It is also important to be aware that individuals who are metabolically unhealthy, regardless of whether they have a healthy BMI, so could be a normal weight, are at increased risk of cardiovascular disease [17].

Although not all the evidence is absolutely clear, the picture is emerging that an individual can be overweight and fit but still have an increased risk of many chronic diseases.

- It is advisable that anyone wishing to lose weight should consult their healthcare provider, who may have access to local resources such as exercise referral schemes, may be able to refer the individual to a dietician and, if necessary, can discuss options such as medication and surgery.

- It is also vital to be aware that obesity is related to social and environmental factors and that managing weight loss must take into consideration an individual's level of education, social and financial background, health literacy and access to healthy foods.
- Taking this into consideration, there are several lifestyle modifications an individual can make to achieve a healthy weight. Simple and effective strategies are outlined in the following:
- **Set realistic goals.** If an individual is overweight, it is very important that weight is lost in a healthy and sustainable manner. It can be very useful to set goals which are SMART — specific, measurable, achievable, realistic and time bound. So, rather than say: 'I want to lose weight by Christmas', it is much more effective to say: 'I want to lose 5 kg in 5 weeks. I will have achieved my target weight by December 25'. It is recommended that weight is monitored at specific time intervals, every week for instance, so that the individual can see their achievements, which provides motivating feedback.
- **Aim for slow and sustainable weight loss.** There is not a safe and healthy quick fix to losing weight. It requires continued effort over time and modifications to diet and activity levels. The maximum amount of weight an individual should aim to lose is 1 kg per week.
- **Eat a healthy diet.** Eating a healthy diet is absolutely necessary, as has been described throughout this book. In Western society, there is easy access to a huge range of cheap, processed foods with a high sugar content. It is really important to avoid these and to focus on fresh foods — encourage your patients to eat fresh fruit and vegetables, to prepare their meals from scratch so they know exactly what they contain and try to minimise intake of processed foods and refined sugar.
- **Watch what is drunk.** Fruit juices, sugar sweetened drinks, many fizzy drinks and alcohol are all relatively high sources of calories. If these are swapped for water, this is a way of cutting energy intake with little effort.
- **Eat slowly.** There is evidence to suggest that eating slowly can lower the energy consumed during a meal and can reduce feelings of hunger [18]. Every time an individual eats, encourage them to make a conscious effort to eat more slowly. One way of doing this is to eat mindfully, really paying attention to what is put into the mouth. Focus on the sight, smell and textures of the food, concentrate on how it tastes and feels, its temperature and what sounds it makes as it is chewed. Other techniques include taking smaller mouthfuls with each bite and using a timer to help chew for longer.
- **Avoid situations where overeating is a possibility.** Some situations can encourage overindulgence, for example all-you-can-eat buffets. Support your patients in avoiding these if at all possible. Similarly, eating in front of the television or while working can prompt an individual to eat more, as they are distracted and not paying attention to their plate [19].
- **Smaller portion sizes.** Some people find that they have difficulty losing weight even though they eat the right sorts of foods. This could be due to portion sizes. Some ways of reducing portion sizes include using a smaller plate, filling half of the plate with vegetables or salad so there is less room for the higher energy components of the meal and, once a meal has been served, immediately putting away any leftover food so there is not the temptation to go back for a second helping.
- **Use an app to track calories.** It is often very hard to know exactly how much energy is in the foods that is being consumed. There are many free apps that can be downloaded which will calculate this — the individual types in what has been eaten, for example an apple, and the app will record its calorie content. As food is logged throughout the day, the app will add up the total

calorie content of what has been consumed, so the individual can understand when they have reached their limit.

- **Exercise regularly.** Weight is a result of how much energy has been consumed and how much has been burnt. Exercise will increase energy output, which will help with weight loss as well as improving health.

Bone health

Claire Stansfield

General Practitioner, National Health Service, West Yorkshire, United Kingdom

- Most actions undertaken to improve general health will have a positive impact on the skeleton.
- There are 212 bones in the human body, which serve to maintain an upright posture, allow movement, provide protection for the internal organs and act as a source of calcium. Furthermore, the bone marrow they contain is involved in the production of blood cells.
- Bone is a living tissue made from proteins, such as collagen, and minerals including calcium.
- Formation of bones begins in utero, and they continue to grow and mature through childhood and adolescence. As adults, bones do not grow in size or length, but they are constantly renewed and reshaped.
- Bones start to lose their density from the age of around 35 years, and this may result in osteoporosis. Osteoporosis is associated with an increased risk of bone fractures. Three and a half million people in the United Kingdom (UK) have osteoporosis, and it accounts for half a million broken bones every year. Half of all females and one in five males will break a bone during their lifetime. In 2017, it was estimated that osteoporosis cost the National Health Service (NHS) £4.5 billion pounds, with this figure predicted to rise by 30% by 2030.
- Osteoporosis is usually undiagnosed until it results in a fracture. Common sites for fractures include the wrists, hips and spine. The force needed to break an osteoporotic bone can be very small, so injuries can be minor, sometimes just twisting or bending is sufficient. Fractures of the spine often occur without any acute symptoms but can result in long-term back pain. 42% of individuals who develop a fracture due to osteoporosis will experience chronic pain, and a third report the pain to be unbearable. Hip fractures can result in significant disability, immobility and social isolation. Older adults who have a hip fracture are between five and eight times more likely to die in the 3 months following the fracture than those without fractures [20].
- Some individuals have an increased risk of developing osteoporosis. This can be due to genetics, early menopause, certain medications such as steroids and some long-term illnesses, including Coeliac disease, Crohn's disease, ulcerative colitis, rheumatoid arthritis, diabetes, renal disease and liver disease.
- The most important period for bone strengthening is the first 30 years of life, as this allows maximisation of bone density. However, a 'bone-friendly diet' and lifestyle factors are important throughout life, as this will minimise the bone loss that occurs with age
- Advice that can be given to patients to enhance their bone health is detailed in the following:
  - **Diet.** Diet is vitally important for bone health. Consuming enough calcium is probably the most important factor, with vitamin D also playing a role.

- **Calcium.** It is well known that dairy foods, including milk, yogurt and cheese, are the best sources of calcium. Other calcium-rich products include calcium-enriched non-dairy milks, Tofu, calcium-enriched orange juice, fortified cereals and bread. Sardines, pilchards, tinned salmon, whitebait, oranges, broccoli and spring greens are examples of nondairy sources.

It is worth noting that some foods which contain calcium also contain chemicals such as oxalates and/or phylates, which will impair the absorption of calcium. These include spinach, dried fruits, beans, nuts, seeds and wheatbran [21].

Adults need between 700 and 1000 mg of calcium a day — 700 mg equates to around 200 mL of milk, a matchbox-sized piece of cheese and a yogurt [22].

- **Vitamin D.** Vitamin D aids the absorption of calcium, and the best source of sunlight features in Chapter 23. Supplements may be beneficial during the winter months.
- **Other foods.** Eating a wide variety of fruits and vegetables helps to ensure an adequate intake of the other vitamins and minerals that contribute to healthy bones, including vitamin K, magnesium, zinc and phosphorus. It is important to avoid consuming too much salt, as this can lead to calcium loss from bones.
- **Drinks.** It is well known that too much alcohol is harmful. One of its many effects is to increase the risk of osteoporosis developing. However, there is some evidence that small volumes may actually help to protect bones [23].

Another chemical we discussed in this book is caffeine. Caffeine may reduce calcium absorption, leading to bone loss. Cola drinks which contain phosphoric acid can cause calcium to leech out of bones. This is not seen not with lighter coloured fizzy drinks or carbonated water [24].

- **Cigarettes** are dangerous for bones. It is unclear whether smokers are more likely to have osteoporosis because of a direct effect of cigarettes on bone health or because smokers are generally thinner, exercise less, drink more alcohol and eat poorer diets. Whatever the reason, the advice is to quit [25]!
- **Exercise** is another major factor in developing good bone health. Bones are like muscles, they respond to exercise by becoming stronger. Weight-bearing exercise is best for bones. This includes activities such as running, walking, dancing or even marching while watching the television.
- **Weight.** Generally, when talking about health, we talk about the need to avoid being overweight. Ironically, for bones, being underweight is more harmful. This is probably because people who are underweight having lower levels of oestrogen, which is important for healthy bones.
- **Supplements.** It is unclear whether calcium supplements can reduce the risk of developing osteoporosis, and it is uncertain that taking them is entirely without risk. It is therefore preferable to obtain calcium from a healthy calcium-rich diet.
- In summary, it is vital to maximise bone health so that individuals remain active, pain-free and mobile for as long as possible. To achieve this, a sensible strategy includes consuming calcium-rich foods with every meal, eating plenty of fruits and vegetables, avoiding excessive amounts of salt and reducing alcohol and caffeine intake. People should also aim to spend some time outside every day, ideally while doing weight-bearing exercises, and consideration should be given to taking a vitamin D supplement, especially during the winter months.

References

Mindfulness references

[1] The mindfulness initiative. Mindful nation UK; n.d. https://themindfulnessinitiative.org.uk/images/reports/Mindfulness-APPG-Report_Mindful-Nation-UK_Oct2015.pdf2015.

[2] Kabat-Zinn J. Mindfulness-based interventions in context: past, present, and future. Am Psychol Assoc Clin Psychol Sci & Pract 2003:145. https://doi.org/10.1093/clipsy/bpg016. University of Massachusetts Medical School.

[3] Etienneblondiaux. Essay on mindfulness; n.d. https://etienneblondiaux.wordpress.com/2015/12/05/essay-on-mindfulness/2015.

[4] Monash University. What is mindfulness; n.d. https://www.monash.edu/__data/assets/pdf_file/0006/233898/what-is-mindfulness.pdf.

[5] Scientific American. What does mindfulness meditation do to your brain; n.d. https://blogs.scientificamerican.com/guest-blog/what-does-mindfulness-meditation-do-to-your-brain/2014.

[6] Killingsworth M, Gilbert D. A wandering mind is an unhappy mind. Science 2010;330(6006):932. https://doi.org/10.1126/science.1192439.

[7] Monteiro, et al. In: Practitioner's guide to ethics and mindfulness based interventions, mindfulness in behavioural health. Springer International Publishing AG; 2017. https://doi.org/10.1007/978-3-319-64924-5_2.

[8] Chambers RH, Lo BCY, Allen NB. The impact of intensive mindfulness training on attentional control, cognitive style, and affect. Cognit Ther Res 2008;32(3):303−22.

[9] Slagter HA, Lutz A, Greischar LL. Mental training affects distribution of limited brain resources. PLoS Biol 2007;5(6).

[10] Hofmann SG, Sawyer AT, Witt AA, Oh D. The effect of mindfulness-based therapy on anxiety and depression: a meta-analytic review. J Consult Clin Psychol 2010;78(2):169.

[11] Neff KD, Kirkpatrick KL, Rude SS. Self-compassion and adaptive psychological functioning. J Res Pers 2007;41(1):139−54.

[12] Horan R. The neuropsychological connection between creativity and meditation. Creativ Res J 2009; 21(2−3):199−222.

Healthy weight references

[13] https://www.nhs.uk/conditions/obesity/.

[14] https://www.cdc.gov/obesity/data/adult.html.

[15] Alberti KGMM, et al. Harmonizing the metabolic syndrome A joint interim statement of the International Diabetes Federation Task Force on Epidemiology and Prevention; National Heart, Lung, and Blood Institute; American Heart Association; World Heart Federation; International Atherosclerosis Society; and International Association for the Study of Obesity. Circulation 2009;120:1640−5.

[16] Ortega FB, et al. The intriguing metabolically healthy but obese phenotype: cardiovascular prognosis and role of fitness. Eur Heart J 2013;34:389−97.

[17] Caleyachetty R, et al. Metabolically healthy obese and incident cardiovascular disease events among 3.5 million men and women. J Am Coll Cardiol 2017;70(12):1429−37.

[18] Von Seck P, et al. Persistent weight loss with a non invasive novel medical device to change eating behaviour in obese individuals with high risk cardiovascular risk profile. PLoS One 2017;12(4):e017458.

[19] https://www.health.harvard.edu/blog/distracted-eating-may-add-to-weight-gain-201303296037.

Bone references

[20] The international osteoporosis foundation. Broken bones, broken lives: a roadmap to solve the fragility fracture crisis in the United Kingdom. 2018. http://share.iofbonehealth.org/EU-6-Material/Reports/IOF_report_UK.pdf.

[21] National osteoporosis foundation. 2018.www.nof.org/patients/treatment/nutrition/.

[22] The association of UK dieticians food fact sheet: calcium. 2017. www.bda.uk.com/foodfacts/Calcium.pdf.

[23] The association of UK dieticians food fact sheet: osteoporosis. 2016. https://www.bda.uk.com/foodfacts/osteoporosis.pdf.

[24] Tucker KL, et al. Colas, but not other carbonated beverages, are associated with low bone mineral density in older women: The Framingham Osteoporosis Study. Am J Clin Nutr 2006;84(4):936−42.

[25] National Institute of Health Osteoporosis and Bone Diseases National Resource Centre; Smoking and Bone Health. 2016. https://www.bones.nih.gov/health-info/bone/osteoporosis/conditions- behaviors/bone-smoking.

Conclusions and how to access reliable health information

Emma Short

Department of Cellular Pathology, Division of Cancer and Genetics, Cardiff University, University Hospital of Wales, Cardiff, United Kingdom

Good health is not merely the absence of disease. It describes a state of physical, mental and social well-being which allows an individual to thrive. An individual's health is determined by a complex interaction of genetic, environmental and socioeconomic factors. Good health is a basic human right, as stated in the WHO Constitution.

The developed world is currently facing a major health crisis, with dangerously high rates of conditions such as obesity, diabetes, cardiovascular disease and cancer, many of which are preventable. While disease prevention and health promotion require support, funding and input from multiple agencies, including the government, education system, social care system and industry, there are many lifestyle modifications which individuals can make to take charge of their well-being and to improve their health.

It is important that all healthcare professionals support and encourage their patients to make positive behavioural changes. The most commonly cited model of behavioural change is the *transtheoretical model*, proposed by Prochaska and colleagues in the 1970s. It describes the six stages of change as **precontemplation**, during which individuals have not yet considered making a behaviour change, **contemplation** when it is identified that behaviour change may be necessary, **preparation** when planning and work towards change is initiated, **action** when the change in behaviour is elicited, **maintenance** during which the new behaviour is continually performed and then potentially **relapse** when old behaviour patterns emerge.

The role of the healthcare professional therefore varies depending on which stage their patients are at. In some instances, practitioners may need to encourage their patients to move from a state of precontemplation to contemplation, prompting them to identify that lifestyle modifications may be desirable. At the other times, the healthcare professional might be involved in ensuring that a beneficial behaviour is maintained. This can often be aided by positive feedback and through the delivery of quantifiable measures of improved health, for example informing a patient who has increased their exercise levels that their resting heart rate has decreased. Whenever a practitioner is involved in supporting a patient to implement lifestyle changes, it is important that they consider the use of tools such as goal setting, developing action plans and contingency plans, boosting sustainable motivation through the selection of appropriate target behaviours and providing advice on how to establish habitual actions.

Lifestyle modifications and behavioural changes do not need to be complicated, time-consuming or expensive. Significant health gains can be achieved through simple measures such as being physically active, minimizing the time spent sitting, eating a healthy and balanced diet, maintaining a healthy weight, not smoking, moderating alcohol intake and maintaining social relationships. This approach to healthcare has recently gained global popularity under the umbrella term, 'Lifestyle Medicine'.

It is the duty of healthcare providers to ensure that the information they share with their patients regarding lifestyle medicine is unbiased and evidence-based. The medical community can obtain

305

relevant information from the medical and scientific literature, and when critically appraising any publications, it is always important to consider factors such as:

- What is the nature of the study? For example, large, randomised double-blinded placebo-controlled trials have greater reliability than single-person case studies.
- Is there a clear research question which the study can answer? Is the study design appropriate?
- What is the sample size?
- How have the outcomes been measured?
- Has rigorous statistical analysis been performed, and are the conclusions drawn meaningful?
- Have potential sources of bias been accounted for?
- Have confounding factors been considered?
- Who carried out the study − do they have a good research reputation, and do they have any conflicts of interest?
- Where was the study published? Has the paper been peer reviewed? What is the impact factor of the publishing journal?
- Was the research sponsored/funded, and if so, who by?

It can be difficult for patients to access reliable health information, especially with the widespread use of the Internet and social media, which is sometimes responsible for sharing 'fake news' disseminating inaccurate and confusing health information. Generally, resources from the government, academic institutions, hospitals and health services should be evidence-based and driven by national or international guidelines. For patients who use the Internet as their major source of information, useful points to be considered are outlined in Box 1. It is important that patients are aware that accessing health information from online sources does not replace talking to their healthcare provider to discuss any concerns they have.

We hope that this book has inspired and encouraged you to empower your patients to make simple behavioral changes which will have a large impact on their physical and mental well-being. Let your patients know that they are in charge of their future and they can help themselves to change it for the better.

Box 1 Points for the general public and patients to consider when accessing health information

Using the Internet for health information: points to consider

- The Internet is not regulated. The information published may not always be correct.
- Who has produced the information? If it has come from an academic institution such as a university, or a hospital/health organisation, it may be more reliable than information from elsewhere, as these institutions usually have strict guidelines addressing what can be published online.
- Are the authors listed? If so, consider whether they are experts in the field and who they work for? Do they have any potential conflicts of interest, for example is an author writing about nutrition also advertising their weight loss programme?
- Has the information been written by a company/individual trying to sell something?
- Where has the information come from? Different countries have different healthcare policies, so the information presented may not be relevant to everyone.
- How old is the information? Medical knowledge is continually evolving, and data from several years ago may have changed in recent times.
- Are statements backed up by evidence? Some websites include bold statements about health, but it is always important to consider whether these are supported by evidence-based, non-biased research.

Index

Printed in the United States
By Bookmasters